MW01505627

Orthogonal Functions

REVISED ENGLISH EDITION

by G. Sansone

Translated From the Italian by
Ainsley H. Diamond

With a Foreword by Einar Hille

DOVER PUBLICATIONS, INC.
New York

This Dover edition, first published in 1991, is an unabridged republication of the edition first published by Interscience Publishers, Inc., New York, in 1959 as Volume IX of the series *Pure and Applied Mathematics*.

Manufactured in the United States of America

Dover Publications, Inc.
31 East 2nd Street
Mineola, New York 11501

Library of Congress Cataloging-in-Publication Data

Sansone, Giovanni, 1888–
[Sviluppi in serie di funzioni ortogonali. English]
Orthogonal functions / by G. Sansone; translated from the Italian by Ainsley H. Diamond ; with a foreword by Einar Hille.—Rev. English ed.
p. cm.
Translation of: Sviluppi in serie di funzioni ortogonali.
Reprint. Originally published: Rev. English ed. New York : Interscience, 1959. (Pure and applied mathematics : vol. 9)
Includes bibliographical references and index.
ISBN 0-486-66730-8 (pbk.)
1. Functions, Orthogonal. I. Title. II. Series: Pure and applied mathematics (Interscience Publishers) ; v. 9.
QA404.5.S313 1991
515′.55—dc 20 91-12235
CIP

Foreword

Modes come and go in Mathematics as in other fields of human endeavor. New modes first have a hard time breaking through until they are greeted with acclaim, copied and multiplied by the debutants, and then become oldfashioned and are disdained by the still younger set. In the meantime the old modes show great reluctance to disappearing, there are always some faithful souls who prefer the Paris mode of the twenties to, say the Princeton mode of the fifties. It is never safe to say that a particular field in Mathematics is dead or has outlived its usefulness. I have heard many pronouncements to that effect, often made with malice toward some; they have usually been belied by later events. A mathematician can and does choose the mathematics he prefers to do, not so the theoretical physicist who often cannot tell what type of mathematics his problem will lead to. We have seen over and over again during the last fifty years how one obsolescent mathematical theory after another had to be dug out of oblivion to meet the varying needs of physics. And there have always been young mathematicians willing to aid in the process who found rich reward in so doing.

The theory of orthogonal expansions had its origin in the debate concerning the vibrating string which animated the mathematical world two hundred years ago. The theory has had a place in the sun ever since, though naturally it meant different things to different times. How much it meant to the mathematicians of the eighteenth and nineteenth centuries can be read off from the monumental, 1800-page report on oscillating functions published by H. Burkhardt in 1908 covering the period 1727–1890. This was before the days of Fejér, Hardy, Hilbert, Lebesgue, Plancherel, F. and M. Riesz, Weyl, and Wiener, to mention only a few of the men whose work completely transformed the theory during the first third of this century. The new quantum mechanics found a rich storehouse to draw from; we mathematicians have been amply paid for the borrowings.

What is nowadays called the classical theory of orthogonal series is ably presented in G. Sansone's treatise which here appears in careful English translation. Since the term "classical" is often used in a derogatory sense by our vigorous youngsters—if lucky, their work will also become "classical"—I hasten to add that the term is here used as praise. The treatise contains what the student needs to know concerning the field, set in proper perspective, rigorously presented with due attention to detail and appropriate technique but nevertheless easy to read. It contains a wealth of information, factual and bibliographical, of use to the working analyst. If the reader should find the book short on maximal ideals, group characters, and other adjuncts of "modern" Fourier analysis, let that be a challenge to him to write a book in which these concepts are placed in the foreground. Such books are also needed.

American mathematics is deeply indebted to Professor Diamond for making this valuable text available to our students. Personally I am glad that my praise of the original had such beneficial effects. I can only hope that the translation will enjoy the same popularity in this country as the original has had and has in Italy.

EINAR HILLE

Preface

G. Vitali's monograph *Moderna Teoria delle Teoria delle Funzioni di Variabili Reale* finds here its natural sequel. The work should prove especially valuable to the applied mathematician who very often does not have access to the original papers. Its purpose is therefore to present general results and convenient criteria concerning Fourier series, Legendre series, Laguerre and Hermite polynomials.

Care has been taken to keep the necessary connections between this work and Vitali's monograph and it is hoped that the reader will be inspired to look further into a number of interesting questions concerning which only the essential points have been covered here.

The first chapter "Expansion in Series of Orthogonal Functions and Preliminary Notions on Hilbert Space" based on G. Vitali's *Geometria nello Spazio Hilbertiano,* contains the most pertinent results on expansion in series of orthogonal functions and includes some theorems on functions summable L^p and on convergence in the mean of order p.

The second chapter, "Expansion in Fourier Series" includes a treatment of the Gibbs phenomenon and a detailed discussion of the classical problem of Fourier on the distribution of heat in a plane. Also the validity of the theorems of Fejér and Lebesgue for (C, k) summability of Fourier series is extended to values of $k > 0$.

In the third chapter, "Expansion in Series of Legendre Polynomials," the representation of Legendre polynomials by the classical formulas of Mehler is established; the elegant expansion in series of Stieltjes-Neumann of $(1-x)^w$ is given prominence; and Bruns' inequalities for the zeros of $P_n(x)$ are established by the method of Szegö. In view of the great importance in applied mathematics of the expansion of functions in series of spherical harmonics in two variables, this chapter has been further amplified by a brief study of the spherical harmonics of Laplace and theorems on (C, k) and Poisson summability of such series.

Chapter four contains sufficient material to provide an introduction to the theory of representation of functions by series of Laguerre and Hermite polynomials. The reader who wishes to acquire a more complete knowledge of the theory is referred to G. Szegö's recent treatise *Orthogonal Polynomials* [107a] and the extensive bibliography of J. H. Shohat, E. Hille, and J. L. Walsh in their *Bibliography on Orthogonal Polynomials* [100b].

Formulas for the asymptotic approximation of Hermite polynomials are obtained by the method of Liouville-Stekloff and are used to obtain bounds on the orthogonal functions for complex arguments. The sixth article give a more precise form of Uspensky's formulas for the asymptotic approximation of the Laguerre polynomials.

GIOVANNI SANSONE

Translator's Note

In 1952 the third edition of Giovanni Sansone's treatise on Series Expansions of Orthogonal Functions appeared as Part II of *Moderna Teoria delle Funzioni di Variable Reale* in the series of Monografie di Matematica Applicata published by the Consiglio Nazionale delle Ricerche. Part I of this treatise was written by G. Vitali.

The enthusiastic reviews by Einar Hille of Sansone's work and the fact that much of the material, especially the chapters on series expansions in terms of Legendre polynomials and Laguerre and Hermite polynomials, was not readily available in English, suggested the desirability of translating the book.

The present volume comprises the first four chapters of Sansone's *Series Expansions of Orthogonal Functions*. The remaining two chapters, "Approximation and Interpolation" and "The Stieltjes Integral," which have no essential connection with the first four chapters, were omitted. Chapter II has been extended to include a section written by Sansone on the Fourier transform.

For the most part the translation is literal. The only essential departure from this procedure is in Chapter I, Sec. 5, "Convergence in the Mean," where the brief translator's note on the concept of internal convergence (*convergenza completa in media*) was added to the text.

In view of the numerous references in the Italian text to theorems in Part I of *Moderna teoria delle funzioni di variable reale*, it was decided to add an appendix listing those definitions and theorems from Part I to which references are made in the translation.

I am especially grateful to Professor Sansone for his encouragement and assistance in correcting the manuscript and selecting the material for the appendix. I also wish to acknowledge the contribution of the administration of Stevens Institute of Technology in making available the time and secretarial assistance necessary to carry out the translation; the personal assistance of Frank Babina, one of my students at Stevens; the patient efforts of the secretary, Katherine Melis; and, finally, the encouragement and cooperation of Interscience Publishers, without which the present volume would not be possible.

AINSLEY H. DIAMOND

November, 1958.

Contents

CHAPTER I

Expansion in Series of Orthogonal Functions and Preliminary Notions of Hilbert Space

1. Square Integrable Functions

1. DEFINITION 1. A function will be said to have a property *almost everywhere* in a set g if it has that property for all points of g except for a set of measure zero. In future the abbreviation "a.e." will denote "almost everywhere."

DEFINITION 2. If g is a measurable set of points of a straight line, $f(t)$ is a function defined a.e. in g and $f^2(t)$ is integrable, then $f(t)$ is called *square integrable* in g.

THEOREM 1. If f_1 and f_2 are two functions measurable and square integrable in a set g, then the product $f_1 f_2$ is also integrable in g.

Proof. It is sufficient to show that $|f_1 f_2|$ is integrable in g. Clearly, $|f_1 f_2| \leqq \frac{1}{2}f_1^2 + \frac{1}{2}f_2^2$ a.e. in g and therefore $|f_1 f_2|$ admits a majorant which is integrable.

Henceforth, we shall tacitly assume that the functions under consideration are measurable.

COROLLARY. If f is a function square integrable in a set g of finite measure, then f is integrable in g.

Proof. In fact the constant 1 is square integrable in g, precisely because g has finite measure, and therefore the product $1 \cdot f$ is integrable in g.

THEOREM 2. If $f_1, f_2, \ldots, f_n$ are square integrable functions in g, and $c_1, c_2, \ldots, c_n$ are constants, then $c_1 f_1 + c_2 f_2 + \cdots + c_n f_n$ is a square integrable function in g.

Proof. $f_1, f_2, \ldots, f_n$ are finite a.e. in g, and since

$$(c_1 f_1 + c_2 f_2 + \ldots + c_n f_n)^2 = \sum_{i,k=1}^{n} c_i c_k f_i f_k$$

and the second member is integrable, it follows that the first member is integrable.

2. Linearly Independent Functions

1. DEFINITION 3. n functions $f_1, f_2, \ldots, f_n$ are called *linearly dependent* in g if there exist n constants $c_1, c_2, \ldots, c_n$, not all zero, for which the function $c_1 f_1 + c_2 f_2 + \ldots + c_n f_n$ is zero a.e. If such constants do not exist, the n functions are called *linearly independent.*

In a set g of measure zero, n functions are always linearly dependent. Accordingly, the discussion which follows is limited to sets of non-zero measure.

2. THEOREM 3. A necessary and sufficient condition for n functions

$$f_1, f_2, \ldots, f_n \tag{1}$$

square integrable in g, to be linearly dependent in g is the vanishing of Gram's determinant

$$G(f_1, f_2, \ldots, f_n) = \begin{vmatrix} \int_g f_1^2\,dt, & \int_g f_1 f_2\,dt, \ldots, & \int_g f_1 f_n\,dt \\ \int_g f_2 f_1\,dt, & \int_g f_2^2\,dt, \ldots, & \int_g f_2 f_n\,dt \\ \cdots & \cdots & \cdots \\ \int_g f_n f_1\,dt, & \int_g f_n f_2\,dt, \ldots, & \int_g f_n^2\,dt \end{vmatrix}$$

(Gram [42]; Kowaleski [59]).

Proof. The condition is sufficient. In fact, if $G(f_1, f_2, \ldots, f_n) = 0$ it is possible to find n constants $\lambda_1, \lambda_2, \ldots, \lambda_n$ not all zero for which

$$\sum_{s=1}^{n} \lambda_s c_{r,s} = 0, \qquad (r = 1, 2, \ldots, n)$$

where

$$c_{r,s} = \int_g f_r f_s dt.$$

If F is defined by $F = \sum_{s=1}^n \lambda_s f_s$, then $\int_g F f_r dt = \sum_{s=1}^n \lambda_s c_{r,s} = 0$, whence

$$0 = \sum_{r=1}^n \lambda_r \int_g F f_r dt = \int_g \sum_{r=1}^n \lambda_r f_r (\sum_{s=1}^n \lambda_s f_s) dt = \int_g F^2 dt.$$

Now, since $\mu(g) \neq 0$ (where $\mu(g)$ denotes the measure of g), then F^2 and therefore F, is zero a.e. in g.

Conversely, if the functions (1) are linearly dependent in g, there will exist a system of constants $\lambda_1, \lambda_2, \ldots, \lambda_n$ not all zero such that

$$\sum_{s=1}^n \lambda_s f_s = 0$$

a.e. in g. From this follows

$$0 = \int_g f_r (\sum_{s=1}^n \lambda_s f_s) dt = \sum_{s=1}^n \lambda_s \int_g f_r f_s dt = \sum_{s=1}^n \lambda_s c_{r,s} \quad (r = 1, 2, \ldots, n)$$

and therefore $G(f_1, f_2, \ldots, f_n) = 0$.

THEOREM 4. The rank of the matrix corresponding to the determinant $G(f_1, f_2, \ldots, f_n)$ gives the maximum number of linearly independent functions $f_1, f_2, \ldots, f_n$. If the rank is r, then r of the functions are linearly independent, and the other $n - r$ functions are linearly dependent on these.

Proof. Let r be the rank of the matrix correponding to the determinant G. Since G is symmetric it contains a non-vanishing principal minor of order r. Without loss of generality, we may suppose that this minor is formed from the first r rows and the first r columns of G. It follows that $G(f_1, f_2, \ldots, f_r) \neq 0$. Therefore the functions $f_1, f_2, \ldots, f_r$ are linearly independent.

From the fact that $G(f_1, f_2, \ldots, f_r, f_{r+j}) = 0, j = 1, 2, \ldots, n-r$, it follows from theorem 3 that every function f_{r+j} is linearly dependent on $f_1, f_2, \ldots, f_r$. In particular, we have

$$\lambda_1 f_1 + \lambda_2 f_2 + \cdots + \lambda_r f_r + \lambda_{r+j} f_{r+j} = 0$$

where $\lambda_{r+j} \neq 0$, and therefore f_{r+j} is a linear, homogenous combination of $f_1, f_2, \ldots, f_r$.

THEOREM 5. If the functions $f_1, f_2, \ldots, f_n$, square integrable in g, are linearly independent, then $G(f_1, f_2, \ldots, f_n) > 0$.

Proof. For n constants $c_1, c_2, \ldots, c_n$, not all zero $(c_1 f_1 + c_2 f_2 + \cdots + c_n f_n)^2 > 0$, a.e. in g. Therefore $\int_g (c_1 f_1 + c_2 f_2 + \cdots + c_n f_n)^2 dt > 0$, and finally the quadratic form in the $c_1, c_2, \ldots, c_n$

$$\sum_{i,k=1}^{n} c_i c_k \int_g f_i f_k \, dt$$

is positive definite, and its discriminant $G(f_1, f_2, \ldots, f_n) > 0$.

COROLLARY 1. If $f_1, f_2, \ldots, f_n$ are square integrable in g, then

$$G(f_1, f_2, \ldots, f_n) \geqq 0$$

and the equality sign holds only in case the given functions are linearly dependent in g.

COROLLARY 2. If f_1 and f_2 are two square integrable functions then

$$G(f_1, f_2) = \begin{vmatrix} \int_g f_1^2 dt, & \int_g f_1 f_2 dt \\ \int_g f_1 f_2 dt, & \int_g f_2^2 dt \end{vmatrix} = \int_g f_1^2 dt \int_g f_2^2 dt - \left(\int_g f_1 f_2 dt\right)^2 \geqq 0$$

and therefore

$$\left(\int_g f_1 f_2 dt\right)^2 \leqq \int_g f_1^2 dt \int_g f_2^2 dt.$$

The equality sign holds only in case f_1 and f_2 are linearly dependent in g. This inequality, usually referred to as the Schwarz inequality, was discovered independently by Bunikowsky in 1861 and Schwarz in 1885. The Schwarz inequality will arise as a particular case of the inequality which we shall establish in Sec. 9 of this chapter.

COROLLARY 3. If a function f defined in a set g of finite measure is square integrable, then

$$\left(\int_g f\,dt\right)^2 \leqq \mu(g) \int_g f^2\,dt.$$

Proof. Put $f_1 = f$, $f_2 = 1$ in the Schwarz inequality.

3. Elementary Notions of Hilbert Space

1. DEFINITION 4. We shall call a *real Hilbert Space*, or simply a *Hilbert Space*, or Space H, the set of all the functions which are square integrable in g. Two functions will be considered as the same element of H if they are equal almost everywhere in g. Any function whatever of the given set will be called a *point* of H; the function which vanishes almost everywhere will be called the *origin* of H.

Two points f_1 and f_2 are two distinct points of H if f_1 and f_2 are not equal almost everywhere.

2. DEFINITION 5. If f_1 and f_2 are two points of H, we shall call the *distance* between these two points the positive square root of $\int_g (f_1 - f_2)^2\,dt$; consequently if d is the distance between two points f_1 and f_2, then $d \geqq 0$ and

$$d^2 = \int_g (f_1 - f_2)^2\,dt.$$

The distance from a point f to the origin of H is given by $\sqrt{\int_g f^2\,dt}$.

3. DEFINITION 6. If f and φ are two distinct points of H and φ is distinct from the origin, we shall call the *straight line* passing through the points f and φ the totality of the points $f + \lambda\varphi$ where λ varies from $-\infty$ to ∞.

There is clearly a one-to-one correspondence between the points of the line and the values of λ.

Consider a line $\lambda\varphi$ passing through the origin; the points of this line which are a unit distance from the origin correspond to the values of λ for which $\lambda^2 \int_g \varphi^2\,dt = 1$; consequently, there exist on the line $\lambda\varphi$ two and only two points $\varphi/\sqrt{\int_g \varphi^2\,dt}$, and $-\varphi/\sqrt{\int_g \varphi^2\,dt}$ which are a unit distance from the origin. These two points will be called normal parameters of the line $\lambda\varphi$.

DEFINITION 7. A function φ square integrable in g will be called *normal* in g if

$$\int_g \varphi^2\,dt = 1.$$

The normal functions in g consist of all the points, and only those points, of H which are a unit distance from the origin.

To *normalize* a function φ square integrable and not vanishing a.e. in g, means to determine a factor c for which $c\varphi$ is normal in g, or, in other words, to find on the line $\lambda\varphi$ the points which are a unit distance from the origin. The required values of c are given by

$$c = 1/\sqrt{\int_g \varphi^2\,dt}, \qquad c = -1/\sqrt{\int_g \varphi^2\,dt}.$$

4. Let λf, $\mu\varphi$ be two lines through the origin; from the Schwarz inequality

$$\left(\int_g f\varphi\,dt\right)^2 \leqq \int_g f^2\,dt \int_g \varphi^2 dt$$

follows the existence of a number ω between 0 and π such that

$$\cos\omega = \left(\int_g f\varphi\,dt\right)/\sqrt{\int_g f^2\,dt\int_g \varphi^2\,dt}. \tag{1}$$

DEFINITION 8. The number ω between 0 and π which satisfies (1) will be called the *angle between the two lines* λf, $\mu\varphi$ *in* H.

In particular $\int_g f\varphi\,dt = 0$ is a necessary and sufficient condition for $\omega = \pi/2$.

We shall say in this case that the two lines λf, $\mu\varphi$ (or the two functions f and φ) are *orthogonal* in H (in g).

A necessary and sufficient condition for $\omega = 0$, π is

$$\left(\int_g f\varphi\,dt\right)^2 = \int_g f^2\,dt \int_g \varphi^2\,dt,$$

which implies that the functions f and φ are linearly dependent (cf. Sec. 2) and the two lines λf, $\mu\varphi$ coincide.

It is easy to show that the functional metric defined in Sec. 3.2 has the properties of distance in ordinary space. For example, if d_1 and d_2 are the distances from the two points f_1

and f_2 to the origin, if ω is the angle between the two lines λf_1, λf_2, and if d denotes the distance between the points f_1 and f_2 then

$$d^2 = d_1^2 + d_2^2 - 2d_1 d_2 \cos \omega.$$

This follows at once from

$$d^2 = \int_g (f_1 - f_2)^2 dt = \int_g f_1^2 dt + \int_g f_2^2 dt - 2\sqrt{\int_g f_1^2 dt \int_g f_2^2 dt}\, \frac{\int_g f_1 f_2 dt}{\sqrt{\int_g f_1^2 dt \int_g f_2^2 dt}}.$$

5. We shall now give the geometric interpretation of Gram's determinant.

THEOREM 6. Let $f_1, f_2, \ldots, f_n$, be n functions square integrable in g and let $\varphi_1, \varphi_2, \ldots, \varphi_n$ be n associated functions defined by the following equations

$$\varphi_i = \sum_{k=1}^{n} b_{i,k} f_k \qquad (i = 1, 2, \ldots, n)$$

where $b_{i,k}$ are constants with $B = \text{det. } ||b_{i,k}|| \neq 0$. We wish to show that

$$G(\varphi_1, \varphi_2, \ldots, \varphi_n) = B^2 G(f_1, f_2, \ldots, f_n).$$

Proof. We first note the following relations

$$c'_{i,l} = \int_g \varphi_i \varphi_l dt = \int_g \sum_{k=1}^{n} b_{i,k} f_k \sum_{m=1}^{n} b_{l,m} f_m dt = \sum_{k,m=1}^{n} b_{i,k} b_{l,m} \int_g f_k f_m dt$$

$$= \sum_{k,m=1}^{n} b_{i,k} b_{l,m} c_{k,m} = \sum_{k=1}^{n} \left(\sum_{m=1}^{n} b_{l,m} c_{k,m}\right) b_{i,k},$$

where, following the notation of Sec. 2 we write $c_{k,m} = \int_g f_k f_m dt$.

If we now define $b'_{l,k} = \sum_{m=1}^{n} b_{l,m} c_{k,m}$ we have $c'_{i,l} = \sum_{k=1}^{n} b'_{l,k} b_{i,k}$ and, finally,

$$BG(f_1, f_2, \ldots, f_n) = \text{det. } ||b_{l,m}|| \text{ det. } ||c_{k,m}|| = \text{det. } ||b'_{l,k}||,$$

$$B^2 G(f_1, f_2, \ldots, f_n) = \text{det.} ||b_{i,k}|| \text{det.} ||b'_{l,k}|| = \text{det.} ||c'_{i,l}|| = G(\varphi_1, \varphi_2, \ldots, \varphi_n).$$

THEOREM 7. If $f_1, f_2, \ldots, f_{n-1}, f_n$ are square integrable and linearly independent in g, there exists a linear combination of them which is orthogonal to $f_1, f_2, \ldots, f_{n-1}$. This is determined except for sign by imposing the condition that it be normal in g.

Proof. If we write

$$\varphi_n = \gamma_1 f_1 + \gamma_2 f_2 + \cdots + \gamma_n f_n \tag{1}$$

and $c_{i,k} = \int_g f_i f_k \, dt$, the orthogonality of φ_n to $f_1, f_2, \ldots, f_n$ implies

$$c_{i,1}\gamma_1 + c_{i,2}\gamma_2 + \cdots + c_{i,n}\gamma_n = 0, \quad (i = 1, 2, \ldots, n-1),$$

and, since in the determinant $||c_{i,k}||$, $(i, k = 1, \ldots, n)$, the minor

$$\begin{vmatrix} c_{1,1}, & \ldots, & c_{1,n-1} \\ \ldots & \ldots & \ldots \\ c_{n-1,1}, & \ldots, & c_{n-1,n-1} \end{vmatrix} = G(f_1, f_2, \ldots, f_{n-1}) \tag{2}$$

is different from zero, the $\gamma_1, \gamma_2, \ldots, \gamma_n$ are determined except for a factor of proportionality. Since the corresponding φ_n is not zero a.e. in g, this factor can be chosen so that φ_n is normalized, (cf. Sec. 3.3).

It follows, moreover, from (2) that the constant γ_n occurring in (1) is different from zero.

THEOREM 8. Let $f_1, f_2, \ldots, f_n$, be n functions linearly independent and square integrable in g; let ϱ_i be the distance of f_i from the origin; φ_3 the linear combination of f_1, f_2, f_3 normal and orthogonal to f_1, f_2; φ_4 the linear combination of f_1, f_2, f_3, f_4 normal and orthogonal to $f_1, f_2, f_3; \ldots; \varphi_n$ the linear combination of $f_1, f_2, \ldots, f_n$ normal and orthogonal to $f_1, f_2, \ldots, f_{n-1}$.

We now establish the following relation

$$\begin{aligned} G(f_1, f_2, \ldots, f_n) \\ = \varrho_1^2 \varrho_2^2 \cdots \varrho_n^2 \sin^2 (f_1, f_2) \cos^2 (f_3, \varphi_3) \ldots \cos^2 (f_n, \varphi_n). \end{aligned} \tag{3}$$

This formula furnishes the basis for the geometric interpretation of Gram's determinant.

Proof. From (1), since $\gamma_n \neq 0$, we have

$$f_n = \alpha_1 f_1 + \alpha_2 f_2 + \ldots + \alpha_{n-1} f_{n-1} + \alpha_n \varphi_n$$

where $\alpha_1, \alpha_2, \ldots, \alpha_{n-1}, \alpha_n$ are constants. Multiplying by φ_n and integrating over g, we have

$$\alpha_n = \int_g f_n \varphi_n \, dt = \sqrt{\int_g f_n^2 \, dt \int_g \varphi_n^2 \, dt} \frac{\int_g f_n \varphi_n \, dt}{\sqrt{\int_g f_n^2 \, dt \int_g \varphi_n^2 \, dt}} = \varrho_n \cos (f_n, \varphi_n).$$

Now, $f_1, f_2, \ldots, f_{n-1}, f_n$ are obtained from $f_1, f_2, \ldots, f_{n-1}$, by the linear transformation

$$f_1 = f_1, \; f_2 = f_2, \ldots, f_{n-1} = f_{n-1},$$
$$f_n = \alpha_1 f_1 + \alpha_2 f_2 + \cdots + \alpha_{n-1} f_{n-1} + \alpha_n \varphi_n$$

where the determinant of the transformation is $\alpha_n = \varrho_n \cos(f_n, \varphi_n)$. Therefore it follows at once from Theorem 6 that

$$G(f_1, f_2, \ldots, f_{n-1}, f_n) = \varrho_n^2 \cos^2 (f_n, \varphi_n) \, G(f_1, f_2, \ldots, f_{n-1}, \varphi_n)$$

and, since φ_n is normal and orthogonal to $f_1, f_2, \ldots, f_{n-1}$, we have

$$G(f_1, f_2, \ldots, f_{n-1}, \varphi_n) = G(f_1, f_2, \ldots, f_{n-1})$$

and, consequently, the recurrence relation

$$G(f_1, f_2, \ldots, f_{n-1}, f_n) = \varrho_n^2 \cos^2 (f_n, \varphi_n) G(f_1, f_2, \ldots, f_{n-1}).$$

This implies

$$G(f_1, f_2, \ldots, f_n) = \varrho_n^2 \varrho_{n-1}^2, \ldots \varrho_3^2 \cos^2 (f_n, \varphi_n) \ldots \cos^2 (f_3, \varphi_3) G(f_1, f_2)$$

from which (3) follows immediately with the help of the relation

$$G(f_1, f_2) = \int_g f_1^2 dt \int_g f_2^2 dt - \left(\int_g f_1 f_2 \, dt\right)^2 = \left[1 - \left(\frac{\int_g f_1 f_2 \, dt}{\sqrt{\int_g f_1^2 dt \int_g f_2^2 dt}}\right)^2\right] \varrho_1^2 \varrho_2^2$$

$$= [1 - \cos^2 (f_1, f_2)] \, \varrho_1^2 \varrho_2^2 = \varrho_1^2 \varrho_2^2 \sin^2 (f_1, f_2).$$

From (3) it follows that if $f_1, f_2, \ldots, f_n$ are normal, then

$$G(f_1, f_2, \ldots, f_n) \leqq 1.$$

In particular, $G = 1$ if

$$\sin^2 (f_1, f_2) = 1, \quad \cos^2 (f_3, \varphi_3) = 1, \ldots, \quad \cos^2 (f_n, \varphi_n) = 1.$$

From the first of these relations it follows that f_1 and f_2 are orthogonal in g.

From $\cos^2 (f_3, \varphi_3) = 1$ follows $(\int_g f_3 \varphi_3 \, dt)^2 = \int_g f_3^2 dt \int_g \varphi_3^2 dt$ and therefore (Sec. 3.2), $f_3 = k\varphi_3$ where k is a constant. But since φ_3 is orthogonal to f_1 and f_2, it follows that f_3 is orthogonal to f_1 and f_2. Continuing in this way we obtain the following theorem:

THEOREM 9. Gram's determinant for n normalized functions is less than or equal to 1, and, if it is equal to 1, the functions are mutually orthogonal.

4. Linear Approximations to Functions

1. DEFINITION 9. A system of functions which are both mutually orthogonal and normal is called an *orthonormal system* and the functions are called *orthonormal.*

THEOREM 10. Given n functions $f_1, f_2, \ldots, f_n$ square integrable and linearly independent in g, there exist n linear combinations of them which are orthonormal.

Proof. By Theorem 7 there exist n normal functions $\varphi_1, \varphi_2, \ldots, \varphi_n$ such that

$$\varphi_i = c_{i,1} f_1 + c_{i,2} f_2 + \cdots + c_{i,i} f_i, \qquad i = 1, 2, \ldots, n; \;\; c_{i,i} \neq 0;$$

$$\int_g \varphi_i f_s \, dt = 0, \;\; s = 1, 2, \ldots, i-1; \;\; \int_g \varphi_i^2 \, dt = 1, \quad i = 1, 2, \ldots, n.$$

The functions $\varphi_1, \varphi_2, \ldots, \varphi_n$ are mutually orthogonal; in fact, for $i < l$

$$\int_g \varphi_i \varphi_l \, dt = \int_g \varphi_l \Big(\sum_{r=1}^{i} c_{i,r} f_r\Big) \, dt = \sum_{r=1}^{i} c_{i,r} \int_g \varphi_l f_r \, dt = 0.$$

Moreover, since $G(\varphi_1, \varphi_2, \ldots, \varphi_n) = 1$, $\varphi_1, \varphi_2, \ldots, \varphi_n$ are linearly independent.

2. We now consider the problem of determining a linear combination $c_1 f_1 + c_2 f_2 + \cdots + c_n f_n$ of n functions $f_1, f_2, \ldots, f_n$, square integrable in g, which shall be a minimum distance from another function F also square integrable in g. Such a linear combination is called *the best approximation in the mean to F with respect to linear combinations of $f_1, f_2, \ldots, f_n$.*

By the preceding theorem, we may suppose that the system $f_1, f_2, \ldots, f_n$ is orthonormal.

Our problem consists, then, in determining the constants $c_1, c_2, \ldots, c_n$ so that

$$(1)\qquad \int_g (F - c_1 f_1 - c_2 f_2 - \dots - c_n f_n)^2 dt$$

shall assume its minimum value.

The integral (1) can be written

$$\int_g F^2 dt + \sum_{i=1}^{n} c_i^2 - 2\sum_{i=1}^{n} c_i \int_g F f_i dt$$

and equating to zero the partial derivatives with respect to $c_1, c_2, \dots, c_n$ we find

$$(2)\qquad c_i = \int_g F f_i dt, \qquad (i = 1, 2, \dots, n).$$

We now prove conversely that the integral (1) assumes its minimum value for constants c_i given by (2). In fact, we have

$$\int_g (F - c_1 f_1 - c_2 f_2 - \dots - c_n f_n)^2 dt$$
$$= \int_g F^2 dt + \sum_{i=1}^{n} \left(c_i - \int_g F f_i dt\right)^2 - \sum_{i=1}^{n} \left(\int_g F f_i dt\right)^2.$$

The only variable term of the second member is $\sum_{i=1}^{n} (c_i - \int_g F f_i dt)^2$ which assumes the value zero for c_i given by (2).

DEFINITION 10. The constants c_i determined by (2) are called the *Fourier coefficients* of F with respect to $f_1, f_2, \dots, f_n$.

THEOREM 11. If F is square integrable in g, and $f_1, f_2, \dots, f_n$ are orthonormal in g, the linear combination $\sum_{i=1}^{n} c_i f_i$, in which the c_i are the Fourier coefficients of F, is the linear combination of $f_1, f_2, \dots, f_n$ which is the best approximation in the mean to F of all possible linear combinations of $f_1, f_2, \dots, f_n$.

DEFINITION 11. If $f_1, f_2, \dots, f_n$; F, are square integrable in g, and $f_1, f_2, \dots, f_n$ are linearly independent in g, the totality of points $F + \lambda_1 f_1 + \lambda_2 f_2 + \dots + \lambda_n f_n$ corresponding to all possible values of λ_i is called a *linear space* or a *Euclidean space of n dimensions*, or simply a S_n.

For $n = 1$, S_1 is called a line (Sec. 3.3); for $n = 2$, S_2 is called a plane.

The problem we have solved can now be stated in geometric

terms: given a linear space, to find which of its points is a minimum distance from the origin.

3. THEOREM 12. If $\{f_n\}$ is a sequence of functions orthonormal in g, F is square integrable in g, $\{c_n\}$ is the sequence of Fourier coefficients with respect to the sequence $\{f_n\}$

$$c_n = \int_g F f_n \, dt, \qquad (n = 1, 2, \ldots)$$

then the series

$$\sum_{n=1}^{\infty} c_n^2$$

is convergent, and, moreover, the c_i satisfy Bessel's inequality:

$$\sum_{n=1}^{\infty} c_n^2 \leqq \int_g F^2 \, dt. \tag{3}$$

Proof. We observe first the relations

$$\int_g \Big[F - \sum_{n=1}^{N} c_n f_n\Big]^2 dt = \int_g F^2 dt + \sum_{n=1}^{N} c_n^2 \int_g f_n^2 \, dt - 2\sum_{n=1}^{N} c_n \int_g F f_n \, dt \tag{4}$$

$$+ 2\sum_{n \neq m}^{N} c_n c_m \int_g f_n f_m \, dt = \int_g F^2 \, dt + \sum_{n=1}^{N} c_n^2 - 2\sum_{n=1}^{N} c_n^2 = \int_g F^2 dt - \sum_{n=1}^{N} c_n^2,$$

and since $[F - \sum_{n=1}^{N} c_n f_n]^2 \geqq 0$, a.e. in g, it follows for every positive integer N, that

$$\sum_{n=1}^{N} c_n^2 \leqq \int_g F^2 \, dt$$

and consequently (3) is established.

5. Convergence in the Mean

1. DEFINITION 12. If g is a measurable set of points of a line, $p > 0$, and $|f(t)|^p$ is integrable in g, then $f(t)$ is called *integrable* L^p *in* g and we write $f(t) \in L^p$ in g.

DEFINITION 13. If $\{f_n\}$ is a sequence of functions integrable L^p in a measurable set g, and $f(t)$ is integrable L^p in g, then $\{f_n\}$ is said *to converge in the mean of order* p *in* g *to* $f(t)$, or *to* *converge strongly L^p in g to $f(t)$,* if

$$\lim_{n\to\infty} \int_g |f(t) - f_n(t)|^p \, dt = 0.$$

If $p = 2$, we shall say merely that $\{f_n(t)\}$ *converges in the mean to* $f(t)$.

Assume $\{f_n\}$ is a sequence of functions orthonormal in g, F is square integrable in g, $\{c_n\}$ is the sequence of Fourier coefficients of F with respect to the sequence $\{f_n\}$ and, finally, that the sign of equality occurring in Bessel's inequality holds, namely

$$\sum_{n=1}^{\infty} c_n^2 = \int_g F^2 \, dt, \tag{1}$$

$$\left[\int_g f_n f_m \, dt = 0, \quad n \neq m; \quad \int_g f_n^2 \, dt = 1; \quad c_n = \int_g F f_n \, dt\right].$$

The first relation in (1) may be written

$$\lim_{N\to\infty} \sum_{n=1}^{N} c_n^2 = \int_g F^2 \, dt.$$

From (4) of the preceding Sec. 4.3 follows

$$\lim_{N\to\infty} \int_g \left[F - \sum_{n=1}^{N} c_n f_n\right]^2 dt = 0$$

whence, by writing

$$\varphi_N = \sum_{n=1}^{N} c_n f_n \tag{2}$$

we have

$$\lim_{N\to\infty} \int_g [F - \varphi_N]^2 \, dt = 0. \tag{3}$$

Conversely from (3), (1) follows by virtue of (4) of the preceding Sec. 4.3. Consequently, we have

THEOREM 13. A necessary and sufficient condition for the equality sign occurring in Bessel's inequality to hold is that the sequence φ_N defined in (2) converges in the mean in g to the function F.

2. To deal with the case in which g may be measurable but of infinite measure we introduce the concept of internal convergence (*convergenza completa*) in the mean of order p by the following definition:

DEFINITION 14. If $\{f_n(t)\}$ and $f(t)$ are defined a.e. in a set g of finite or infinite measure, and are integrable L^p in every subset of g of finite measure, then $\{f_n(t)\}$ is said to *converge internally in the mean of order p in g to $f(t)$* if, for every subset γ of g of finite measure $\mu(\gamma)$,

$$\lim_{n\to\infty} \int_\gamma |f_n - f|^p \, dt = 0 \tag{4}$$

i.e., if $\{f_n(t)\}$ converges strongly L^p to $f(t)$ in every subset γ of g of finite measure. If $p = 2$ we shall say merely that $f_n(t)$ *converges internally in the mean in g to $f(t)$*, and if $p = 1$ we shall say merely that $\{f_n(t)\}$ *converges internally in g to $f(t)$*.

[TRANSLATOR'S REMARK: A consideration of the sequence $\{f_n(x)\}$ where $f_n = 0$ for $-n \leqq x \leqq n$ and $f_n = 1$ for $x > n$ and $x < -n$ will illustrate the difference between the concepts of internal convergence in the mean of order p and convergence in the mean of order p for the interval $g = (-\infty, \infty)$. For, the sequence in question converges internally of order $p > 1$ in $g = (-\infty, \infty)$ since it converges in the mean of order $p > 1$ in every finite subset γ of g, but it is not convergent in the mean of order $p > 1$ in $g = (-\infty, \infty)$.]

THEOREM 14. A sequence $\{\varphi_n\}$ internally convergent in the mean to a function φ is internally convergent to φ.

Proof. If we write $C = \sqrt{\mu(\gamma)}$, then by the Schwarz inequality we have

$$0 \leqq \int_\gamma |\varphi_n - \varphi| \, dt \leqq C \sqrt{\int_\gamma (\varphi_n - \varphi)^2 \, dt} \quad \text{[Sec. 2.1, Cor. 3]}$$

and from (4) by taking $p = 2$, we have

$$\lim_{n\to\infty} \int_\gamma |\varphi_n - \varphi| \, dt = 0 \qquad \text{(Sec. 2.1, Cor. 3)}$$

and the sequence $\{\varphi_n\}$ converges internally to φ.

It follows from this that if the sequence $\{\varphi_n\}$ converges internally in the mean in g to two functions φ and ψ, then these functions are equal a.e. in g. (Appendix, Th. 7).

3. THEOREM 15. Given a sequence $\{\varphi_n\}$ of functions defined a.e. in a measurable set g, and square integrable in every measurable subset of g of finite measure, a necessary and sufficient condition that they should be internally convergent in the mean in g is that for every measurable subset γ of finite measure of g,

$$\lim_{\substack{n\to\infty \\ m\to\infty}} \int_\gamma |\varphi_n - \varphi_m|^2\, dt = 0. \tag{5}$$

Proof. The condition is necessary. If the sequence $\{\varphi_n\}$ is internally convergent in the mean to $\varphi(t)$ then

$$0 \leqq \int_\gamma |\varphi_n - \varphi_m|^2\, dt = \int_\gamma [(\varphi_n - \varphi) - (\varphi_m - \varphi)]^2\, dt$$

$$\leqq 2\int_\gamma |\varphi_n - \varphi|^2\, dt + 2\int_\gamma |\varphi_m - \varphi|^2\, dt$$

and by virtue of

$$\lim_{n\to\infty} \int_\gamma |\varphi_n - \varphi|^2\, dt = 0, \qquad \lim_{m\to\infty} \int_\gamma |\varphi_m - \varphi|^2\, dt = 0$$

(5) follows at once.

The condition is sufficient. In fact, if we write $C = \sqrt{\mu(\gamma)}$, then by the Schwarz inequality we have

$$0 \leqq \int_\gamma |\varphi_n - \varphi_m|\, dt \leqq C\sqrt{\int_\gamma |\varphi_n - \varphi_m|^2\, dt}$$

and from (5) follows

$$\lim_{\substack{n\to\infty \\ m\to\infty}} \int_\gamma |\varphi_m - \varphi_n|\, dt = 0,$$

and consequently the sequence $\{\varphi_n\}$ is internally convergent in g. (Appendix, Th. 8).

If $\{\varphi_n\}$ converges internally to φ, then there exists a subsequence which is convergent a.e. in g to φ. Consequently there exists a subsequence of the sequence $\{\varphi_n^2\}$ which is convergent a.e. in g to φ^2, say,

$$\varphi_{s_1}^2, \quad \varphi_{s_2}^2, \ldots, \quad \varphi_{s_n}^2, \ldots .$$

We now prove that

$$\lim_{n\to\infty} \int_\gamma \varphi_{s_n}^2 \, dt = \int_\gamma \varphi^2 \, dt. \tag{6}$$

For this purpose it will suffice to show that the integrals of the φ_n are equi-absolutely continuous in γ. (Appendix, Def. 13 and Th. 8).

For any positive ε there exists n_0 such that for every integer $p \geqq 0$ we have

$$\int_\gamma |\varphi_{n_0+p} - \varphi_{n_0}|^2 \, dt < \frac{\varepsilon}{4}$$

and consequently for every measurable subset γ' contained in γ

$$\int_{\gamma'} |\varphi_{n_0+p} - \varphi_{n_0}|^2 \, dt < \frac{\varepsilon}{4}.$$

Now take $m > 0$ such that for $\mu(\gamma') < m$ we have

$$\int_{\gamma'} \varphi_{n_0}^2 \, dt < \frac{\varepsilon}{4}.$$

Then, for $\mu(\gamma') < m$ and for every p

$$\begin{aligned} \int_{\gamma'} \varphi_{n_0+p} \, dt &= \int_{\gamma'} [(\varphi_{n_0+p} - \varphi_{n_0}) + \varphi_{n_0}]^2 \, dt \\ &\leqq 2 \int_{\gamma'} |\varphi_{n_0+p} - \varphi_{n_0}|^2 \, dt + 2 \int_{\gamma'} \varphi_{n_0}^2 \, dt < \varepsilon. \end{aligned} \tag{7}$$

If now we take m sufficiently small, we may suppose that for $\mu(\gamma') < m$ we also have

$$\int_{\gamma'} \varphi_{n_0-p}^2 \, dt < \varepsilon, \qquad (p = 1, 2, \ldots, n_0 - 1) \tag{7'}$$

whence, from (7) and (7′), follows that the integrals of the sequence $\{\varphi_n^2\}$ are equi-absolutely continuous.

Finally, we show that

$$\lim_{n\to\infty} \int_\gamma |\varphi_n - \varphi|^2\, dt = 0. \tag{8}$$

For a fixed positive integer n, the same hypothesis we assume for $\{\varphi_n\}$ will also hold with respect to the sequence $\{\varphi_{n+k} - \varphi_n\}$ $k = 1, 2, \ldots$, and, therefore, there exists a subsequence

$$\varphi_{s_1} - \varphi_n,\ \varphi_{s_2} - \varphi_n, \ldots, \varphi_{s_k} - \varphi_n, \ldots$$

which converges internally to $\varphi - \varphi_n$ and such that

$$\lim_{k\to\infty} \int_\gamma |\varphi_{s_k} - \varphi_n|^2 dt = \int_\gamma |\varphi_n - \varphi|^2\, dt,$$

whence, passing to the limit as $n \to \infty$, and taking (5) into account, (8) follows immediately.

The theorem is thus proved.

4. THEOREM 16. If $\{\varphi_n\}$ is a sequence of functions and F is a function defined a.e. in g and square integrable in every measurable subset γ of g of finite measure, and if

$$\lim_{\substack{n\to\infty\\ m\to\infty}} \int_\gamma |\varphi_n - \varphi_m|^2\, dt = 0, \tag{9}$$

where $\{\varphi_n\}$ converges internally to φ, it follows that

$$\lim_{n\to\infty} \int_\gamma F\varphi_n\, dt = \int_\gamma F\varphi\, dt. \tag{10}$$

Proof. The Schwarz inequality implies

$$0 \leqq \int_\gamma |F\varphi_n - F\varphi_m|\, dt \leqq \sqrt{\int_\gamma F^2\, dt \int_\gamma |\varphi_n - \varphi_m|^2\, dt}$$

and consequently by (9)

$$\lim_{\substack{n\to\infty\\ m\to\infty}} \int_\gamma |F\varphi_n - F\varphi_m|\, dt = 0 \tag{11}$$

and, therefore, the sequence $\{F\varphi_n\}$ is internally convergent in g. On the other hand, since φ is the limit of a subsequence of $\{\varphi_n\}$

then $F\varphi$ is the limit of a subsequence of $\{F\varphi_n\}$ and by virtue of (11), $\{F\varphi_n\}$ converges internally to $F\varphi$. (Appendix, Th. 10). Therefore

$$\lim_{n\to\infty} \int_\gamma |F\varphi_n - F\varphi|\, dt = 0$$

and since

$$0 \leqq \left| \int_\gamma (F\varphi_n - F\varphi)\, dt \right| \leqq \int_\gamma |F\varphi_n - F\varphi| dt \quad \text{(Appendix, Th. 3)}$$

we have

$$\lim_{n\to\infty} \int_\gamma (F\varphi_n - F\varphi)\, dt = 0$$

which implies (10).

5. THEOREM 17. If $\{\varphi_n\}$, $\{\psi_n\}$ are two sequences internally convergent in the mean in g to two functions φ and ψ, respectively, then the sequence $\{\varphi_n + \psi_n\}$ converges internally in the mean in g to $\varphi + \psi$.

Proof. Clearly

$$0 \leqq \int_\gamma [(\varphi_n + \psi_n) - (\varphi + \psi)]^2\, dt$$

$$= \int_\gamma [(\varphi_n - \varphi) + (\psi_n - \psi)]^2 dt \leqq 2 \int_\gamma |\varphi_n - \varphi|^2 dt + 2 \int_\gamma |\psi_n - \psi|^2 dt$$

and by hypothesis

$$\lim_{n\to\infty} \int_\gamma |\varphi_n - \varphi|^2\, dt = 0, \quad \lim_{n\to\infty} \int_\gamma |\psi_n - \psi|^2\, dt = 0$$

therefore,

$$\lim_{n\to\infty} \int_\gamma [(\varphi_n + \psi_n) - (\varphi + \psi)]^2\, dt = 0. \qquad \text{Q.E.D.}$$

6. Expansion in Series of Orthogonal Functions

1. DEFINITION 15. If g is a measurable set and r is a rational

number then ψ_r will denote the function which is equal to 1 at all points of g which precede r and which is zero at all remaining points of g.

Evidently, since the rational numbers are denumerable the functions ψ_r are also denumerable.

THEOREM 18. If a function φ, integrable in g, is orthogonal in g to all ψ_r then it vanishes a.e. in g.

Proof. $\int_g \varphi\psi_r \, dt = 0$ implies $\int_{g_r} \varphi \, dt = 0$, where g_r denotes the set of points of g which precede r, and therefore, φ is zero almost everywhere in g. (Appendix, Th. 4).

2. THEOREM 19. A system of orthonormal functions in a set g of finite measure consists of a finite number or a denumerable number of functions.

Proof. Let (φ) denote a system of orthonormal functions and suppose $\varphi_1, \varphi_2, \ldots, \varphi_n$ are n of these. If, now, ψ denotes a ψ_r and we write

$$a_i = \int_g \psi\varphi_i \, dt$$

then

$$\int_g (\psi - a_1\varphi_1 - a_2\varphi_2 - \cdots - a_n\varphi_n)^2 \, dt \geqq 0,$$

whence, by the orthogonality of the φ, we have (cf. Th. 12)

$$\int_g \psi^2 \, dt \geqq a_1^2 + a_2^2 + \cdots + a_n^2$$

and finally

$$a_1^2 + a_2^2 + \cdots + a_n^2 \leqq \mu(g).$$

It follows from this that for every $\lambda > 0$ there exists a finite number of functions of the system (φ) for which we have

$$\left|\int_g \psi\varphi \, dt\right| \geqq \lambda,$$

and, therefore, all the functions of the system (φ) for which the corresponding constants a_i satisfy the inequalities

$$|a_i| > 1, \qquad 1 \geqq |a_i| > 1/2,$$
$$1/2 \geqq |a_i| > 1/3, \ldots, \qquad 1/n \geqq |a_i| > 1/(n+1), \ldots,$$

form a finite or denumerable set. However, since the ψ form a denumerable set, it follows that all φ such that for at least one value of the index r

$$\left| \int_g \psi_r \varphi \, dt \right| > 0$$

form a finite or denumerable set. But for every r the equation $\int_g \psi_r \varphi \, dt = 0$ is satisfied only by functions φ vanishing a.e., which proves the theorem.

3. THEOREM 20. (of Fischer-Riesz) (F. Riesz [93a], Fischer [36]). If g is a set of finite measure, $\{\varphi_n\}$ is a sequence of functions orthonormal in g, $\{a_n\}$ is a sequence of constants for which the series $\sum_{n=1}^{\infty} a_n^2$ is convergent, and finally, if we write

$$f_n = a_1\varphi_1 + a_2\varphi_2 + \cdots + a_n\varphi_n$$

then the sequence $\{f_n\}$ converges in the mean in g to a function f which has for Fourier coefficients the terms of the sequence $\{a_n\}$, that is

$$a_i = \int_g f\varphi_i \, dt, \qquad (i = 1, 2, \ldots).$$

Proof. In fact for $n > m$

$$f_n - f_m = a_{m+1}\varphi_{m+1} + a_{m+2}\varphi_{m+2} + \cdots + a_n\varphi_n$$

and, therefore,

$$\int_g (f_n - f_m)^2 \, dt = a_{m+1}^2 + a_{m+2}^2 + \cdots + a_n^2$$

and by the convergence of the series $\sum_{n=1}^{\infty} a_n^2$

$$\lim_{\substack{n\to\infty \\ m\to\infty}} \int_g |f_n - f_m|^2 \, dt = 0$$

and since g is of finite measure, the sequence $\{f_n\}$ converges in the mean to a function f (Th. 15). Therefore, for all $n > i$ we have

$$\int_g \varphi_i f_n \, dt = a_i$$

and, consequently, by Theorem 16 we have in the limit as $n \to \infty$

$$\int_g \varphi_i f dt = a_i.$$

The theorem is valid even if g is of infinite measure (E. W. Hobson [48a]; II, p. 761).

4. DEFINITION 16. A system of functions orthonormal in g is called *complete in* g when the only square integrable function which is orthogonal in g to all functions of the system is the function which is zero a.e.

It is easy to construct a complete system in a set g. For, consider the system of functions ψ_r defined in 1 of this section. Since they are denumerable they may be arranged in a sequence; now form a new sequence $\psi_1', \psi_2', \ldots, \psi_n', \ldots$ by suppressing from the original sequence the ψ_r which are zero a.e. and the ψ_r which are linearly dependent on the preceding ones. By Theorem 7 we can substitute for every function ψ_n' another ψ_n'' which is a linear combination of $\psi_1', \psi_2', \ldots, \psi_n'$ normal and orthogonal to $\psi_1', \psi_2' \ldots \psi_{n-1}'$. The system $\psi_1'', \psi_2'', \ldots, \psi_n'', \ldots$ is complete in g. In fact every ψ_r can be expressed linearly and homogeneously in terms of the ψ_n'' and a function which is orthogonal to all of the ψ_n'' is orthogonal to all of the ψ_r and, therefore, is zero a.e. (Th. 18).

THEOREM 21. A complete system in a set g of finite measure consists of a denumerable number of functions.

Proof. To prove the theorem it will suffice to show that if $f_1, f_2, \ldots, f_n$ are n functions orthonormal in g then there exists a function, not zero a.e., linearly independent of these. In fact, if φ is such a function it follows by Theorem 7 that there exists a linear combination of $f_1, f_2, \ldots, f_n, \varphi$ normal and orthogonal to $f_1, f_2, \ldots, f_n$.

Now if f is any bounded function, measurable and not zero a.e. in g, which assumes constant values, no two of which are equal, in at least $n + 2$ mutually exclusive subsets of g of non-zero measure, then among the functions $f, f^2, \ldots, f^n, f^{n+1}$ there is at least one which is not a linear, homogeneous combination of the $f_1, f_2, \ldots, f_n$. If this were not the case, then the $n + 1$ relations in g

$$f^i = c_{i,1}f_1 + c_{i,2}f_2 + \cdots + c_{i,n}f_n, \quad (i = 1, 2, \ldots, n + 1)$$

with the $c_{i,k}$ constants, would imply that there exists between the functions $f, f^2, \ldots, f^n, f^{n+1}$ a relation of the type $\lambda_1 f + \lambda_2 f^2 + \cdots + \lambda_{n+1} f^{n+1} = 0$ with $\lambda_1, \lambda_2, \ldots, \lambda_{n+1}$ not all zero, whence f could assume in g at most $n + 1$ distinct values, contrary to hypothesis.

Since a complete system consists of an infinite number of functions by what we have just shown, it follows immediately from Theorem 19 that a complete system is denumerably infinite.

5. THEOREM 22. If $\{\varphi_n\}$ is a complete system of orthonormal function is a set g of finite measure, and f is any function square integrable in g, then the sequence $\{f_n\}$ where

$$f_n = a_1\varphi_1 + a_2\varphi_2 + \cdots + a_n\varphi_n \tag{1}$$

with

$$a_i = \int_g \varphi_i f\,dt \tag{2}$$

converges in the mean to f, that is

$$\lim_{n\to\infty} \int_g |f - f_n|^2\,dt = 0, \tag{3}$$

and therefore (Th. 13)

$$\sum_{i=1}^{\infty} a_i^2 = \int_g f^2\,dt. \tag{4}$$

This formula will be referred to as *Bessel's equality*.

Proof. By Bessel's inequality we have

$$\sum_{i=1}^{\infty} a_i^2 \leq \int_g f^2\,dt$$

and consequently the series $\sum_{i=1}^{\infty} a_i^2$ is convergent. It follows from the Fischer-Riesz Theorem that the sequence $\{f_n\}$ defined in (1) converges in the mean to a function $\bar{f}$. Therefore

$$a_i = \int_g \varphi_i \bar{f} dt \tag{2'}$$

and in view of (2) and (2′)

$$\int_g \varphi_i (f - \bar{f}) dt = 0, \qquad (i = 1, 2, \ldots).$$

But since the system $\{\varphi_n\}$ is complete, the difference $f - \bar{f}$ is zero a.e. in g and (2′) coincides with (2). Q.E.D.

The theorem is valid even if g is of infinite measure (E. W. Hobson, [48a]; II, p. 761).

DEFINITION 17. A series $a_1\varphi_1 + a_2\varphi_2 + \ldots + a_n\varphi_n + \ldots$ is called *convergent in the mean* in g to f if

$$\lim_{n\to\infty} \int_g (f - \sum_{i=1}^{n} a_i\varphi_i)^2 dt = 0.$$

Remark. It should be observed that convergence in the mean to f of a series $\sum_{n=1}^{\infty} a_n\varphi_n$ does not imply pointwise convergence of the series, that is, if t is a point of g it does not follow necessarily that

$$f(t) = a_1\varphi_1(t) + a_2\varphi_2(t) + \cdots + a_n\varphi_n(t) + \cdots.$$

However, by Theorem 15, there are partial sums of the series

$$\sum_{i=1}^{s_1} a_i\varphi_i(t), \quad \sum_{i=1}^{s_2} a_i\varphi_i(t), \ldots, \quad \sum_{i=1}^{s_n} a_i\varphi_i(t), \ldots$$

which converge a.e. to $f(t)$ as $n \to \infty$.

6. THEOREM 23. A necessary and sufficient condition for a sequence $\{\varphi_n\}$ orthonormal in a set g of finite measure to form a complete system is that, for every function f, square integrable in g,

$$\int_g f^2 dt = \sum_{n=1}^{\infty} \left(\int_g \varphi_n f dt\right)^2. \tag{5}$$

Proof. That the condition is necessary follows from Theorem 22.

The condition is also sufficient. Indeed, if the sequence is not complete, there exists a function f square integrable, not zero a.e. and orthogonal to all the φ_n,

$$\int_g \varphi_n f dt = 0, \qquad (n = 1, 2, \ldots).$$

For such a function, (5) implies

$$\int_g f^2 dt = 0,$$

and therefore if g_t is the subset of the points of g which precede t, we have also

$$\int_{g_t} f^2 dt = 0$$

and the f is zero almost everywhere (Appendix, Th. 4) contrary to hypothesis.

The theorem is also valid if g is of infinite measure (E. W. Hobson [48a] II, p. 761).

The following is a well-known theorem of Picone (M. Picone [88d]).

THEOREM 24. If $\theta(t)$ is a non-negative function integrable in g, and $\{\theta^{1/2}(t) f_n(t)\}$ is an orthonormal sequence and f is a function for which θf^2 is integrable in g, then a necessary and sufficient condition for f to satisfy Bessel's equality

$$\sum_{n=1}^{\infty} \left[\int_g \theta f f_n dt\right]^2 = \int_g \theta f^2 dt$$

is that, for every real x

$$\sum_{n=1}^{\infty} \left[\int_g [\theta\,(t)\, e^{itx} f_n(t)\, dt\right]^2 = \int_g \theta(t) dt.$$

But this is just the necessary and sufficient condition that the system $\{\theta^{1/2}(t) f_n(t)\}$ be complete in g.

THEOREM 25. (Lauricella [66]). A necessary and sufficient condition for an orthonormal sequence $\{\varphi_n\}$ in a measurable set

g to form a complete system is that (5) should hold for all functions f which belong to a complete system.

Proof. If $\{f_n\}$ is a complete system then, since (5) holds for every square integrable function f, it also holds for every f_n.

Conversely we show that if (5) holds for every f_n of a complete system then the system $\{\varphi_n\}$ is also complete. Indeed, if this were not the case there would exist a function θ, not zero a.e., normal and orthogonal to all φ_n. But Bessel's inequality when applied to the system $\{\varphi_n\}$ and θ would give

$$\int_g f_n^2 dt \geqq \sum_{m=1}^{\infty} \left(\int_g \varphi_m f_n dt\right)^2 + \left(\int_g f_n \theta dt\right)^2$$

whence by (5)

$$\int_g f_n \theta dt = 0, \qquad (n = 1, 2, \ldots),$$

and consequently θ would be orthogonal to every f_n which is impossible.

THEOREM 26 (Vitali [116c]). Let $\mu(t)$ denote the measure of the subset $g_{a,t}$ of points of g between a fixed point a and a variable point t of the line containing the set g. Then a necessary and sufficient condition for an orthonormal sequence $\{\varphi_n\}$ in a set g of finite measure to form a complete system is

$$\mu(t) = \sum_{n=1}^{\infty} \left[\int_{g_{a,t}} \varphi_n dt\right]^2. \tag{6}$$

Proof. To show that the condition is necessary it is sufficient to consider (5) for the case in which f is the function which is equal to 1 at all points of $g_{a,t}$ and zero at all remaining points.

In order to prove that the condition is sufficient we note that if the sequence $\{\varphi_n\}$ is not complete then there exists a function θ, not zero a.e., normal and orthogonal to all the φ_n. Then, considering the system which results from adjoining θ to the sequence $\{\varphi_n\}$ and the function f which is equal to 1 at points of $g_{a,t}$ and zero at remaining points, we have by Bessel's inequality

$$\mu(t) \geqq \sum_{n=1}^{\infty} \left[\int_{g_{a,t}} \varphi_n\, dt\right]^2 + \left[\int_{g_{a,t}} \theta\, dt\right]^2$$

and by (6) $\int_{g_{a,t}} \theta dt = 0$ for all values of t. From this follows that θ is zero a.e. in any finite segment which has a as an end point and therefore θ is zero a.e., contrary to hypothesis.

If we remove the restriction that the set g be of finite measure the second part of the argument is still valid, and therefore, we have the following theorem.

THEOREM 27. Let $\mu(t)$ denote the measure of the subset $g_{a,t}$ of points of g between a fixed point a and a variable point t of the line containing the set g. Then a sufficient condition for an orthonormal sequence $\{\varphi_n\}$ in a set g (of finite or infinite measure) to form a complete system is (6).

It may be remarked that (5) may be derived directly from (6) even if g is infinite (G. Sansone [98e]).

7. *Orthogonal Cartesian Systems of Hilbert Space*

1. DEFINITION 18. A complete system of orthonormal functions in a set g of finite measure is called an *orthogonal cartesian system in* H, and the lines $\lambda_1\varphi_1, \lambda_2\varphi_2, \ldots, \lambda_n\varphi_n, \ldots$ are called the *axes of the system.*

2. DEFINITION 19. If

(1) $$\varphi_1, \quad \varphi_2, \ldots, \quad \varphi_n, \ldots$$

is an orthogonal cartesian system in H and $f(t)$ is a point of H, then the numbers

(2) $$a_n = \int_g f\varphi_n\, dt, \qquad (n = 1, 2, \ldots)$$

are called the *Fourier coefficients* (cf. Sec. 4. 2) or the *coordinates of the point* f with respect to the system (1) and we shall write

$$f \equiv (a_1, a_2, \ldots, a_n, \ldots); \qquad f \sim \sum_{n=1}^{\infty} a_n \varphi_n.$$

THEOREM 28. If two points f_1 and f_2 have the same coordinates they coincide.

Proof. Indeed the equality

$$\int_g f_1 \varphi_n \, dt = \int_g f_2 \varphi_n \, dt$$

implies

$$\int_g (f_1 - f_2) \varphi_n \, dt = 0,$$

and since $\{\varphi_n\}$ is complete, the difference $f_1 - f_2$ is zero a.e. in g.

THEOREM 29. If $f \equiv (a_1, a_2, \ldots, a_n, \ldots)$, then the series $\sum_{n=1}^{\infty} a_n^2$ is convergent and

$$\sum_{n=1}^{\infty} a_n^2 = \int_g f^2 \, dt. \tag{3}$$

Conversely if $\{a_n\}$ is a sequence of real constants, and the series $\sum_{n=1}^{\infty} a_n^2$ is convergent then there exists one and only one point f such that

$$f \equiv (a_1, a_2, \ldots, a_n, \ldots).$$

Proof. It is sufficient to refer to Theorems 22, 20 (Fischer-Riesz [28]).

3. THEOREM 30. If

$$f \equiv (a_1, a_2, \ldots, a_n, \ldots), \quad f_2 \equiv (b_1, b_2, \ldots, b_n, \ldots)$$

and d is the distance between the two points f_1 and f_2 (cf. Sec. 3. 2) it follows that

$$d^2 = \sum_{n=1}^{\infty} (a_n - b_n)^2. \tag{4}$$

Proof. The point $f_1 - f_2$ has by (2) the coordinates $a_1 - b_1$, $a_2 - b_2, \ldots, a_n - b_n, \ldots$, and applying (3) we obtain (4) which is an extension of the well-known formula of ordinary euclidean space.

In particular if $f \equiv (a_1, a_2, \ldots, a_n, \ldots)$, $\sum_{i=1}^{\infty} a_i^2$ denotes the distance of f from the origin.

4a. THEOREM 31. If

$$f_1 \equiv (a_1, a_2, \ldots, a_n, \ldots), \quad f_2 \equiv (b_1, b_2, \ldots, b_n, \ldots)$$

then

$$\int_g f_1 f_2 dt = \sum_{n=1}^{\infty} a_n b_n. \tag{5}$$

Proof. The sequence

$$\omega_1 = a_1\varphi_1, \ \omega_2 = a_1\varphi_1 + a_2\varphi_2, \ldots, \ \omega_n = a_1\varphi_1 + a_2\varphi_2 + \cdots + a_n\varphi_n, \ldots$$

converges internally to f_1 (Th. 22, Th. 14) and therefore, by Theorem 16,

$$\begin{aligned}\int_g f_1 f_2 dt &= \lim_{n\to\infty} \int_g \omega_n f_2 dt \\ &= \lim_{n\to\infty} \left[a_1 \int_g f_2\varphi_1 dt + a_2 \int_g f_2\varphi_2 dt + \cdots + a_n \int_g f_2\varphi_n dt\right] \\ &= \lim_{n\to\infty} [a_1b_1 + a_2b_2 + \cdots + a_nb_n] = \sum_{n=1}^{\infty} a_n b_n.\end{aligned}$$

REMARK. Consider the two lines $\lambda_1 f_1$, $\lambda_2 f_2$ through the origin where f_1 and f_2 are two points a unit distance from the origin:

$$f_1 \equiv (a_1, a_2, \ldots, a_n, \ldots), \quad f_2 \equiv (b_1, b_2, \ldots, b_n, \ldots);$$

$$\int_g f_1^2 dt = \int_g f_2^2 dt = 1.$$

By Definition 8, the angle ω between the two lines is given by the following formula

$$\cos\omega = \int_g f_1 f_2 dt = \sum_{n=1}^{\infty} a_n b_n, \tag{6}$$

which is a generalization of the analogous formula for ordinary euclidean space.

4b. The following theorem which is a generalization of (5) is due to Picone (M. Picone [88d]) and is of importance in the applications.

THEOREM 32. Suppose $\theta(t)$ is a nonnegative function integrable in a measurable set g (not necessarily of finite measure), $\{\theta^{1/2}(t)\varphi_n(t)\}$ is a system (not necessarily complete) of orthonormal functions in g; $\theta^{1/2}f_1(t)$ is square integrable in g, and if we write

$$a_n = \int_g \theta\varphi_n f_1\, dt, \qquad \theta^{1/2}f_1 \sim \theta^{1/2}\sum_{n=1}^{\infty} a_n\varphi_n, \tag{7}$$

suppose further that

$$\sum_{n=1}^{\infty} a_n^2 = \int_g \theta f_1^2 dt. \tag{8}$$

(If the system $\{\theta^{1/2}\varphi_n\}$ is complete, (8) follows immediately from Theorem 23.) Suppose $f_2(t)$ is measurable in g such that

$$|f_2(t)| \leqq \theta(t) \tag{9}$$

and for any measurable subset γ of g let

$$b_n(\gamma) = \int_\gamma f_2\varphi_n\, dt. \tag{10}$$

Then it follows that

$$\int_\gamma f_1 f_2 dt = \sum_{n=1}^{\infty} a_n b_n(\gamma) \tag{11}$$

and the series on the right converges absolutely and uniformly with respect to γ.

Proof. The integrability in g of $\theta^{1/2}$, $\theta^{1/2}|f_1|$, and $\theta^{1/2}|\varphi_n|$ implies the integrability in g of $\theta|f_1|$ and $\theta|\varphi_n|$, and by (9) also the integrability of f_1f_2 and $f_2\varphi_n$. Therefore the integrals which occur in (10) and (11) exist.

If we write

$$F_n(t) = \sum_{k=1}^{n} a_k\varphi_k(t) \tag{12}$$

then

$$\left|\int_\gamma f_2 F_n\,dt - \int_\gamma f_2 f_1\,dt\right| = \left|\int_\gamma f_2(F_n - f_1)\,dt\right| \leqq \int_\gamma \theta\,|F_n - f_1|\,dt$$

and consequently by the Schwarz inequality

$$\left|\int_\gamma f_2 F_n\,dt - \int_\gamma f_1 f_2\,dt\right| \leqq \sqrt{\int_g \theta dt}\,\sqrt{\int_g \theta\,|F_n - f_1|^2\,dt}. \tag{13}$$

Now by Theorem 13, (8) implies

$$\lim_{n\to\infty}\int_g \theta\,|F_n - f_1|^2\,dt = 0$$

and from (13) follows (11) and the uniform convergence with respect to γ.

Changing the order of the terms of the sequence $\{\theta^{1/2}\varphi_n\}$ does not affect the result, and, consequently, the series on the right side of (11) is also absolutely convergent, and the theorem is proved.

REMARK 1. If we suppose $\theta(t) = \theta_1(t)\theta_2(t)$ such that $\theta_2(t) \geqq \varrho > 0$, as t varies over a measurable subset g' of g, and if, instead of the condition imposed on f_2 in (9), we assume

$$|f_2| \leqq \theta_1, \tag{9'}$$

then (11) is still valid for all measurable subsets γ of g'.

Indeed $\theta_1 = \theta/\theta_2 \leqq \theta/\varrho$ and in place of (13) we have

$$\left|\int_\gamma f_2 F_n\,dt - \int_\gamma f_1 f_2\,dt\right| \leqq \frac{1}{\varrho}\sqrt{\int_g \theta\,dt}\,\sqrt{\int_g \theta\,|F_n - f_1|^2 dt}. \tag{13'}$$

REMARK 2. If $\theta = 1$, g is finite, $\{\varphi_n\}$ is complete, $\gamma = g$, then (11) is a particular case of (5).

5. Suppose that $\{\varphi_n\}$ and $\{\psi_n\}$, are two orthogonal cartesian systems in H, and consider a point f of H expressible in the two systems by the following:

$$f \equiv (x_1, x_2, \ldots, x_n, \ldots), \quad x_n = \int_g f\varphi_n\,dt, \tag{14}$$

$$f \equiv (y_1, y_2, \ldots, y_n, \ldots), \quad y_n = \int_g f\psi_n\,dt. \tag{14'}$$

We seek the relations between the coordinates $(x_1, x_2, \ldots, x_n, \ldots)$, $(y_1, y_2, \ldots, y_n, \ldots)$ of the point f.

Since the series $\sum_{n=1}^{\infty} x_n \varphi_n$ converges in the mean to f (Def. 18), we have by Theorem 16

$$\lim_{m\to\infty} \int_g (x_1\varphi_1 + x_2\varphi_2 + \cdots + x_m\varphi_m)\,\psi_n\,dt = \int_g f\psi_n\,dt = y_n.$$

If now we write

$$a_{m,n} = \int_g \varphi_m \psi_n\,dt, \qquad (m, n = 1, 2, \ldots) \tag{15}$$

then

$$\sum_{m=1}^{\infty} a_{m,n} x_m = y_n, \qquad (n = 1, 2, \ldots) \tag{16}$$

and, similarly, by interchanging φ_n and ψ_n

$$\sum_{n=1}^{\infty} a_{m,n} y_n = x_m, \qquad (m = 1, 2, \ldots) \tag{16'}$$

(16) and (16′) are the desired formulas.

A set of fundamental relations between the constants $a_{m,n}$ follows immediately.

If we write $f = \psi_r$ then $x_m = \int_g \psi_r \varphi_m\,dt = a_{m,r}$; $y_1 = 0, y_2 = 0, \ldots,$ $y_{r-1} = 0,\ y_r = 1,\ y_{r+1} = 0, \ldots$ and from (15) follows

$$\sum_{m=1}^{\infty} a_{m,n}^2 = 1; \qquad \sum_{m=1}^{\infty} a_{m,n} a_{m,r} = 0 \quad \text{if} \quad n \neq r. \tag{17}$$

Similarly, if we write $f = \varphi_r$, we have

$$\sum_{n=1}^{\infty} a_{m,n}^2 = 1; \qquad \sum_{n=1}^{\infty} a_{m,n} a_{r,n} = 0 \quad \text{if} \quad m \neq r. \tag{17'}$$

8. *L^p Integrability. The Holder-Riesz and the Minkowski Inequalities*

1. This section and the succeeding one are devoted to some theorems of frequent use in analysis and extend various results of preceding sections.

We recall that f is called integrable L^p in g, and we write $f(t) \in L^p$, if $|f|^p$ is integrable in g (Def. 12).

THEOREM 33. If $f_1(t)$, $f_2(t)$ are integrable L^p, $(p > 0)$ in g and

c_1 and c_2 are constants, then $c_1f_1(t) + c_2f_2(t)$ is also integrable L^p in g.

Proof. Clearly

$$|c_1f_1 + c_2f_2|/2 \leqq \max\,[|c_1|\,|f_1|,\ |c_2|\,|f_2|],$$
$$|c_1f_1 + c_2f_2|^p \leqq 2^p \max\,[|c_1|^p\,|f_1|^p,\ |c_2|^p\,|f_2|^p],$$
$$|c_1f_1 + c_2f_2|^p \leqq 2^p[|c_1|^p\,|f_1|^p + |c_2|^p\,|f_2|^p]$$

and the last inequality implies $c_1f_2 + c_2f_2 \in L^p$.

2a. THEOREM 34. If p and p' are two positive conjugate numbers such that

$$\frac{1}{p} + \frac{1}{p'} = 1, \tag{1}$$

and if $f_1(t)\in L^p$, $f_2(t)\in L^{p'}$, then $f_1(t)f_2(t)$ is integrable in g, and Hölder's inequality (Hölder [49], F. Riesz [93d, e]) holds

$$\left|\int_g f_1f_2\,dt\right| \leqq \left[\int_g |f_1|^p dt\right]^{1/p}\left[\int_g |f_2|^{p'} dt\right]^{1/p'}. \tag{2}$$

Proof. Consider the function

$$\frac{x^p}{p} + \frac{1}{p'} - x$$

for $x \geqq 0$. Its derivative $x^{p-1} - 1$ vanishes for $x = 1$, is negative for $x < 1$ and positive for $x > 1$. It follows that $(x^p/p) + (1/p') - x$ attains its minimum at $x = 1$ and therefore vanishes there. Finally

$$\frac{x^p}{p} + \frac{1}{p'} \geqq x.$$

If we write $x = |\alpha|\,|\beta|^{-1/(p-1)} = |\alpha|\,|\beta|^{-p'/p}$ we have

$$\frac{|\alpha|^p\,|\beta|^{-p'}}{p} + \frac{1}{p'} \geqq \alpha\,|\beta|^{-p'/p}$$

whence multiplying by $|\beta|^{p'}$, $[p' - p'/p = 1]$, follows

$$|\alpha\beta| \leqq \frac{1}{p}\,|\alpha|^p + \frac{1}{p'}\,|\beta|^{p'}.$$

If we now write

$$\alpha = f_1(t)\left[\int_g |f_1|^p dt\right]^{-1/p}, \qquad \beta = f_2(t)\left[\int_g |f_2|^{p'} dt\right]^{-1/p'},$$

then

$$\frac{|f_1(t)f_2(t)|}{\left[\int_g |f_1|^p dt\right]^{1/p}\left[\int_g |f_2|^{p'} dt\right]^{1/p'}} \leqq \frac{1}{p}\frac{|f_1(t)|^p}{\int_g |f_1|^p dt} + \frac{1}{p'}\frac{|f_2(t)|^{p'}}{\int_g |f_2|^{p'} dt}, \tag{3}$$

and since the second member is integrable in g it follows that the first member is also, and therefore $|f_1(t)f_2(t)| \in L$.

Integrating in g the second member of (3) we have

$$\frac{\int_g |f_1(t)f_2(t)|\,dt}{\left[\int_g |f_1|^p dt\right]^{1/p}\left[\int_g |f_2|^{p'} dt\right]^{1/p'}} \leqq \frac{1}{p} + \frac{1}{p'} = 1, \quad \text{Q.E.D.}$$

2b. Clearly if f is integrable L^p in a set g of finite measure, and $p > 1$ then f is integrable L (cf. Th. 1).

Indeed $|f| = |f| \cdot 1$ and 1 is integrable $L^{p'}$.

3. THEOREM 35. If $f_1(t)$ and $f_2(t)$ are integrable L^p in g and $p \geqq 1$, then the Minkowski inequality (Minkowski [76]) holds

$$\left[\int_g |f_1 + f_2|^p dt\right]^{1/p} \leqq \left[\int_g |f_1|^p dt\right]^{1/p} + \left[\int_g |f_2|^p dt\right]^{1/p}. \tag{4}$$

Indeed

$$\int_g |f_1 + f_2|^p dt \leqq \int_g |f_1 + f_2|^{p-1}|f_1|\,dt + \int_g |f_1 + f_2|^{p-1}|f_2|\,dt$$

and applying the Hölder inequality to obtain a majorant of the second member where $p' = p/(p-1)$ we have

$$\int_g |f_1 + f_2|^p dt \leqq \left[\int_g |f_1 + f_2|^p dt\right]^{1/p'}\left[\int_g |f_1|^p dt\right]^{1/p}$$
$$+ \left[\int_g |f_1 + f_2|^p dt\right]^{1/p'}\left[\int_g |f_2|^p dt\right]^{1/p}.$$

Upon dividing thet wo members by $[\int_g |f_1 + f_2|^p dt]^{1/p'}$ then (4) follows immediately.

9. *Generalized Convergence in the Mean of Order p*

1. THEOREM 36. If a sequence $\{f_n\}$ integrable L^p in a measurable set g, converges strongly L^p (Def. 13) to a function $f(t)$ integrable L^p in g,

$$\lim_{n\to\infty}\int_g |f(t)-f_n(t)|^p\,dt = 0 \tag{1}$$

if $p > 1$, and if $\varphi(t)$ is integrable $L^{p'}$ in g, where p' is the conjugate of p, that is $1/p + 1/p' = 1$, then

$$\lim_{n\to\infty}\int_g f_n(t)\,\varphi(t)\,dt = \int_g f(t)\,\varphi(t)\,dt. \tag{2}$$

Proof. Clearly

$$0 \leqq \int_g |f - f_n|\,|\varphi|\,dt \leqq \Big[\int_g |f - f_n|^p\,dt\Big]^{1/p'}\Big[\int_g |\varphi|^p\,dt\Big]^{1/p'},$$

and by (1)

$$\lim_{n\to\infty}\int_g |f - f_n|\,|\varphi|\,dt = 0\,, \tag{3}$$

however,

$$0 \leqq \left|\int_g f\varphi\,dt - \int_g f_n\varphi\,dt\right| \leqq \int_g |f - f_n|\,|\varphi|\,dt,$$

and, therefore, (2) follows from (3).

THEOREM 37. If a sequence $\{f_n\}$ of functions integrable L^p in g converges strongly L^p to a function $f(t)$ integrable L^p in g, and if $p > 1$, then

$$\lim_{n\to\infty}\int_g |f_n|^p\,dt = \int_g |f|^p\,dt. \tag{4}$$

Proof. From Minkowski's inequality (Th. 35) follows

$$0 \leqq \Big[\int_g |f|^p\,dt\Big]^{1/p} = \Big[\int_g |(f - f_n) + f_n|^p\,dt\Big]^{1/p}$$
$$\leqq \Big[\int_g |f - f_n|^p\,dt\Big]^{1/p} + \Big[\int_g |f_n|^p\,dt\Big]^{1/p},$$

and by (1)

$$\left[\int_g |f|^p\, dt\right]^{1/p} \leqq \varliminf_{n\to\infty} \left[\int_g |f_n|^p\, dt\right]^{1/p},$$

whence also

$$\int_g |f|^p\, dt \leqq \varliminf_{n\to\infty} \int_g |f_n|^p\, dt. \tag{5}$$

Analogously

$$0 \leqq \left[\int_g |f_n|^p\, dt\right]^{1/p} \leqq \left[\int_g |f_n - f|^p dt\right]^{1/p} + \left[\int_g |f|^p\, dt\right]^{1/p}$$

$$\varlimsup_{n\to\infty} \int_g |f_n|^p\, dt \leqq \int_g |f|^p\, dt. \tag{6}$$

Clearly (4) follows from (5) and (6).

REMARK. From (4) follows that if a sequence $\{f_n\}$ of functions integrable L^p in g converges strongly L^p to a function $f(t)$ integrable L^p, then the integrals $\int_g |f_n|^p dt$, $(n = 1, 2, \ldots)$ are uniformly bounded.

THEOREM 38. If $\{f_n(t)\}$ is a sequence of functions integrable L^p, $(p > 0)$, in a set g of finite measure, if the integrals of the sequence $\{|f_n|^p\}$ are equi-absolutely continuous in g, and if

$$\lim_{n\to\infty} f_n(t) = f(t) \tag{7}$$

a.e. in g, then $f(t) \in L^p$ and

$$\lim_{n\to\infty} \int_g |f(t) - f_n(t)|^p\, dt = 0,$$

i.e., $f_n(t)$ converges strongly L^p to $f(t)$ in g.

Proof. From (7), $\lim_{n\to\infty} |f_n|^p = |f|^p$ a.e. in g, and the equi-absolute continuity of the integrals of the sequence $\{f_n^p(t)\}$, follows (Appendix, Th. 8)

$$\lim_{n\to\infty} \int_g |f_n|^p\, dt = \int_g |f|^p\, dt.$$

Since, moreover $|f| \in L^p$ in g, for every positive ε there exists an $\eta > 0$ such that the integrals of $|f_n|^p$ and $|f|^p$ in every subset of g with $\mu(\gamma) < \eta$ are less than ε. By the theorem of Egoroff-

Severini (Appendix, Th. 1) there exists a subset g^* of g and an integer N such that $\mu(g - g^*) < \eta$, and $|f(t) - f_n(t)| < \varepsilon$ for t in g^* and $n > N$.

It follows, then, that

$$\int_g |f(t) - f_n(t)|^p \, dt = \int_{g^*} |f - f_n|^p \, dt + \int_{g-g^*} |f - f_n|^p \, dt$$

$$\leqq \int_{g^*} |f - f_n|^p \, dt + 2^p \int_{g-g^*} |f_n|^p \, dt + 2^p \int_{g-g^*} |f|^p \, dt \leqq \varepsilon^p \mu(g) + 2^{p+1}\varepsilon,$$

and since ε is arbitrary the theorem is proved.

2. THEOREM 39. If $\{f_n(t)\}$ is a sequence of functions measurable and finite a.e. in a set g of finite measure, if

$$\lim_{n\to\infty} f_n(t) = f(t), \tag{8}$$

a.e. in g and if $p > 0$ then

$$\varliminf_{n\to\infty} \int_g |f_n(t)|^p \, dt \geqq \int_g |f(t)|^p \, dt, \tag{9}$$

and therefore, if the first member is finite, $f(t) \in L^p$ in g.

Proof. Since $\lim_{n\to\infty} |f_n(t)|^p = |f(t)|^p$ a.e. in g, then, by the Egoroff-Severini theorem, for any $\eta > 0$, there exists a subset g^* of g such that $\mu(g - g^*) < \eta$, and $\{|f^p|\}$ converges uniformly to $|f|^p$ in g^*.

It follows that if $\varepsilon > 0$, there exists an n_0 such that for $n > n_0$ and t in g^* we have

$$|f_n|^p \geqq |f|^p - \varepsilon$$

and from this follows

$$\int_g |f_n|^p \, dt \geqq \int_{g^*} |f_n|^p \, dt \geqq \int_{g^*} |f|^p \, dt - \varepsilon\mu\,(g^*) \geqq \int_{g^*} |f|^p \, dt - \varepsilon\mu(g),$$

and, therefore,

$$\varliminf_{n\to\infty} \int_g |f_n|^p \, dt \geqq \int_{g^*} |f|^p dt - \varepsilon\mu(g). \tag{10}$$

If we assume $f(t) \in L^p$ in g, and let $\eta \to 0$, and consequently $g^* \to g$ then by the absolute continuity of $\int_g |f|^p \, dt$, we have

$$\varliminf_{n\to\infty} \int_g |f_n|^p \, dt \geqq \int_g |f|^p \, dt - \varepsilon\mu(g),$$

and, since ε is arbitrary, (9) follows.

On the other hand, if we assume $\int_g |f(t)|^p dt = +\infty$, there exists a sequence of sets $g_1^*, g_2^*, \ldots, g_m^*, \ldots$, each contained in the one following, such that $\{|f_n|^p\}$ converges uniformly in every g_m^* to $|f|^p$, and, moreover, such that

$$\lim_{m\to\infty} \mu(g - g_m^*) = 0.$$

Then

$$\varliminf_{n\to\infty} \int_g |f_n|^p\, dt \geqq \int_{g_m^*} |f|^p\, dt - \varepsilon\mu(g) \tag{11}$$

and since $\lim_{m\to\infty} \int_{g_m^*} |f|^p\, dt = +\infty$ (otherwise $f(t)$ would be integrable L^p in g (Appendix, Th. 2)), (9) follows from (11).

THEOREM 40. If $\{f_n(t)\}$ is a sequence of functions integrable L^p, ($p > 0$), in a set g of finite measure, if

$$\lim_{n\to\infty} f_n(t) = f(t),$$

a.e. in g, and if

$$\int_g |f_n|^p dt \leqq A < +\infty, \qquad (n = 1, 2, \ldots),$$

then $f(t)$ is integrable L^p in g, and

$$\int_g |f(t)|^p\, dt \leqq A.$$

Proof. Clearly $\varliminf_{n\to\infty} \int_g |f_n(t)|^p\, dt \leqq A$, and the preceding theorem is applicable.

3. THEOREM 41. A sequence $\{f_n\}$ internally convergent of order $p > 1$ (Def. 14) to a function f, is internally convergent to f.

Proof. The reasoning is similar to that of Theorem 14. Indeed, if p' denotes the conjugate of p, and if we write $C = [\mu(\gamma)]^{1/p'}$, then by the Hölder-Riesz inequality we have

$$0 \leqq \int_\gamma |f_n - f|\, dt \leqq C \left[\int_g |f_n - f|^p\, dt\right]^{1/p}.$$

REMARK 1. It is known that if $\{f_n\}$ converges internally to two functions f_1 and f_2, then $f_1 = f_2$ a.e. in g (Appendix, Th. 7). From this result and Theorem 41 it follows that if the sequence $\{f_n\}$ converges internally in the mean of order $p > 1$ in g to two functions f_1 and f_2, then $f_1 = f_2$ a.e. in g.

REMARK 2. By a corollary of Vitali's theorem (Appendix, Th. 10) it follows that if $\{f_n(t)\}$ converges internally in the mean of order $p > 1$ to $f(t)$, then there exists a subsequence of $f_n(t)$ which converges to $f(t)$ a.e. in g.

4. For internal convergence in the mean of order $p > 1$ we have the following heorem of which Theorem 15 is a particular case.

THEOREM 42. Given a sequence $\{f_n(t)\}$ of functions defined a.e. in a measurable set g, and integrable L^p, $p > 1$, in every subset of g of finite measure, a necessary and sufficient condition that $\{f_n(t)\}$ be internally convergent in the mean in g is that for every subset γ of g, of finite measure,

$$\lim_{\substack{n\to\infty \\ m\to\infty}} \int_\gamma |f_n - f_m|^p \, dt = 0. \tag{12}$$

Proof. From the last inequality in the proof of Theorem 33, it follows that

$$0 \leqq \int_\gamma |f_n - f_m|^p \, dt = \int_\gamma |(f_n - f) - (f_m - f)|^p \, dt$$
$$\leqq 2^p \int_\gamma |f_n - f|^p \, dt + 2^p \int_\gamma |f_m - f|^p \, dt$$

and, therefore, that (12) is necessary.

To prove the condition is sufficient we proceed as in the proof of Theorem 15, taking into account the inequality

$$0 \leqq \int_\gamma |f_n - f_m| \, dt \leqq C \left[\int_\gamma |f_n - f_m|^p \, dt\right]^{1/p}$$

where $C = [\mu(\gamma)]^{1/p'}$, $(1/p' + 1/p = 1)$, and observe that after establishing the equi-absolute continuity of the integrals of $|f_n|^p$ in γ, Theorem 38 is applicable.

CHAPTER II

Expansions in Fourier Series

1. Approximation in the Mean of a Function by a Trigonometric Polynomial of Order n

1. DEFINITION 1. A polynomial of the form

$$\tfrac{1}{2}a_0 + a_1 \cos x + a_2 \cos 2x + \cdots + a_n \cos nx$$
$$+ b_1 \sin x + b_2 \sin 2x + \cdots + b_n \sin nx$$

where the a_k and b_k are constants such that $a_n^2 + b_n^2 \neq 0$, is called a trigonometric polynomial of order n.

Equating real and imaginary parts on the two sides of De Moivre's formula

$$(\cos x + i \sin x)^k = \cos kx + i \sin kx$$

we have

$$\cos kx = \binom{k}{0} \cos^k x - \binom{k}{2} \cos^{k-2} x \sin^2 x + \cdots,$$
$$\sin kx = \binom{k}{1} \cos^{k-1} x \sin x - \binom{k}{3} \cos^{k-3} x \sin x + \cdots.$$

Consequently, every trigonometric polynomial of order n is also a polynomial of degree n in $\cos x$ and $\sin x$. Conversely, equating real and imaginary parts on the two sides of the formula

$$\cos^m x = [(e^{ix} + e^{-ix})/2]^m, \qquad \sin^m x = [(e^{ix} - e^{-ix})/2i]^m$$

we have

$$2^{2k-1}\cos^{2k}x = \binom{2k}{0}\cos 2kx + \binom{2k}{1}\cos(2k-2)x + \cdots + \binom{2k}{k-1}\cos 2x + \frac{1}{2}\binom{2k}{k};$$

$$2^{2k}\cos^{2k+1}x = \binom{2k+1}{0}\cos(2k+1)x + \binom{2k+1}{1}\cos(2k-1)x + \cdots + \binom{2k+1}{k}\cos x;$$

$$(-1)^k 2^{2k-1}\sin^{2k}x = \binom{2k}{0}\cos 2kx - \binom{2k}{1}\cos(2k-2)x + \binom{2k}{2}\cos(2k-4)x - \cdots + (-1)^k\frac{1}{2}\binom{2k}{k};$$

$$(-1)^k 2^{2k}\sin^{2k+1}x = \binom{2k+1}{0}\sin(2k+1)x - \binom{2k+1}{1}\sin(2k-1)x + \binom{2k+1}{2}\sin(2k-3)x + \cdots + (-1)^k\binom{2k+1}{k}\sin x;$$

and therefore every trigonometric polynomial of degree n in $\cos x$ and $\sin x$ is a polynomial of order not higher than n.

We now seek the trigonometric polynomial of order not exceeding n which is a minimum distance from a given function $f(x)$ square integrable in $[-\pi, \pi]$ (Ch. I, Sec. 4. 2).

The following system is clearly an orthonormal system in $[-\pi, \pi]$:

$$\frac{1}{\sqrt{2\pi}};\quad \frac{\cos kx}{\sqrt{\pi}};\quad \frac{\sin kx}{\sqrt{\pi}};\qquad (k = 1, 2, \ldots). \tag{1}$$

Consequently, if $\frac{1}{2}A_0$; A_k, B_k $(k = 1, 2, \ldots)$ denote the Fourier coefficients of $f(x)$ with respect to the system (1), i.e. if

$$\begin{cases} \frac{1}{2}A_0 = \dfrac{1}{\sqrt{2\pi}}\displaystyle\int_{-\pi}^{\pi} f(x)dx; \\ A_k = \dfrac{1}{\sqrt{\pi}}\displaystyle\int_{-\pi}^{\pi} f(x)\cos kx\,dx, \quad B_k = \dfrac{1}{\sqrt{\pi}}\displaystyle\int_{-\pi}^{\pi} f(x)\sin kx\,dx, \end{cases} \tag{2}$$

then the polynomial

$$\tfrac{1}{2}A_0 \frac{1}{\sqrt{2\pi}} + \frac{1}{\sqrt{\pi}} A_1 \cos x + \frac{1}{\sqrt{\pi}} A_2 \cos 2x + \cdots + \frac{1}{\sqrt{\pi}} A_n \cos nx$$
$$+ \frac{1}{\sqrt{\pi}} B_1 \sin x + \frac{1}{\sqrt{\pi}} B_2 \sin 2x + \cdots + \frac{1}{\sqrt{\pi}} B_n \sin nx$$

represents the trigonometric polynomial of order not exceeding n which is a minimum distance from $f(x)$. Therefore if we write

$$(3) \quad \begin{cases} a_0 = \dfrac{1}{\pi}\displaystyle\int_{-\pi}^{\pi} f(x)\,dx; \\ a_k = \dfrac{1}{\pi}\displaystyle\int_{-\pi}^{\pi} f(x)\cos kx dx, \quad b_k = \dfrac{1}{\pi}\displaystyle\int_{-\pi}^{\pi} f(x)\sin kx\,dx, \end{cases} \quad (k = 1, 2, \ldots$$

the polynomial

$$(4) \quad \begin{aligned} &\tfrac{1}{2}a_0 + a_1 \cos x + a_2 \cos 2x + \cdots + a_n \cos nx \\ &\quad + b_1 \sin x + b_2 \sin 2x + \cdots + b_n \sin nx \end{aligned}$$

is the required polynomial.

This establishes the theorem:

THEOREM 1. Among all trigonometric polynomials of order n, the polynomial (4), whose coefficients are given by (3), is the polynomial which is a minimum distance from $f(x)$ in $[-\pi, \pi]$.

2. Convergence in the Mean of the Fourier Series of a Square Integrable Function

1. DEFINITION 2. Given an integrable function $f(x)$ of period 2π, then the constants given by the formulas

$$(1) \quad \begin{cases} a_0 = \dfrac{1}{\pi}\displaystyle\int_{-\pi}^{\pi} f(x)\,dx; \\ a_k = \dfrac{1}{\pi}\displaystyle\int_{-\pi}^{\pi} f(x)\cos kx\,dx, \quad b_k = \dfrac{1}{\pi}\displaystyle\int_{-\pi}^{\pi} f(x)\sin kx\,dx, \end{cases} \quad (k = 1, 2, \ldots$$

are called the Fourier constants of $f(x)$, and the series

$$\frac{1}{2}a_0 + \sum_{k=1}^{\infty} (a_k \cos kx + b_k \sin kx) \tag{2}$$

is called the Fourier series of $f(x)$.

By virtue of (2) of Sec. 1, there exist the following relations between the Fourier coefficients and the Fourier constants of $f(x)$:

$$\frac{1}{2}A_0 = \sqrt{2\pi}\,(\tfrac{1}{2}a_0); \quad A_k = \sqrt{\pi}\,a_k, \quad B_k = \sqrt{\pi}\,b_k, \ (k = 1, 2, \ldots). \tag{3}$$

Without loss of generality we may assume that $f(x)$ is periodic with period 2π. For if $f(x)$ is defined in $(-\infty, \infty)$ with period ω, then the function $F(x) = f(\omega x/2\pi)$ is periodic with period 2π.

If $f(x)$ is periodic with period 2π, then the Fourier constants are also given by the following formulas:

$$(1') \begin{cases} a_0 = \dfrac{1}{\pi}\displaystyle\int_a^{a+2\pi} f(x)\,dx; \\ a_k = \dfrac{1}{\pi}\displaystyle\int_a^{a+2\pi} f(x)\cos kx\,dx, \quad b_k = \dfrac{1}{\pi}\displaystyle\int_a^{a+2\pi} f(x)\sin kx\,dx, \\ \qquad (k = 1, 2, \ldots), \end{cases}$$

where a is an arbitrary constant.

To indicate that the series (2) is the Fourier series of $f(x)$ we write

$$f(x) \sim \frac{1}{2}a_0 + \sum_{k=1}^{\infty} (a_k \cos kx + b_k \sin kx),$$

in accordance with a notation due to Hurwitz [50b]. When we write this, it should be understood that we affirm neither that the series represents a convergent series nor that, if the series does converge, it converges to $f(x)$.

2. THEOREM 2. Given $f(x)$ is integrable in $(-\pi, \pi)$. If $f(x)$ satisfies the condition $f(x) = f(-x)$, i.e. $f(x)$ is an even function, then its Fourier series reduces to a cosine series. If $f(x)$ satisfies the condition $f(x) = -f(-x)$, i.e. $f(x)$ is an odd function, then its Fourier series reduces to a sine series.

Proof. If $f(x)$ is even, then

$$f(x) \cos kx = f(-x) \cos(-kx);\ f(x) \sin kx = -f(-x) \sin(-kx),$$

whence

$$a_k = \frac{1}{\pi}\int_{-\pi}^{\pi} f(x) \cos kx\, dx = \frac{2}{\pi}\int_0^{\pi} f(x) \cos kx dx;\ b_k = 0, \qquad (k = 0, 1, 2 \ldots),$$

and, therefore,

$$(5)\quad f(x) \sim \tfrac{1}{2} a_0 + \sum_{k=1}^{\infty} a_k \cos kx, \qquad a_k = \frac{2}{\pi}\int_0^{\pi} f(x) \cos kx dx.$$

Similarly, if $f(x)$ is odd, it follows that

$$(5')\qquad f(x) \sim \sum_{k=1}^{\infty} b_k \sin kx, \qquad b_k = \frac{2}{\pi}\int_0^{\pi} f(x) \sin kx dx.$$

It is therefore clear that a function integrable in $[0, \pi]$ can be expanded either into a Fourier cosine or sine series.

3. In order to apply to the series the results of Chapter 1, the following theorem is established:

THEOREM 3. The system of functions

$$(6)\qquad \frac{1}{\sqrt{2\pi}}; \qquad \frac{\cos kx}{\sqrt{\pi}}, \qquad \frac{\sin kx}{\sqrt{\pi}}, \qquad (k = 1, 2, \ldots)$$

is complete in $[0, 2\pi]$, i.e. if $\theta(x)$ is square integrable in $[0, 2\pi]$ such that

$$(7)\quad \int_0^{2\pi} \theta(x) dx = 0;\ \int_0^{2\pi} \theta(x) \cos kx dx = 0,\ \int_0^{2\pi} \theta(x) \sin kx dx = 0, \qquad (k = 1, 2, \ldots),$$

then $\theta(x)$ vanishes a.e. in $[0, 2\pi]$.

Proof. By Vitali's theorem, it will suffice to prove that (Ch. 1, Th. 26, cf. A. Tonolo [112])

$$x = \left(\frac{1}{\sqrt{2\pi}}\int_0^x dx\right)^2 + \sum_{n=1}^{\infty}\frac{1}{\pi}\left[\left(\int_0^x \cos nx\,dx\right)^2 + \left(\int_0^x \sin nx\,dx\right)^2\right]$$

$$= \frac{x^2}{2\pi} + \frac{2}{\pi}\sum_{n=1}^{\infty}\frac{1}{n^2} - \frac{2}{\pi}\sum_{n=1}^{\infty}\frac{\cos nx}{n^2},$$

or, since $\sum_{n=1}^{\infty} 1/n^2 = \pi^2/6$ (cf. W. Rogosinski, Fourier Series, p. 15) it will suffice to prove that

$$\sum_{n=1}^{\infty}\frac{2\cos nx}{n^2} = \frac{x^2}{2} - \pi x + \frac{\pi^2}{3}, \qquad (0 \leqq x \leqq 2\pi). \tag{8}$$

This identity, due to Bernoulli (Hobson [48a] II, p. 480) will be proved directly as follows: Let

$$f(x) = \frac{x^2}{2} - \pi x + \frac{\pi^2}{3}$$

then

$$\int_0^{2\pi} f^2(x)dx = \frac{2\pi^5}{45}. \tag{9}$$

The Fourier coefficients of $f(x)$ with respect to the orthonormal system (6) are

$$A_0 = 0, \quad A_n = \frac{2\sqrt{\pi}}{n^2}, \quad B_n = 0, \qquad (n = 1, 2, \ldots), \tag{10}$$

and thus we have established that the first member of (8) represents the Fourier series of the second member.

Now

$$\sum_{n=1}^{\infty} A_n^2 = 4\pi \sum_{n=1}^{\infty} 1/n^4 = 2\pi^5/45$$

and, in view of (9),

$$\int_0^{2\pi} f^2(x)\,dx = \sum_{n=1}^{\infty} A_n^2$$

Consequently (Ch. 1. Th. 13) the series

$$\sum_{n=1}^{\infty} (2 \cos nx)/n^2 \tag{11}$$

converges in the mean in $[0, 2\pi]$ to $f(x)$. It follows that there exists a sequence of partial sums of the series which converges to $f(x)$ a.e. (Ch. I. Sec. 6. 5, Remark).

On the other hand, the terms of the series of absolute values of (11), whose terms are continuous functions of x, are less than the corresponding terms of the series $2 \sum_{n=1}^{\infty} 1/n^2$. The series (11) is therefore absolutely and uniformly convergent to a continuous function $\psi(x)$. The difference $f(x) - \psi(x)$ vanishes a.e. in $[0, 2\pi]$ and therefore, by the continuity of $f(x)$ and $\psi(x)$, vanishes everywhere.

We can complete the preceding theorem by

THEOREM 4. If $\theta(x)$ is integrable in $[0, 2\pi]$ and satisfies (7) then $\theta(x)$ vanishes a.e. in $(0, 2\pi)$.[1]

Proof. Suppose $\theta(x)$ satisifes (7) and define the absolutely continuous function $\omega(x)$ by the relation

$$\omega(x) = c + \int_0^x \theta(x)dx$$

---

[1] If a function $\theta(x)$, integrable in $[0, \pi]$ and defined in $[\pi, 2\pi]$ by the relation $\theta(x) = \theta(2\pi - x)$ has the property that

$$\int_0^\pi \theta(x)dx = 0, \quad \int_0^\pi \theta(x) \cos kx \, dx = 0, \qquad (k = 1, 2, \ldots),$$

then the relations (7) are satisfied and it follows, therefore, from theorem 4 that $\theta(x)$ vanishes a.e. and consequently the system

$$\frac{1}{\sqrt{\pi}}; \quad \sqrt{\frac{2}{\pi}} \cos kx, \qquad (k = 1, 2, \ldots)$$

is complete in $[0, \pi]$.

Similarly if $\theta(x)$, integrable in $[0, \pi]$ and defined in $[\pi, 2\pi]$ by the relation $\theta(x) = -\theta(2\pi - x)$, has the property that

$$\int_0^\pi \theta(x) \sin kx \, dx = 0,$$

then the relations (7) are satisfied and it follows, therefore, that the system

$$\sqrt{\frac{2}{\pi}} \sin kx, \qquad (k = 1, 2, \ldots),$$

is complete in $[0, \pi]$ (cf. 2).

where

$$c = -\frac{1}{2\pi}\int_0^{2\pi} dx \int_0^x \theta(x)\,dx.$$

Then

$$\int_0^{2\pi} \omega(x)dx = 0, \quad \omega(0) = c, \quad \omega(2\pi) = c + \int_0^{2\pi} \theta(x)dx = c,$$

whence, integrating by parts (Appendix, Th. 17)

$$0 = \int_0^{2\pi} \theta(x) \cos nx\,dx = [\omega(x) \cos nx]_0^{2\pi}$$
$$+ n\int_0^{2\pi} \omega(x) \sin nx\,dx = n\int_0^{2\pi} \omega(x) \sin nx\,dx,$$
$$0 = \int_0^{2\pi} \theta(x) \sin nx\,dx = [\omega(x) \sin nx]_0^{2\pi}$$
$$- n\int_0^{2\pi} \omega(x) \cos nx\,dx = -n\int_0^{2\pi} \omega(x) \cos nx\,dx,$$

and consequently $\omega(x)$ satisfies (7). Since $\omega(x)$ is continuous, and therefore square integrable, by the preceding theorem it vanishes identically. However $\omega'(x) = \theta(x)$ a.e. (Appendix Th. 11). Therefore $\theta(x)$ vanishes a.e. in $[0,2\pi]$. Q.E.D.

COROLLARY. If two integrable functions have the same Fourier constants, they are equal a.e., and conversely.

From theorem 3, by virtue of Theorem 22 of Ch. I, follows

THEOREM 5. If $f(x)$ is square integrable in $[-\pi, \pi]$ its Fourier series

$$\tfrac{1}{2}a_0 + \sum_{k=1}^{\infty} (a_k \cos kx + b_k \sin kx)$$

converges in the mean in $[-\pi, \pi]$ to $f(x)$. That is

$$\lim_{n\to\infty}\int_{-\pi}^{\pi} \left[f(x) - \tfrac{1}{2}a_0 - \sum_{k=1}^{n} (a_k \cos kx + b_k \sin kx)\right]^2 dx = 0.$$

If we now write

$$S_n(x) = \tfrac{1}{2}a_0 + \sum_{k=1}^{n} (a_k \cos kx + b_k \sin kx)$$

it follows (Ch. I, Sec. 6. 5, Remark) that there exists a sequence of partial sums

$$S_{n_1}(x), \quad S_{n_2}(x), \ldots, \quad S_{n_m}(x), \ldots$$

such that

$$\lim_{m\to\infty} S_{n_m}(x) = f(x)$$

a.e. in $[-\pi, \pi]$.

The following paragraphs deal with pointwise convergence of Fourier series.

4. If $f(x)$ is square integrable in $[-\pi, \pi]$ then the Fourier coefficients of $f(x)$ satisfy Bessel's equality (Ch. I, Th. 22)

$$\tfrac{1}{4}A_0^2 + \sum_{k=1}^{\infty} (A_k^2 + B_k^2) = \int_{-\pi}^{\pi} f^2(x)dx,$$

it follows from (3) of Sec. 9. 1 that

$$\tfrac{1}{2}a_0^2 + \sum_{k=1}^{\infty} (a_k^2 + b_k^2) = \frac{1}{\pi}\int_{-\pi}^{\pi} f^2(x)dx, \tag{12}$$

(Parseval's formula (Parseval [86])), and therefore

THEOREM 6. If $f(x)$ is square integrable in $[-\pi, \pi]$, then its Fourier constants satisfy (12).

More generally, if $f(x)$ and $g(x)$ are two functions square integrable in $[-\pi, \pi]$, and we write

$$a_k = \frac{1}{\pi}\int_{-\pi}^{\pi} f(x)\cos kx\,dx, \quad b_k = \frac{1}{\pi}\int_{-\pi}^{\pi} f(x)\sin kx\,dx \tag{13}$$

$$\alpha_k = \frac{1}{\pi}\int_{-\pi}^{\pi} g(x)\cos kx\,dx, \quad \beta_k = \frac{1}{\pi}\int_{-\pi}^{\pi} g(x)\sin kx\,dx \tag{13'}$$

$$(k = 0, 1, 2, \ldots),$$

then, by Theorem 31 of Ch. I and (3) of Sec. 9.1 it follows that

$$\tfrac{1}{2}a_0\alpha_0 + \sum_{k}^{\infty} (a_k\alpha_k + b_k\beta_k) = \frac{1}{\pi}\int_{-\pi}^{\pi} f(x)g(x)dx, \tag{14}$$

(generalized Parseval's formula), and, therefore

THEOREM 6[1]. If $f(x)$ and $g(x)$ are two functions square integrable in $[-\pi, \pi]$, then their Fourier constants, given by (13) and (13′), satisfy (14).

THEOREM 7. If $f(x)$ is square integrable in the finite interval $g = (a, b)$, and σ is any arbitrary positive constant, then there exists a continuous function $P(x)$ defined in (a, b) such that

$$\int_g |f(x) - P(x)|\, dx < \sigma.$$

Proof. Suppose first that g coincides with the interval $(-l\pi, l\pi)$. If $f(x)$ is bounded and therefore square integrable, the proposition to be proved is a consequence of the preceding remark and the remark of Ch. I, sec. 6. 5. In this case it suffices to take $P(x)$ equal to

$$\tfrac{1}{2}a_0 + \sum_{k=1}^{n} \left(a_k \cos \frac{k}{l}\, x + b_k \sin \frac{k}{l}\, x\right)$$

for n sufficiently large.

Suppose $f(x)$ is integrable in g and denote by g_n the set of points of g for which $|f(x)| \leqq n$. If $\delta_n = g - g_n$, then every δ_n contains the succeeding one, their intersection is of measure zero, and therefore (Appendix, Th. 3a) $\lim_{n\to\infty} \int_{\delta_n} |f|\, dx = 0$. Consequently, for $\sigma > 0$, there exists a δ_n such that

$$\int_{\delta_n} |f|\, dx < \sigma/2.$$

Let $f_1(x)$ denote the function such that $f_1(x) = f(x)$ in g_n and $f_1(x) = 0$ in δ_n and $f_2(x)$ denote the function such that $f_2(x) = f(x)$ in δ_n and $f_2(x) = 0$ in g_n. Then $f_1(x)$ is bounded in g and therefore there exists a continuous function $P(x)$ such that

$$\int_g |f_1(x) - P(x)|\, dx < \sigma/2.$$

Consequently

$$\int_g |f - P|\, dx = \int_g |f_1 + f_2 - P|\, dx \leqq \int_g |f_1 - P|\, dx + \int_{\delta_n} |f|\, dx < \sigma.$$

If g does not coincide with $(-l\pi, l\pi)$, determine l so that $-l\pi \leqq a < b \leqq l\pi$ and let $F(x) = f(x)$ in g and $F(x) = 0$ in

$(-l\pi, a)$ and $(b, l\pi)$. By what we have just proved there exists a continuous function $P(x)$ which has the desired property with respect to $F(x)$ in $(-l\pi, l\pi)$ and therefore with respect to $f(x)$ in (a, b).

3. Continuous Functions: Sufficient Conditions for Pointwise Convergence

1. THEOREM 8. If $f(x)$ is continuous in $[-\pi, \pi]$, and its Fourier series

$$\tfrac{1}{2}a_0 + \sum_{k=1}^{\infty} (a_k \cos kx + b_k \sin kx)$$

is uniformly convergent in $[-\pi, \pi]$, the series converges to $f(x)$.

Proof. Let

$$\varphi(x) = \tfrac{1}{2}a_0 + \sum_{k=1}^{\infty} (a_k \cos kx + b_k \sin kx)$$

$\varphi(x)$ is continuous in $[-\pi\ \pi]$. Moreover the series obtained by multiplying the right side of (1) by $\cos nx$ and $\sin nx$ respectively are uniformly convergent. Therefore we may integrate term by term between $-\pi$ and π to obtain

$$a_n = \frac{1}{\pi}\int_{-\pi}^{\pi} \varphi(x) \cos nx\, dx, \qquad b_n = \frac{1}{\pi}\int_{-\pi}^{\pi} \varphi(x) \sin nx\, dx.$$

Consequently $f(x)$ and $\varphi(x)$ have the same Fourier constants and the difference $f(x) - \varphi(x)$ vanishes a.e. (Cor. Th. 4). But $f(x)$ and $\varphi(x)$ are continuous and therefore $f(x) \equiv \varphi(x)$ in $[-\pi, \pi]$.

COROLLARY. If $f(x)$ is continuous in $[-\pi, \pi]$ and its Fourier constants are such that the series $\sum_{k=1}^{\infty}\{|a_k| + |b_k|\}$ is convergent, then the Fourier series of $f(x)$ converges uniformly to $f(x)$ in $[-\pi, \pi]$.

Proof. The absolute values of the terms of the series $\sum_{k=1}^{\infty} |a_k \cos kx + b_k \sin kx|$ are less than the corresponding terms of the series $\sum_{k=1}^{\infty}\{|a_k| + |b_k|\}$. Therefore, the original series is everywhere uniformly convergent.

Examples:

i. Let

$$f(x) = -\tfrac{1}{4}\pi(\pi + x) \quad \text{for} \quad -\pi \leqq x \leqq -\tfrac{1}{2}\pi;$$
$$f(x) = \tfrac{1}{4}\pi x \quad \text{for} \quad -\tfrac{1}{2}\pi \leqq x \leqq \tfrac{1}{2}\pi;$$
$$f(x) = \tfrac{1}{4}\pi(\pi - x) \quad \text{for} \quad \tfrac{1}{2}\pi \leqq x \leqq \pi.$$

Then

$$f(x) \sim \sin x - \frac{\sin 3x}{3^2} + \frac{\sin 5x}{5^2} - \dots$$

and since the series on the right side is uniformly convergent

$$f(x) = \sin x - \frac{\sin 3x}{3^2} + \frac{\sin 5x}{5^2} - \dots$$

ii. Let

$$f(x) = \tfrac{1}{4}\pi(\pi + x) \quad \text{for} \quad -\pi \leqq x \leqq -\tfrac{1}{2}\pi;$$
$$f(x) = -\tfrac{1}{4}\pi x \quad \text{for} \quad -\tfrac{1}{2}\pi \leqq x \leqq 0;$$
$$f(x) = \tfrac{1}{4}\pi x \quad \text{for} \quad 0 \leqq x \leqq \tfrac{1}{2}\pi;$$
$$f(x) = \tfrac{1}{4}\pi(\pi - x) \quad \text{for} \quad \tfrac{1}{2}\pi \leqq x \leqq \pi.$$

Then

$$f(x) = \tfrac{1}{16}\pi^2 - 2\left[\frac{\cos 2x}{2^2} + \frac{\cos 6x}{6^2} + \frac{\cos 10x}{10^2} + \dots\right].$$

iii. Let

$$f(x) = \tfrac{1}{12}\pi^2 - \tfrac{1}{4}x^2 \quad \text{for} \quad -\pi \leqq x \leqq \pi;$$

then

$$\tfrac{1}{12}\pi^2 - \tfrac{1}{4}x^2 = \cos x - \frac{\cos 2x}{2^2} + \frac{\cos 3x}{3^2} - \dots. \qquad \text{(Euler)}$$

2. DEFINITION 3. If $f(x)$ is continuous in the finite interval $[a, b]$, the function $\omega(\delta)$ defined in the interval $[0, b - a]$ which assumes for every value of δ the maximum value of $|f(x'') - f(x')|$

for all x', x'' in $[a, b]$ such that $0 < x'' - x' \leqq \delta$ is called the *modulus of continuity* of $f(x)$ in $[a, b]$. (De la Vallée-Poussin [27b] pp. 7—9).

The function $\omega(\delta)$ has the following properties:

i. $\omega(\delta) \geqq 0$

(2) ii. $\omega(\delta)$ is a non-decreasing function of δ.

iii. $\omega(\delta + h) \leqq \omega(\delta) + \omega(h)$.

For, suppose x', x'' are two points of $[a, b]$ such that $a \leqq x' \leqq x'' \leqq x' + \delta + h$. If $a \leqq x' \leqq x'' \leqq x' + \delta$ then $|f(x'') - f(x')| \leqq \omega(\delta) \leqq \omega(\delta) + \omega(h)$. On the other hand, if $a \leqq x' \leqq x' + \delta < x'' \leqq x' + \delta + h$ then again $|f(x'') - f(x')| \leqq |f(x'') - f(x' + \delta)| + |f(x' + \delta) - f(x')| \leqq \omega(h) + \omega(\delta)$.

iv. From (2) by induction it follows that for every positive integer n

$$\omega(n\delta) \leqq n\omega(\delta). \tag{3}$$

v. If $\lambda > 0$ then

$$\omega(\lambda\delta) \leqq (\lambda + 1)\omega(\delta). \tag{4}$$

If λ is an integer (4) follows from (3). If λ is not an integer and $\lambda + \varepsilon$ is the first integer greater than λ, $0 < \varepsilon < 1$, then

$$\omega(\lambda\delta) \leqq \omega((\lambda + \varepsilon)\delta) \leqq (\lambda + \varepsilon)\omega(\delta) \leqq (\lambda + 1)\omega(\delta).$$

vi. $\omega(\delta)$ is a continuous function of δ. Since $f(x)$ is continuous in the interval $[a, b]$, it is uniformly continuous there. It follows that $\lim_{h\to 0} \omega(h) = 0$. Since, morever,

$$0 \leqq \omega(\delta + h) - \omega(\delta) \leqq \omega(h)$$

it also follows that

$$\lim_{h\to 0} \omega(\delta + h) = \omega(\delta).$$

vii. If there exists a $\delta \neq 0$ such that $\omega(\delta) = 0$, then $f(x)$ is a constant.

Indeed, if $0 \leqq \delta_1 \leqq \delta$ then $\omega(\delta_1) = 0$ and therefore $\omega(x)$ vanishes in every interval of length $\leqq \delta$. Consequently $f(x)$ is constant.

viii. If $f(x)$ is a periodic function then its modulus of continuity can be defined for any value of δ. It should be observed that if $f(x)$ has period 2π, for example, then the maximum value of $\omega(\delta)$ is $\omega(\pi)$. Indeed, two values x', x'' of the variable can be replaced by any two values which are congruent to them modulo 2π, and such that the absolute value of their difference does not exceed π.

3. DEFINITION 4. If $f(x)$ is defined in $[a, b]$ it is called *lipschitzian of order* $\alpha > 0$ in $[a, b]$ if there exists a constant L such that if x_1, x_2 are any two points of $[a, b]$, then

$$| f(x_2) - f(x_1) | \leqq L | x_2 - x_1 |^{\alpha}, \ (\alpha > 0).$$

It follows from the definition that lipschitzian functions are continuous.

Every lipschitzian function of order $\alpha > 1$ is a constant. Indeed

$$0 \leqq | [f(x+h) - f(x)]/h | \leqq Lh^{\alpha-1}$$

whence $f'(x) = 0$ for x in $[a, b]$ and therefore $f(x)$ is constant.

The modulus of continuity $\omega(\delta)$ of a lipschitzian function of order α satisfies the formula $\omega(\delta) \leqq L\delta^{\alpha}$.

Conversely if the modulus of continuity $\omega(\delta)$ of a function $f(x)$ continuous in $[a, b]$ satisfies the inequality $\omega(\delta) \leqq L\delta^{\alpha}$ for $0 \leqq \delta \leqq b - a$ then $f(x)$ is lipschitzian of order α. Indeed, for $0 < x'' - x' = \delta$ follows

$$| f(x'') - f(x') | \leqq \omega(\delta) \leqq L\delta^{\alpha} = L | x'' - x' |^{\alpha}.$$

4. THEOREM 9. If $f(x)$ is continuous with period 2π and $\omega(x)$ is its modulus of continuity, then its Fourier constants satisfy the following inequalities

$$(5) \qquad | a_n | \leqq \omega\left(\frac{\pi}{n}\right), \quad | b_n | \leqq \omega\left(\frac{\pi}{n}\right), \qquad (n = 1, 2, \ldots).$$

Proof. Clearly

$$a_n = \frac{1}{\pi}\int_0^{2\pi} f(x) \cos nx \, dx = \frac{1}{\pi}\int_{-\pi/n}^{2\pi-\pi/n} f\left(\xi + \frac{\pi}{n}\right) \cos (n\xi + \pi) d\xi$$

$$= -\frac{1}{\pi}\int_{-\pi/n}^{2\pi-\pi/n} f\left(x+\frac{\pi}{n}\right)\cos nx\,dx.$$

Also, since $f(x)$ has period 2π

$$a_n = -\frac{1}{\pi}\int_0^{2\pi} f\left(x+\frac{\pi}{n}\right)\cos nx\,dx,$$

whence

$$a_n = \frac{1}{2\pi}\int_0^{2\pi}\left[f(x) - f\left(x+\frac{\pi}{n}\right)\right]\cos nx\,dx,$$

and therefore

$$|\,a_n\,| \leqq \frac{1}{2\pi}\int_0^{2\pi}\omega\left(\frac{\pi}{n}\right)dx = \omega\left(\frac{\pi}{n}\right).$$

The second inequality in (5) can be proved in the same way. From (5) follows the

COROLLARY. If $f(x)$ is continuous with period 2π, then

$$\lim_{n\to\infty} a_n = 0, \quad \lim_{n\to\infty} b_n = 0. \tag{6}$$

In Sec. 4. Th. 13, (6) will be established for any integrable function.

THEOREM 10. If $f(x)$ is periodic with period 2π and has a continuous derivative of order r, and $\omega_r(x)$ denotes the modulus of continuity of $f^{(r)}(x)$, then the Fourier constants satisfy

$$|\,a_n\,| \leqq \frac{1}{n^r}\omega_r\left(\frac{\pi}{n}\right), \quad |\,b_n\,| \leqq \frac{1}{n^r}\omega_r\left(\frac{\pi}{n}\right), \quad (n = 1, 2, \ldots). \tag{7}$$

Proof. Clearly $f'(x), f''(x), \ldots, f^{(r)}(x)$ are periodic with period 2π.

If $\frac{1}{2}a_0'$; a_n', b_n' denote the Fourier constants of $f'(x)$, then

$$a_n' = \frac{1}{\pi}\int_{-\pi}^{\pi} f'(x)\cos nx\,dx$$

$$= \frac{1}{\pi}\left[f(x)\cos nx\right]_{-\pi}^{\pi} + \frac{n}{\pi}\int_{-\pi}^{\pi} f(x)\sin nx\,dx = nb_n,$$

and similarly

$$b'_n = -na_n.$$

By induction it is easily proved that the Fourier constants of the rth derivative except for sign are $n^r a_n$, $n^r b_n$ (or these expressions in reverse order). Consequently by Theorem 9

$$n^r |a_n| \leqq \omega_r \left(\frac{\pi}{n}\right), \quad n^r |b_n| \leqq \omega_r \left(\frac{\pi}{n}\right)$$

which is equivalent to (7).

5. THEOREM 11. If $f(x)$ is continuous with period 2π and has a continuous derivative which satisfies a Lipschitz condition of order $\alpha > 0$, then the Fourier series of $f(x)$ converges uniformly to $f(x)$.

Proof. From Theorem 10 follows immediately

$$|a_n| \leqq \frac{1}{n} \omega_1 \left(\frac{\pi}{n}\right), \quad |b_n| \leqq \frac{1}{n} \omega_1 \left(\frac{\pi}{n}\right).$$

Moreover, from Sec. 3.3

$$\omega_1 \left(\frac{\pi}{n}\right) \leqq L \left(\frac{\pi}{n}\right)^\alpha,$$

whence

$$\sum_{n=1}^{\infty} \{|a_n| + |b_n|\} \leqq 2L\pi^\alpha \sum_{n=1}^{\infty} \frac{1}{n^{1+\alpha}}.$$

Now from the convergence of the series $\sum_{n=1}^{\infty} 1/n^{1+\alpha}$, it follows that the series $\sum_{n=1}^{\infty} \{|a_n| + |b_n|\}$ also converges and consequently Theorem 11 follows by the corollary of Theorem 8.

Theorem 11 is a particular case of Corollary 2 of Theorem 17 which will be proved in the next section.

4. *Criteria for Pointwise Convergence*

1. THEOREM 12. If $f(x)$ is integrable in the finite closed interval $[A, B]$, a and b are any two numbers such that $A \leqq a \leqq b \leqq B$, and μ is any real number, then

$$\lim_{\mu\to\infty} \int_a^b f(x) \cos \mu x \, dx = 0, \qquad \lim_{\mu\to\infty} \int_a^b f(x) \sin \mu x \, dx = 0.$$

Moreover, the integrals tend to zero uniformly with respect to a and b, that is, for any $\varepsilon > 0$ there exists a number $k > 0$ independent of a and b such that for all $\mu > k$ it follows

$$\left| \int_a^b f(x) \cos \mu x \, dx \right| < \varepsilon, \qquad \left| \int_a^b f(x) \sin \mu x \, dx \right| < \varepsilon$$

(H. Lebesgue [67a], p. 61). For infinite intervals cf. Sec 9.3: The Fourier Integral.

Proof. Suppose first that $f(x)$ is continuous in $[A, B]$ and take μ so that

$$a < a + \frac{\pi}{\mu} < b - \frac{\pi}{\mu} < b.$$

Then

$$\int_a^b f(x) \cos \mu x \, dx = \int_a^{a+\pi/\mu} f(x) \cos \mu x \, dx + \int_{a+\pi/\mu}^b f(x) \cos \mu x \, dx$$

$$= \int_a^{a+\pi/\mu} f(x) \cos \mu x \, dx - \int_a^{b-\pi/\mu} f\left(x + \frac{\pi}{\mu}\right) \cos \mu x \, dx,$$

$$\int_a^b f(x) \cos \mu x \, dx = \int_a^{b-\pi/\mu} f(x) \cos \mu x \, dx + \int_{b-\pi/\mu}^b f(x) \cos \mu x \, dx.$$

Therefore

$$2\int_a^b f(x) \cos \mu x \, dx = \int_a^{a+\pi/\mu} f(x) \cos \mu x \, dx \tag{1}$$

$$+ \int_{b-\pi/\mu}^b f(x) \cos \mu x dx + \int_a^{b-\pi/\mu} \left[f(x) - f\left(x + \frac{\pi}{\mu}\right)\right] \cos \mu x \, dx.$$

Since $f(x)$ is continuous in $[A, B]$, there exists a positive constant

M such that $|f(x)| < M$, and therefore the first two integrals of (1) are in absolute value less than $M\pi/\mu$. Moreover, by the uniform continuity of $f(x)$ in $[A, B]$, for any $\sigma > 0$, there exists a number $k > 0$ such that for all $\mu > k$ and for all x in $[A, B]$

$$\left| f(x) - f\left(x + \frac{\pi}{\mu}\right) \right| < \sigma. \tag{2}$$

Consequently from (1) follows

$$\left| \int_a^b f(x) \cos \mu x \, dx \right| < M \frac{\pi}{\mu} + \frac{B - A}{2} \sigma. \tag{3}$$

Now for $\varepsilon > 0$, choose σ so that $\sigma < \varepsilon/(B - A)$ and determine k so that simultaneously (2) holds for $\mu > k$ and $M\pi/k < \varepsilon/2$, then for all $\mu > k$ (where k is independent of a and b) it follows from (3) that

$$\left| \int_a^b f(x) \cos \mu x \, dx \right| < \varepsilon.$$

The theorem will now be proved for $f(x)$ integrable in $[A, B]$. By Theorem 7 there exists a function $P(x)$ continuous in $[A, B]$ such that

$$\int_A^B | f(x) - P(x) | \, dx < \frac{\varepsilon}{2}.$$

Clearly

$$\left| \int_a^b f(x) \cos \mu x \, dx \right| = \left| \int_a^b [f(x) - P(x) + P(x)] \cos \mu x \, dx \right|$$
$$\leqq \int_a^b |f(x) - P(x)| \, dx + \left| \int_a^b P(x) \cos \mu x \, dx \right|$$

whence

$$\left| \int_a^b f(x) \cos \mu x \, dx \right| < \frac{\varepsilon}{2} + \left| \int_a^b P(x) \cos \mu x \, dx \right|. \tag{4}$$

Since $P(x)$ is continuous in $[A, B]$, there exists a k independent of a and b such that for $\mu > k$

$$\left| \int_a^b P(x) \cos \mu x \, dx \right| < \frac{\varepsilon}{2}. \tag{5}$$

From (4) and (5) the theorem follows for the first integral in question. The reasoning for the second integral is similar.

COROLLARY 1. If $f(x)$ is integrable in $[A, B]$, α is any point of $[A, B]$, and a and b satisfy

$$A \leqq \alpha + a \leqq \alpha + b \leqq B, \tag{6}$$

then

$$\lim_{\mu \to \infty} \int_a^b f(\alpha + x) \cos \mu x \, dx = 0, \quad \lim_{\mu \to \infty} \int_a^b f(\alpha + x) \sin \mu x \, dx = 0$$

and the integrals tend to zero uniformly with respect to α, a, b satisfying (6).

Proof. Let $\alpha + x = t$. Then

$$\int_a^b f(\alpha + x) \cos \mu x \, dx = \int_{a+\alpha}^{b+\alpha} f(t) \cos \mu (t - \alpha) \, dt$$
$$= \cos \mu\alpha \int_{a+\alpha}^{b+\alpha} f(t) \cos \mu t \, dt + \sin \mu\alpha \int_{a+\alpha}^{b+\alpha} f(t) \sin \mu t \, dt.$$

Consequently

$$\left| \int_a^b f(\alpha+x) \cos \mu x \, dx \right| \leqq \left| \int_{a+\alpha}^{b+\alpha} f(t) \cos \mu t \, dt \right| + \left| \int_{a+\alpha}^{b+\alpha} f(t) \sin \mu t \, dt \right|.$$

The corollary follows immediately for the first integral from Theorem 12. The proof for the second integral is similar.

COROLLARY 2. If $f(x)$ is integrable in $[A, B]$, $g(x)$ is absolutely continuous in $[A - B, B - A]$, and α is in $[A, B]$, then for α, a, b satisfying (6)

$$\lim_{\mu \to \infty} \int_a^b f(\alpha + x) g(x) \cos \mu x \, dx = 0, \quad \lim_{\mu \to \infty} \int_a^b f(\alpha + x) g(x) \sin \mu x \, dx = 0$$

and the integrals tend to zero uniformly with respect to α, a, b satisfying (6).

Proof. Let

$$F(x, \alpha) = \int_a^x f(\alpha + \beta) \cos \mu\beta \, d\beta.$$

Integrating by parts (Appendix, Th. 17) gives

$$\int_a^b f(\alpha + x)g(x) \cos \mu x \, dx = [g(x)F(x, \alpha)]_{x=a}^{x=b} - \int_a^b F(x, \alpha)g'(x)dx$$

$$= g(b) \int_a^b f(\alpha + \beta) \cos \mu\beta \, d\beta - \int_a^b F(x, \alpha) g'(x) \, dx.$$

By Corollary 1, for $\varepsilon > 0$, there exists a $k > 0$ independent of a, α, x, such that for $\mu > k$

$$\left| \int_a^x f(\alpha + \beta) \cos \mu\beta \, d\beta \right| = \left| F(x, \alpha) \right| < \varepsilon$$

and if $|g(x)| < L$ for all x in $[A - B, B - A]$, then

$$\left| \int_a^b f(\alpha + x)g(x) \cos \mu x \, dx \right| < \varepsilon \left[L + \int_{A-B}^{B-A} | g'(x) | \, dx \right]$$

and since ε is arbitrary the corollary follows.

THEOREM 13. If $f(x)$ is integrable in $(-\pi, \pi)$ the sequence $\{a_n, b_n\}$ of its Fourier constants tends to zero as $n \to \infty$, (Lebesgue [67a], p. 45; cf. also Cor. Th. 9).

Proof. Since

$$a_n = \frac{1}{\pi} \int_{-\pi}^{\pi} f(x) \cos nx \, dx, \qquad b_n = \frac{1}{\pi} \int_{-\pi}^{\pi} f(x) \sin nx \, dx$$

the proposition follows immediately from Theorem 12.

2. The sum of the first n terms of the Fourier series of a function integrable in $(-\pi, \pi)$ can be expressed as an integral (Dirichlet's formula).

Let

$$f(x) \sim \frac{a_0}{2} + \sum_{k=1}^{\infty} (a_k \cos kx + b_k \sin kx)$$

and write

$$S_n(x) = \frac{a_0}{2} + \sum_{k=1}^{n} (a_k \cos kx + b_k \sin kx). \tag{1}$$

Throughout this section it will be assumed that $f(x)$ is periodic in $(-\infty, +\infty)$ with period 2π and in particular that $f(\pi) = f(-\pi)$. Then by Sec. 2.1,

$$\begin{aligned} S_n(x) &= \frac{1}{2\pi}\int_{x-\pi}^{x+\pi} f(\alpha)d\alpha \\ &\quad + \sum_{k=1}^{n} \frac{1}{\pi}\int_{x-\pi}^{x+\pi} f(\alpha)[\cos k\alpha \cos kx + \sin k\alpha \sin kx]\,d\alpha \\ &= \frac{1}{\pi}\int_{x-\pi}^{x+\pi} f(\alpha)\left[\tfrac{1}{2} + \sum_{k=1}^{n} \cos k(\alpha - x)\right] d\alpha. \end{aligned}$$

Substituting $\alpha - x = \beta$ gives

$$S_n(x) = \frac{1}{\pi}\int_{-\pi}^{\pi} f(x+\beta)\left[\tfrac{1}{2} + \sum_{k=1}^{n} \cos k\beta\right] d\beta. \tag{2}$$

Equating imaginary parts of both sides of the following identity

$$\begin{aligned} \tfrac{1}{2} + \sum_{k=1}^{n} e^{k\beta i} &= \frac{e^{(n+1)\beta i} - 1}{e^{\beta i} - 1} - \frac{1}{2} = \frac{e^{(n+\frac{1}{2})\beta i} - e^{-\frac{1}{2}\beta i}}{e^{\frac{\beta i}{2}} - e^{-\frac{\beta i}{2}}} - \frac{1}{2} \\ &= \frac{1}{2}\left[\frac{e^{(n+\frac{1}{2})\beta i} - e^{-\frac{1}{2}\beta i}}{i \sin \frac{\beta}{2}} - 1\right] = \frac{-ie^{(n+\frac{1}{2})\beta i} + i\cos\frac{\beta}{2}}{2\sin\frac{\beta}{2}} \end{aligned}$$

gives

$$\tfrac{1}{2} + \sum_{k=1}^{n} \cos k\beta = \frac{\sin(n+\frac{1}{2})\beta}{2\sin\frac{\beta}{2}}, \quad \sum_{k=1}^{n} \sin k\beta = \frac{\cos\frac{\beta}{2} - \cos(n+\frac{1}{2})\beta}{2\sin\frac{\beta}{2}}. \tag{3}$$

Substituting this result in (2) yields the Dirichlet formula for $S_n(x)$:

(4) $$S_n(x) = \frac{1}{\pi}\int_{-\pi}^{\pi} f(x+\alpha)\frac{\sin(n+\frac{1}{2})\alpha}{2\sin\frac{\alpha}{2}}\,d\alpha.$$

3. THEOREM 14 (Riemann [92]). The behaviour of the Fourier series at a point x depends on the values of $f(x)$ in an arbitrarily small neighborhood of x.

Proof. Take ε so that $0 < \varepsilon < \pi$ and divide the interval of integration in Dirichlet's formula into the three intervals $(-\pi, -\varepsilon)$, $(-\varepsilon, \varepsilon)$, (ε, π).
Then

$$\int_{-\pi}^{\pi} f(x+\alpha)\frac{\sin(n+\frac{1}{2})\alpha}{2\sin(\alpha/2)}\,d\alpha = \int_{-\pi}^{-\varepsilon} f(x+\alpha)\frac{1}{2\sin(\alpha/2)}\sin(n+\tfrac{1}{2})\alpha\,d\alpha$$

$$+\int_{\varepsilon}^{\pi} f(x+\alpha)\frac{1}{2\sin(\alpha/2)}\sin(n+\tfrac{1}{2})\alpha\,d\alpha + \int_{-\varepsilon}^{\varepsilon} f(x+\alpha)\frac{\sin(n+\frac{1}{2})\alpha}{2\sin(\alpha/2)}\,d\alpha.$$

The integrand of the first integral on the right side is the product of $f(x+\alpha)$, $1/(2\sin\alpha/2)$ [with continuous derivative in $(-\pi, -\varepsilon)$] and $\sin(n+\frac{1}{2})\alpha$. Therefore, reasoning as in the proof of Theorem 12, Cor. 2, we find

$$\lim_{n\to\infty}\int_{-\pi}^{-\varepsilon} f(x+\alpha)\frac{1}{2\sin\frac{\alpha}{2}}\sin(n+\tfrac{1}{2})\alpha\,d\alpha = 0,$$

and

$$\lim_{n\to\infty}\int_{\varepsilon}^{\pi} f(x+\alpha)\frac{1}{2\sin\frac{\alpha}{2}}\sin(n+\tfrac{1}{2})\alpha\,d\alpha = 0.$$

Both of these integrals converge to zero with respect to x and ε varying in (ε_0, π), $0 < \varepsilon_0 < \pi$. Therefore, the determination of the limit of $S_n(x)$ as $n \to \infty$ reduces to determining the limit as $n \to \infty$ of

$$\int_{-\varepsilon}^{\varepsilon} f(x+\alpha)\frac{\sin(n+\frac{1}{2})\alpha}{2\sin\frac{\alpha}{2}}\,d\alpha.$$

But the value of this integral depends only on the values which $f(x)$ assumes in $(x-\varepsilon, x+\varepsilon)$, and the value of the integral remains unchanged even if the values $f(x)$ assumes in $(x-\varepsilon, x+\varepsilon)$ are changed in a set of measure zero.

Riemann's theorem can be restated thus: If $f(x)$ and $g(x)$ are two functions integrable in $(-\pi, \pi)$ and whose values coincide in an interval $[a, b]$ contained in $(-\pi, \pi)$, or differ in $[a, b]$ for a set of measure zero, then their Fourier series converge or diverge for the same values of x in $[a, b]$ and for those values of x for which they converge they have the same sum. Moreover, if one of the series converges uniformly in a set g interior to $[a, b]$ then the other series converges uniformly in the same set g.

4. We now establish a necessary and sufficient condition for the sum $S_n(x)$ of the Fourier series of $f(x)$ to converge to a finite value $S(x)$ for a given x, where $S(x)$ need not be equal to $f(x)$.

From (3) of Sec. 9. 2

$$\tfrac{1}{2}+\sum_{k=1}^{n}\cos k\alpha=\frac{\sin(n+\frac{1}{2})\alpha}{2\sin\frac{\alpha}{2}}.$$

Integrating both sides from $-\pi$ to π gives

$$1=\frac{1}{\pi}\int_{-\pi}^{\pi}\frac{\sin(n+\frac{1}{2})\alpha}{2\sin\frac{\alpha}{2}}\,d\alpha,\qquad S(x)=\frac{1}{\pi}\int_{-\pi}^{\pi}S(x)\frac{\sin(n+\frac{1}{2})\alpha}{2\sin\frac{\alpha}{2}}\,dx.$$

Therefore, by (4) of Sec. 3. 2

$$(1)\quad S_n(x)-S(x)=\frac{1}{\pi}\int_{-\pi}^{\pi}[f(x+\alpha)-S(x)]\frac{\sin(n+\frac{1}{2})\alpha}{2\sin\frac{\alpha}{2}}\,d\alpha.$$

This establishes

THEOREM 15. A necessary and sufficient condition for the Fourier series of an integrable function $f(x)$ to converge at a point

x to the value $S(x)$ is that, given any positive number σ, there exists an integer $N(\sigma)$ such that for every integer $n > N(\sigma)$

$$\frac{1}{\pi}\left|\int_{-\pi}^{\pi} [f(x+\alpha) - S(x)] \frac{\sin (n+\frac{1}{2})\alpha}{2 \sin \frac{\alpha}{2}} \, d\alpha\right| < \sigma.$$

If $-\pi \leqq A \leqq B \leqq \pi$ and $S(x)$ is continuous in $[A, B]$, the convergence of the Fourier series of $f(x)$ to $S(x)$ is uniform for all x in $[A, B]$, if for every σ, the $N(\sigma)$ is independent of x. (When $A = -\pi$ and $B = \pi$, it will be assumed here and subsequently that $S(-\pi) = S(\pi)$.

An equivalent form of Theorem 15 will now be derived.

From (1) follows

$$\text{(2)} \qquad S_n(x) - S(x) = \frac{1}{\pi}\int_{-\pi}^{\pi} [f(x+\alpha) - S(x)] \frac{\sin (n+\frac{1}{2})\alpha}{\alpha} \, d\alpha$$

$$+ \frac{1}{\pi}\int_{-\pi}^{\pi} [f(x+\alpha) - S(x)] \sin (n+\tfrac{1}{2})\alpha \left[\frac{1}{2 \sin \frac{\alpha}{2}} - \frac{1}{\alpha}\right] d\alpha.$$

If we let $g(\alpha) = 1/(2 \sin \alpha/2) - 1/\alpha$ for $\alpha \neq 0$ and $g(0) = 0$, then $g(\alpha)$ and $g'(\alpha)$ are clearly continuous in $[-\pi, \pi]$. Therefore, by Corollaries 1 and 2 of Theorem 12, the second integral on the right of (2) converges to zero as $n \to \infty$ (uniformly with respect to x if $S(x)$ is continuous and therefore bounded in $[-\pi, \pi]$). Consequently

$$\lim_{n\to\infty} [S_n(x) - S(x)] = \lim_{n\to\infty} \frac{1}{\pi}\int_{-\pi}^{\pi} [f(x+\alpha) - S(x)] \frac{\sin (n+\frac{1}{2})\alpha}{\alpha} \, d\alpha.$$

If we divide the interval of convergence of the integral on the right side into the three intervals $(-\pi, -\varepsilon)$, $(-\varepsilon, \varepsilon)$, (ε, π), and reason as in the proof of Theorem 14, we obtain

$$\text{(3)} \quad \lim_{n\to\infty} [S_n(x) - S(x)] = \lim_{n\to\infty} \frac{1}{\pi}\int_{-\varepsilon}^{\varepsilon} [f(x+\alpha) - S(x)] \frac{\sin (n+\frac{1}{2})\alpha}{\alpha} \, d\alpha,$$

$$0 < \varepsilon \leqq \pi;$$

which gives

THEOREM 15[1]. A necessary and sufficient condition for the Fourier series of an integrable function $f(x)$ to converge at a point x to the value $S(x)$ is that for any $\sigma > 0$ there should exist a positive $\varepsilon \leqq \pi$ and an integer $N(\sigma)$ such that for every integer $n > N(\sigma)$

$$\left| \frac{1}{\pi} \int_{-\varepsilon}^{\varepsilon} [f(x + \alpha) - S(x)] \frac{\sin (n + \frac{1}{2})\alpha}{\alpha} d\alpha \right| < \sigma.$$

If $-\pi \leqq A \leqq B \leqq \pi$ and $S(x)$ is continuous in $[A, B]$, the convergence of the Fourier series of $f(x)$ to $S(x)$ is uniform for all x in $[A, B]$ if ε and $N(\sigma)$ are independent of x.

Replacing α by 2α we obtain

$$\frac{1}{\pi} \int_{-\varepsilon}^{\varepsilon} [f(x + \alpha) - S(x)] \frac{\sin (n + \frac{1}{2})\alpha}{\alpha} d\alpha$$

$$= \frac{1}{\pi} \int_{-\varepsilon/2}^{\varepsilon/2} [f(x + 2\alpha) - S(x)] \frac{\sin (2n + 1)\alpha}{\alpha} d\alpha$$

$$= \frac{1}{\pi} \int_{0}^{\varepsilon/2} [f(x + 2\alpha) - S(x)] \frac{\sin (2n + 1)\alpha}{\alpha} d\alpha$$

$$+ \frac{1}{\pi} \int_{-\varepsilon/2}^{0} [f(x + 2\alpha) - S(x)] \frac{\sin (2n + 1)\alpha}{\alpha} d\alpha$$

$$= \frac{1}{\pi} \int_{0}^{\varepsilon/2} [f(x + 2\alpha) + f(x - 2\alpha) - 2S(x)] \frac{\sin (2n + 1)\alpha}{\alpha} d\alpha,$$

so that if we let

$$f(x + 2\alpha) + f(x - 2\alpha) - 2S(x) = \varphi(x, \alpha) \tag{4}$$

and replace ε by 2ε, then (3) becomes

$$\lim_{n \to \infty} [S_n(x) - S(x)] = \lim_{n \to \infty} \frac{1}{\pi} \int_{0}^{\varepsilon} \varphi(x, \alpha) \frac{\sin (2n + 1)\alpha}{\alpha} d\alpha, \tag{5}$$

which gives

THEOREM 15^2. A necessary and sufficient condition for the Fourier series of an integrable function $f(x)$ to converge at a point x to the value $S(x)$ is that for every $\sigma > 0$ there should exist a positive $\varepsilon \leqq \pi/2$, and an integer $N(\sigma)$ such that for every integer $n > N(\sigma)$

$$\left| \frac{1}{\pi} \int_0^{\varepsilon} \varphi(x, \alpha) \frac{\sin (2n+1)\alpha}{\alpha} \, d\alpha \right| < \sigma, \quad 0 < \varepsilon \leqq \pi/2.$$

It should be observed that by (4) $\varphi(x, \alpha)$ is integrable and $[\sin (2n + 1)\alpha]/\alpha$ and its derivative can be defined at $\alpha = 0$ so that they are both continuous in $[0, \varepsilon]$.

If $-\pi \leqq A \leqq B \leqq \pi$ and $S(x)$ is continuous in $[A, B]$, the convergence of the Fourier series of $f(x)$ to $S(x)$ is uniform for all x in $[A, B]$ if ε and $N(\sigma)$ are independent of x.

THEOREM 15^3. A necessary and sufficient condition for the Fourier series of an integrable function $f(x)$ to converge at a point x to the value $S(x)$ is that for every $\sigma > 0$ there should exist a positive $\varepsilon \leqq \pi/2$ and a number $M(\sigma) > 0$, such that for all numbers $m > M(\sigma)$

$$\left| \frac{1}{\pi} \int_0^{\varepsilon} \varphi(x, \alpha) \frac{\sin m\alpha}{\alpha} \, d\alpha \right| < \sigma.$$

Proof. Suppose the series is convergent. By theorem 15^2 and (4), for any $\sigma > 0$, there exists a positive $\varepsilon \leqq \pi/2$ and an integer $N(\sigma)$ such that for every $n > N(\sigma)$

$$\left| \frac{1}{\pi} \int_0^{\varepsilon} \varphi(x, \alpha) \frac{\sin (2n+1)\alpha}{\alpha} \, d\alpha \right| < \frac{\sigma}{2}$$

and

$$\frac{1}{\pi} \int_0^{\varepsilon} |\varphi(x, \alpha)| \, d\alpha < \sigma/4. \tag{5'}$$

Now let $M(\sigma) = 2N(\sigma) + 3$ and let m be any number such that $m > M(\sigma)$. If $2n + 1$ denotes the largest odd number $\leqq m$ and δ is such that $m = 2n + 1 + 2\delta$ then $0 \leqq \delta < 1$ and $n > N(\sigma)$. It follows that

$$\left|\frac{1}{\pi}\int_0^\varepsilon \varphi(x, \alpha)\frac{\sin m\alpha}{\alpha}d\alpha\right| \leqq \left|\frac{1}{\pi}\int_0^\varepsilon \varphi(x, \alpha)\frac{1}{\alpha}[\sin m\alpha - \sin (2n+1)\alpha]\,d\alpha\right|$$

$$+\left|\frac{1}{\pi}\int_0^\varepsilon \varphi(x, \alpha)\frac{\sin (2n+1)\alpha}{\alpha}d\alpha\right|$$

$$< \left|\frac{1}{\pi}\int_0^\varepsilon \varphi(x, \alpha)2\delta\frac{\sin \delta\alpha}{\delta\alpha}\cos (2n+1+\delta)\alpha\, d\alpha\right| + \frac{\sigma}{2}$$

$$\leqq \frac{2}{\pi}\int_0 |\varphi(x, \alpha)|\, d\alpha + \frac{\sigma}{2} < \sigma,$$

which proves the condition is necessary. That the condition is sufficient is obvious from Theorem 15^2.

As in the previous theorems, if ε and $M(\sigma)$ are independent of x for x in $[A, B]$, then the convergence of the Fourier series of $f(x)$ to $S(x)$ is uniform in $[A, B]$. In fact, $S(x)$ is continuous in $[A, B]$ and by (4) the inequality (5′) can be satisfied for all x in $[A, B]$.

It will be observed that Theorem 15^3 has removed the restriction of Theorem 15^2 that m should be a positive odd number. This result will be of importance later in the proof of Theorem 51.

5. THEOREM 16 (Dini's theorem). A sufficient condition for the Fourier series of a periodic function with period 2π integrable in $[-\pi, \pi]$ to converge at a point x to the value $S(x)$ is that there should exist a $\delta > 0$ such that $\varphi(x, \alpha)/\alpha$ is integrable with respect to α in $[0, \delta]$. (Dini, [28 b] p. 102).

Proof. Since $\varphi(x, \alpha)/\alpha$ is integrable in $[0, \delta]$ by hypothesis, then by a theorem on absolute continuity of indefinite integrals of integrable functions (Appendix, Th. 6) it follows that

$$\lim_{\varepsilon\to 0}\int_0^\varepsilon \frac{|\varphi(x, \alpha)|}{\alpha}d\alpha = 0$$

whence, for any $\sigma > 0$, there exists an $\varepsilon > 0$ such that

$$\frac{1}{\pi}\int_0^\varepsilon \frac{|\varphi(x, \alpha)|}{\alpha}d\alpha < \sigma,$$

but clearly for any integer n

$$\frac{1}{\pi}\left|\int_0^\varepsilon \varphi(x, \alpha)\frac{\sin(2n+1)\alpha}{\alpha}\,d\alpha\right| \leqq \frac{1}{\pi}\int_0^\varepsilon \frac{|\varphi(x, \alpha)|}{\alpha}\,d\alpha < \sigma,$$

and the theorem follows at once.

If $-\pi \leqq A \leqq B \leqq \pi$ and $S(x)$ is continuous in $[A, B]$, the convergence of the Fourier series of $f(x)$ to $S(x)$ is uniform for all x in $[A, B]$ if ε is independent of x.

COROLLARY 1. If for a given x there exists a $\delta > 0$ such that for all values of δ for which $0 \leqq \alpha \leqq \delta$ it follows that

$$|\varphi(x, \alpha)| \leqq L\alpha^r \quad (r > 0, L \text{ constant}) \tag{6}$$

then the condition of Dini's theorem is satisfied.

Proof. Indeed

$$\int_0^\varepsilon \frac{|\varphi(x, \alpha)|}{\alpha}\,d\alpha \leqq L\int_0^\varepsilon \frac{1}{\alpha^{1-r}}\,d\alpha = L\,\frac{\varepsilon^r}{r}.$$

If (6) is valid for all x in $[A, B]$ and for $0 < \alpha \leqq \delta$ and L is independent of x, then the convergence of the Fourier series of $f(x)$ to $S(x)$ is uniform in $[A, B]$.

Sufficient conditions for the convergence of the Fourier series of $f(x)$ at a point x to $f(x)$ can obviously be obtained by writing $f(x)$ in place of $S(x)$ in the preceding theorems and considering the function

$$\varphi(x, \alpha) = f(x + 2\alpha) + f(x - 2\alpha) - 2f(x).$$

Thus we obtain

COROLLARY 2. If for a given x, there exists a $\delta > 0$ such that

$$\frac{f(x + 2\alpha) + f(x - 2\alpha) - 2f(x)}{\alpha}$$

is integrable with respect to α in $[0, \delta]$ and in particular if for α in $[0, \delta]$

$$|f(x + 2\alpha) + f(x - 2\alpha) - 2f(x)| < L\alpha^r, \quad (r > 0, L \text{ constant}) \tag{7}$$

then the Fourier series of $f(x)$ is convergent at the point x to $f(x)$.

If $-\pi \leqq A \leqq B \leqq \pi$ and $f(x)$ is continuous in $[A, B]$ (we suppose $f(-\pi) = f(\pi)$ in case $A = -\pi$ and $B = \pi$), and for all x in $[A, B]$ there exists a $\delta > 0$ and a constant L independent of x such that for α in $[0, \delta]$, (7) is valid, then the Fourier series of $f(x)$ converges uniformly in $[A, B]$ to $f(x)$.

At points where $f(x)$ is continuous or has finite discontinuities, the following theorem results from Dini's theorem by writing $S(x) = [f(x+) + f(x-)]/2$. It will be assumed that $f[(-\pi)-] = f(\pi-)$, $f[(-\pi)+] = f(\pi+)$.

THEOREM 17. If $f(x)$ is periodic with period 2π, $f(x)$ is integrable in $[-\pi, \pi]$, and corresponding to any point x where $f(x)$ is continuous or has a finite discontinuity, there exists a $\delta > 0$ such that for all h in $[0, \delta]$, the two ratios

$$\frac{f(x+h) - f(x+)}{h}, \quad \frac{f(x-h) - f(x-)}{-h}$$

are integrable in $[0, \delta]$, then the Fourier series of $f(x)$ converges at x to $[f(x+) + f(x-)]/2$. (The last part of the hypothesis is clearly satisfied if the two ratios

$$[f(x+h) - f(x+)]/h, \quad [f(x-h) - f(x-)]/(-h),$$

are integrable in $[0, \delta]$).

At a point where $f(x)$ is regular (Appendix, Def. 15) and the conditions of this theorem are verified, the Fourier series converges to $f(x)$.

COROLLARY 1. If $f(x)$ is periodic with period 2π and is integrable in $[-\pi, \pi]$, then at a point x where $f(x)$ is continuous, or has a finite discontinuity, the Fourier series of $f(x)$ converges to $[f(x+) + f(x-)]/2$ provided that there exists a $\delta > 0$ such that for $0 < h < \delta$

$$(8) \quad |f(x+h) - f(x+)| < Lh^r, \quad |f(x-h) - f(x-)| < Lh^r, \quad (r > 0, L \text{ constant}).$$

REMARK. Later in Sec. 6, Theorem 37, Corollaries 3 and 4, it will appear that at points x where $f(x)$ is continuous or has finite discontinuities and for which its Fourier series converges, then the sum of the series is $[f(x+) + f(x-)]/2$.

COROLLARY 2. If $f(x)$ is periodic with period 2π and is integrable in $[-\pi, \pi]$, then at a point x where $f(x)$ is continuous and has finite left-hand and right-hand derivatives the Fourier series of $f(x)$ converges to $f(x)$.

This follows immediately from Corollary 1 since (8) is clearly satisfied for $r = 1$.

If $-\pi \leqq A \leqq B \leqq \pi$, $f(x)$ is continuous in $[A, B]$ and $f'(x)$ exists and is bounded in $[A, B]$, then the Fourier series of $f(x)$ converges uniformly in $[A, B]$ to $f(x)$.

6. The preceding theorems and corollaries will be illustrated by the following examples.

1) Let $f(x) = (1/12)(\pi^2 x - x^3)$ for $-\pi \leqq x \leqq \pi$. The Fourier series of $f(x)$ is a sine series Sec. 2.2], and it can be easily verified that

$$b_k = (-1)^{k+1}/k^3.$$

Therefore the following relation holds uniformly in $(-\pi, \pi)$ (Th. 17, Cor. 2)

$$(1) \quad \frac{1}{12}(\pi^2 x - x^3) = \sin x - \frac{\sin 2x}{2^3} + \frac{\sin 3x}{3^3} - \ldots, [-\pi \leqq x \leqq \pi].$$

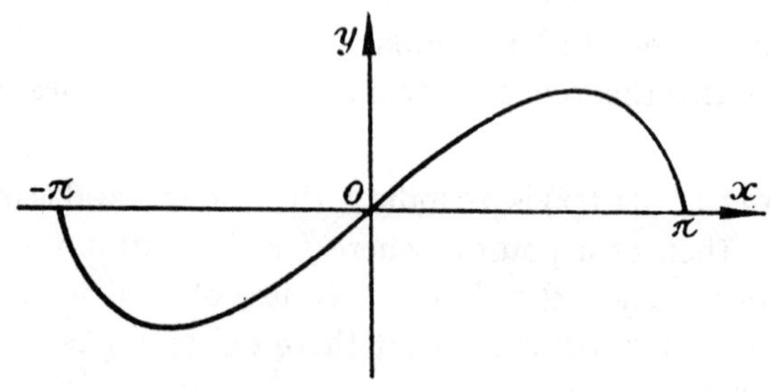

Fig. 1.

2) Let $f(x) = x$ for $-\pi < x < \pi$; $f(-\pi) = f(\pi) = 0$. Then

$$f(x) \sim 2\left[\frac{\sin x}{1} - \frac{\sin 2x}{2} + \frac{\sin 3x}{3} - \ldots\right].$$

$f(x)$ has a bounded derivative (equal to 1) in every interval interior to $[-\pi, \pi]$, at $-\pi$ and $+\pi$ the increment ratios

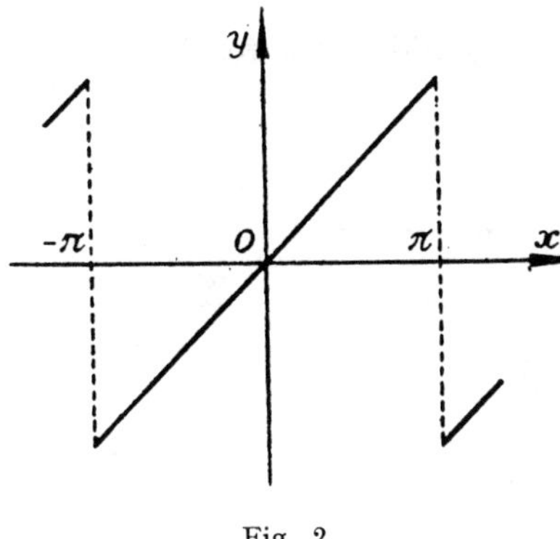

Fig. 2.

$$\frac{f(-\pi \pm h) - f(-\pi \pm)}{\pm h},$$

$$\frac{f(\pi \pm h) - f(\pi \pm)}{\pm h}$$

are equal to 1, and since

$$f(-\pi +) + f(-\pi -) = 0$$

$$[f(\pi +) + f(\pi -) = 0],$$

it follows that for $-\pi \leqq x \leqq \pi$ (Euler [31 a)]

$$(2) \quad f(x) = 2\left[\frac{\sin x}{1} - \frac{\sin 2x}{2} + \frac{\sin 3x}{3} - \dots\right], \; (-\pi \leqq x \leqq \pi),$$

and the convergence of the series is uniform in every interval interior to $[-\pi, \pi]$.

3) Let $f(x) = (\pi - x)/2$ for $0 < x \leqq \pi$, $f(0) = 0$, $f(x) = -f(-x)$ for $-\pi \leqq x < 0$.

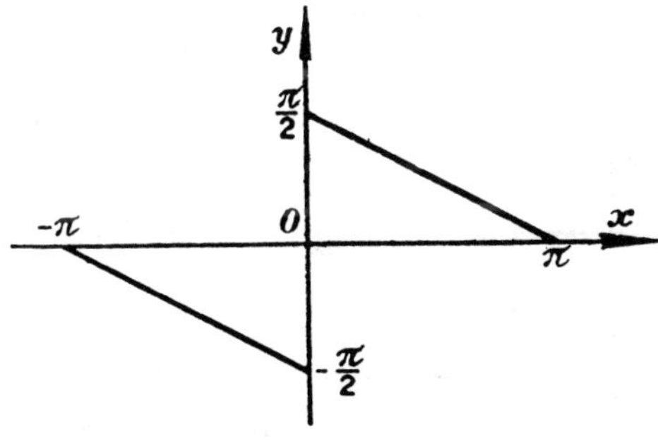

Fig. 3.

The Fourier series of $f(x)$ is a sine series (Sec. 2.2), namely

$$(3) \qquad \frac{\pi - x}{2} = \sin x + \frac{\sin 2x}{2} + \frac{\sin 3x}{3} + \dots, \qquad 0 < x \leqq \pi$$

and the right side is uniformly convergent in every interval interior to $[0, \pi]$.

4) Dividing through (2) by 2 and adding the results to (3) we obtain

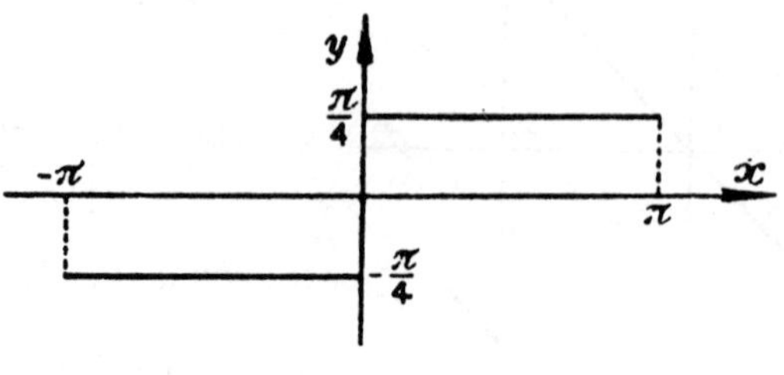

Fig. 4.

$$\frac{\pi}{4} = \sin x + \frac{\sin 3x}{3} + \frac{\sin 5x}{5} + \dots, \qquad 0 < x < \pi, \tag{4}$$

and the right side is uniformly convergent in every interval interior to $[0, \pi]$.

5) If $f(x) = c$ for $-\pi/2 < x < \pi/2$, $f(x) = -c$ for $-\pi \leqq x < -\pi/2$, $\pi/2 < x \leqq \pi$, and $f(-\pi/2) = f(\pi/2) = 0$.

The Fourier series of $f(x)$ is a cosine series (Sec. 2.2) for which $a_0 = 0$, $a_k = \dfrac{4c}{k} \sin (k\pi/2)$, whence

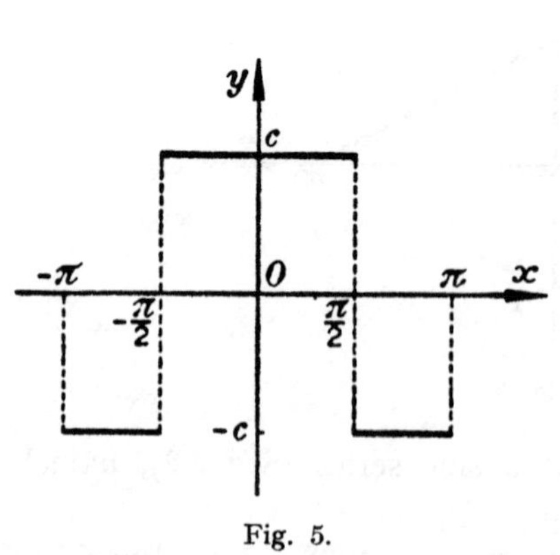

Fig. 5.

$$\frac{\pi}{4} = \cos x - \frac{\cos 3x}{3} + \frac{\cos 5x}{5} - \dots, \qquad -\frac{\pi}{2} < x < \frac{\pi}{2}; \tag{5}$$

$$-\frac{\pi}{4} = \cos x - \frac{\cos 3x}{3} + \frac{\cos 5x}{5} - \dots, \qquad -\pi \leqq x < -\frac{\pi}{2}, \frac{\pi}{2} < x \leqq \pi. \tag{5^1}$$

The series on the right are uniformly convergent in every interval interior to

$$\left(-3\frac{\pi}{2}, -\frac{\pi}{2}\right), \qquad \left(-\frac{\pi}{2}, \frac{\pi}{2}\right), \qquad \left(\frac{\pi}{2}, 3\frac{\pi}{2}\right).$$

(5) follows from (4) by the transformation $x = t + \pi/2$.

6) Let

$$f(x) = c \text{ for } 0 < x < \frac{\pi}{2};$$

$$f(x) = -c \text{ for } \frac{\pi}{2} < x < \pi;$$

$$f(0) = f\left(\frac{\pi}{2}\right) = f(\pi) = 0$$

$$f(x) = -f(-x)$$

for $-\pi \leqq x \leqq 0$.

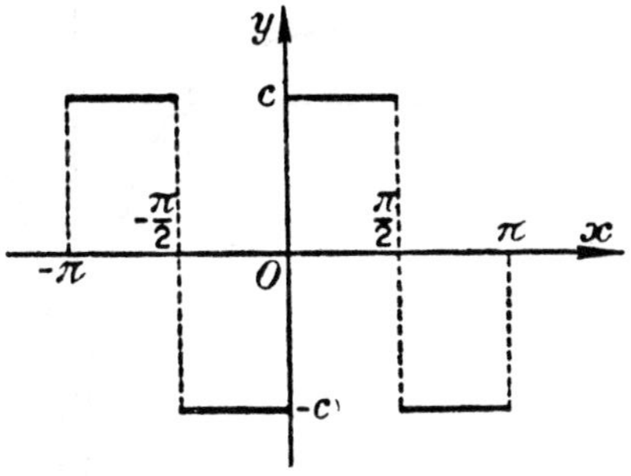

Fig. 6.

The Fourier series of $f(x)$ is a sine series for which

$$b_k = \frac{2c}{k\pi}\left[1 - 2\cos k\frac{\pi}{2} + \cos k\pi\right].$$

Therefore

$$\frac{\pi}{8} = \frac{\sin 2x}{2} + \frac{\sin 6x}{6} + \frac{\sin 10x}{10} + \ldots, \text{ for } 0 < x < \frac{\pi}{2}. \tag{6}$$

$$-\frac{\pi}{8} = \frac{\sin 2x}{2} + \frac{\sin 6x}{6} + \frac{\sin 10x}{10} + \ldots, \text{ for } \frac{\pi}{2} < x < \pi. \tag{6'}$$

The series on the right are uniformly convergent in every closed interval which does not contain points with coordinate $k\pi/2$ where k is an integer.

7) Let

$$f(x) = \log\left(2 \sin \frac{x}{2}\right) \qquad \text{for } 0 \leqq x \leqq \pi, \ [f(0) = -\infty]$$

$$f(x) = f(-x) \qquad \text{for } -\pi \leqq x \leqq 0.$$

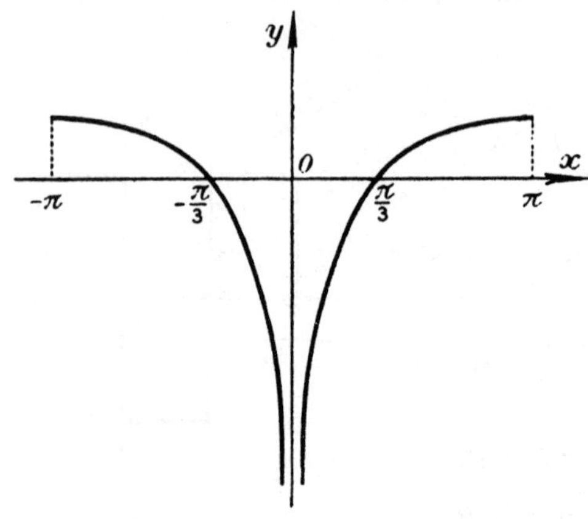

Fig. 7.

The Fourier series of $f(x)$ is a cosine series [Sec. 2. 2] for which

$$a_0 = \frac{2}{\pi}\int_0^\pi \log\left(2 \sin \frac{\alpha}{2}\right) d\alpha = 2 \log 2 + \frac{2}{\pi}\int_0^\pi \log\left(\sin \frac{\alpha}{2}\right) d\alpha$$

$$= 2 \log 2 + \frac{4}{\pi}\int_0^{\pi/2} \log (\sin x) dx.$$

Clearly

$$\int_0^\pi \log (\sin x) dx = \int_0^{\pi/2} \log (\sin x)\, dx + \int_{\pi/2}^\pi \log(\sin x)\, dx$$

$$= 2\int_0^{\pi/2} \log(\sin x) dx,$$

writing $x = \pi - z$ in the second integral. Similarly,

$$\int_0^\pi \log (\sin x)\, dx = \int_0^\pi \log \left(2 \sin \frac{x}{2} \cos \frac{x}{2}\right) dx$$

$$= \pi \log 2 + \int_0^\pi \log \left(\sin \frac{x}{2}\right) dx + \int_0^\pi \log \left(\cos \frac{x}{2}\right) dx$$

$$= \pi \log 2 + 2\int_0^{\pi/2} \log (\sin x)\, dx + 2\int_0^{\pi/2} \log (\cos x)\, dx$$

$$= \pi \log 2 + 4 \int_0^{\pi/2} \log (\sin x)\, dx,$$

writing $x = \pi/2 - t$ in the second integral. Therefore

$$4\int_0^{\pi/2} \log (\sin x) dx + \pi \log 2 = 2\int_0^{\pi/2} \log (\sin x)\, dx$$

whence

$$2\int_0^{\pi/2} \log (\sin x)\, dx = - \pi \log 2,$$

and, consequently

$$a_0 = 0.$$

Now
$$a_n = \frac{2}{\pi}\int_0^\pi \cos n\alpha \log \left(2 \sin \frac{\alpha}{2}\right) d\alpha$$

$$= \frac{2}{\pi}\left[\frac{\sin n\alpha}{n} \log \left(2 \sin \frac{\alpha}{2}\right)\right]_{\alpha=0}^{\alpha=\pi} - \frac{1}{n\pi}\int_0^\pi \frac{\sin n\alpha \cos \dfrac{\alpha}{2}}{\sin \dfrac{\alpha}{2}}\, d\alpha$$

$$= - \frac{1}{n\pi}\int_0^\pi \frac{\sin (n + \frac{1}{2})\alpha}{2 \sin \dfrac{\alpha}{2}}\, d\alpha - \frac{1}{n\pi}\int_0^\pi \frac{\sin (n - \frac{1}{2})\alpha}{2 \sin \dfrac{\alpha}{2}}\, d\alpha = - \frac{1}{n}$$

(cf. Sec. 4. 4) and therefore for any x in $[0, \pi]$

$$(7) \qquad - \log \left(2 \sin \frac{x}{2}\right) = \cos x + \frac{\cos 2x}{2} + \frac{\cos 3x}{3} + \dots,$$

and the series on the right side converges uniformly in every interval interior to $[0, 2\pi]$ and diverges for $x = 2k\pi$ for integral values of k.

8) An example will now be given of a bounded function $f(x)$ defined in $[-\pi, \pi]$ with an infinite number of discontinuities of the first kind, and one discontinuity of the second kind, whose Fourier series converges to $f(x)$ at all points.

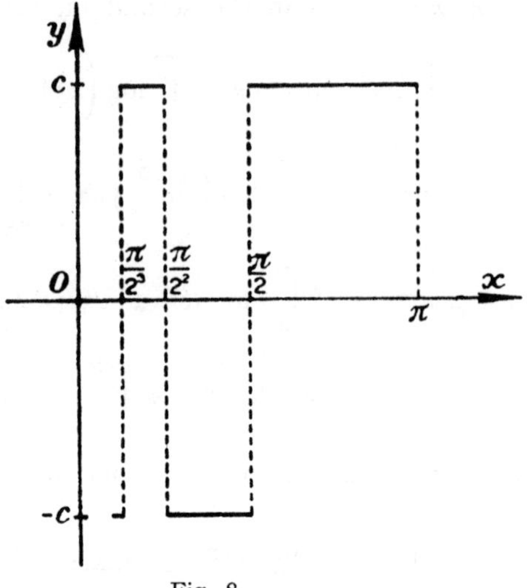

Fig. 8.

Consider the function $f(x)$ defined in $[-\pi, \pi]$ as follows:

i) $f(x) + f(-x) = 0;$

ii) $f\left(\frac{\pi}{2^p}\right) = 0, \qquad (p = 0, 1, 2, \ldots);$

iii) $f(x) = c > 0$ for $\frac{\pi}{2^{2k+1}} < x < \frac{\pi}{2^{2k}};$

iv) $f(x) = -c$ for $\frac{\pi}{2^{2k}} < x < \frac{\pi}{2^{2k-1}} \qquad (k = 0, 1, 2, \ldots).$

$f(x)$ is integrable since its absolute value is constant and equal to c except for a set of measure zero. It has discontinuities of the first kind at the points $\pm \pi/2^p$ and a discontinuity of the second kind at the origin.

Its Fourier series is a sine series, which therefore vanishes at the origin. At a point of continuity of $f(x)$ for α sufficiently small we have

$$\varphi(x, \alpha) = f(x + 2\alpha) + f(x - 2\alpha) - 2f(x) = \pm (c + c - 2c) = 0.$$

At a point of discontinuity of the first kind, for α sufficiently small we have

$$\varphi(x, \alpha) = (+c) + (-c) - 2 \cdot 0 = 0.$$

Consequently at every point the Fourier series of $f(x)$ converges to $f(x)$ and converges uniformly in every interval interior to $[\pi/2, \pi]$, $[\pi/4, \pi/2]$, . . . and in intervals symmetric to these with respect to the origin.

9) Finally we give an example of an unbounded function defined in $[-\pi, \pi]$ with an infinite number of discontinuities of the first kind, and one discontinuity of the second kind, whose Fourier series converges to $f(x)$ at all points.

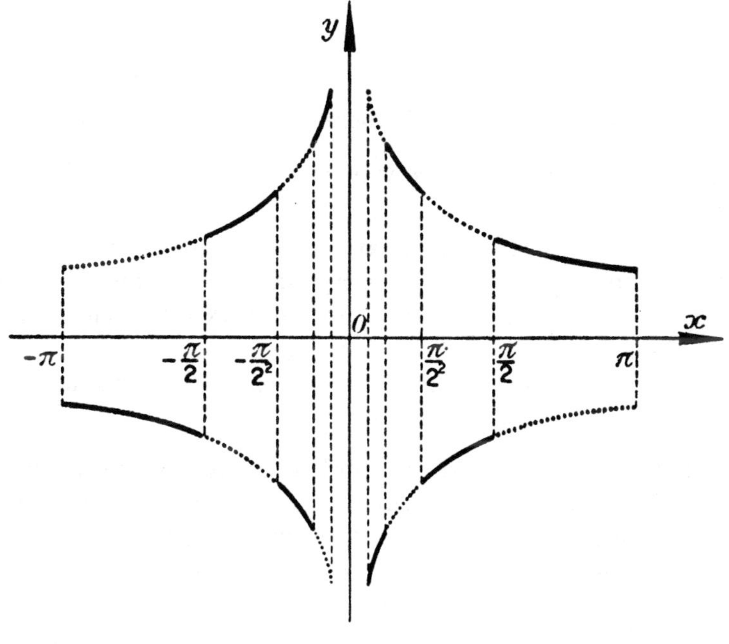

Fig. 9.

$f(x)$ is defined in $[-\pi, \pi]$ as follows:

i) $f(x) + f(-x) = 0$;

ii) $f\left(\frac{\pi}{2^p}\right) = 0,$ $(p = 0, 1, 2, \ldots);$

iii) $f(x) = \frac{1}{\sqrt{|x|}}$ for $\frac{\pi}{2^{2k+1}} < x < \frac{\pi}{2^{2k}},$

$(k = 0, 1, 2, \ldots).$

iv) $f(x) = -\frac{1}{\sqrt{|x|}}$ for $\frac{\pi}{2^{2k}} < x < \frac{\pi}{2^{2k-1}},$

$|f(x)| = x^{-1/2}$ a.e. in $[0, \pi]$ and is therefore integrable. Therefore $f(x)$ is integrable and its Fourier series is a sine series.

$f(x)$ has a finite derivative at every point of continuity. At the points of discontinuity of the first kind for which $x > 0$, we have, for α sufficiently small,

$$\varphi(x, \alpha) = f(x + 2\alpha) + f(x - 2\alpha) - 2f(x)$$
$$= \pm [(x - 2\alpha)^{-1/2} - (x + 2\alpha)^{-1/2}],$$

$\lim_{\alpha \to 0} \frac{\varphi(x, \alpha)}{\alpha} = \pm 2x^{-3/2}$, and therefore the Fourier series of $f(x)$ converges to $f(x)$ at every point x (Th. 17, Cor. 2; Th. 16, Cor. 2).

7. The following theorems, stated without proof, give the most noted criteria for pointwise convergence. (For the proofs of the theorems, the reader is referred to the work of Tonelli [111b] pp. 282—297).

It is tacitly assumed in the statements of the theorems that $f(x)$ is periodic with period 2π and integrable.

THEOREM 18. (The Dirichlet-Jordan criterion of convergence.) (C. Jordan [55a]). (An alternative proof of this theorem will be given under the remark to Theorem 41′ in Sec. 6. 8). The Fourier series of $f(x)$ converges at every point x interior to an interval in which $f(x)$ is of bounded variation. The series converges to $f(x)$ or $[f(x+) + f(x-)]/2$ according as x is a point of continuity or discontinuity.

If $f(x)$ is continuous and of bounded variation in an interval $[A, B]$ of $[-\pi, \pi]$ then the Fourier series of $f(x)$ converges uniformly to $f(x)$ in $[A_1, B_1]$, $A < A_1 < B_1 < B$.

THEOREM 19. (De la Vallée-Poussin's criterion of convergence [27e].) At every point x, for which the following function of h

$$G(h) = \frac{1}{2h}\int_{-h}^{h} f(x + \alpha)d\alpha$$

is of bounded variation in an interval $[0, \delta]$, $\delta > 0$, the Fourier series of $f(x)$ converges to $G(0+)$.

THEOREM 20. (Lebesgue's criterion of convergence [67c]). If for a given x, $S(x)$ satisfies the relations

$$\lim_{\delta\to 0} \frac{1}{\delta}\int_{0}^{\delta} | \varphi(x, \alpha) | d\alpha = 0,$$

$$[\varphi(x, \alpha) = f(x + 2\alpha) + f(x - 2\alpha) - 2S(x)],$$

and $(\delta > 0)$

$$\lim_{\delta\to 0}\int_{\delta}^{\pi/2} \frac{\varphi(x, \alpha + \delta) - \varphi(x, \alpha)}{\alpha} \sin\frac{\pi\alpha}{\delta}\, d\alpha = 0$$

then the Fourier series of $f(x)$ converges at x to $S(x)$.

From this theorem follows immediately

THEOREM 21. (Dini's criterion of convergence [28b], p. 102.) If $f(x)$ is continuous in $[A, B]$ and $| f(x + \delta) - f(x) | \log | \delta |$ tends uniformly to zero as $\delta \to 0$, in every interval $[A_1, B_1]$ interior to $[A, B]$, then the Fourier series of $f(x)$ converges uniformly to $f(x)$ in $[A_1, B_1]$.

THEOREM 22. (Tonelli's criterion of convergence [111f].) If $\varphi(x, \alpha) \to 0$ as $\alpha \to 0$ for a given point x; if there exists a $\delta > 0$ such that for $0 < \alpha < \delta$, $\varphi(x, \alpha)$ is absolutely continuous in the interval $[\alpha, \delta]$; and, finally, if, as $\alpha \to 0$, $\underline{\lim}_{\alpha\to+0}\, \alpha\varphi'_\alpha(x, \alpha) \geqq 0$ or $\overline{\lim}_{\alpha\to+0}\, \alpha\varphi'_\alpha(x, \alpha) \leqq 0$, for those values of α for which $\varphi'_\alpha(x, \alpha)$ exists and is finite (cf. Appendix Ths. 13 and 14); then the Fourier series of $f(x)$ converges to $S(x)$ at the given point x.

5. Term by Term Integration of the Fourier Series: The Hardy-Littlewood Criterion for Pointwise Convergence

1. THEOREM 23. If $f(x)$ is a periodic function with period 2π, integrable in $[0, 2\pi]$ and

$$f(x) \sim \tfrac{1}{2}a_0 + \sum_{k=1}^{\infty} \left[a_k \cos kx + b_k \sin kx \right], \tag{1}$$

then, for any x,

$$\int_0^x f(x)dx = \tfrac{1}{2}\, a_0 x + \sum_{k=1}^{\infty} \left[-\frac{b_k}{k}(\cos kx - 1) + \frac{a_k}{k} \sin kx \right], \tag{2}$$

and the series on the right, obtained by term by term integration of (1), is uniformly convergent for all x in any finite interval (U. Dini, p. 199; Lebesgue [67a] p. 103).

Proof. If $f(x)$ is square integrable for x in $[0, 2\pi]$, the theorem follows as a particular case of Theorem 32 of Ch. I. To show this, it suffices to set $\theta = 1$, $f_1 = f$, $f_2 = 1$ in (11) of the theorem mentioned, identify the orthogonal system $\{ \varphi_k \}$ with

$$\left\{ \frac{1}{\sqrt{2\pi}},\ \frac{1}{\sqrt{\pi}} \cos kx,\ \frac{1}{\sqrt{\pi}} \sin kx \right\}$$

and take $\gamma = [0, x]$.

Suppose more generally that $f(x)$ is merely integrable in $[0, 2\pi]$ and consider the function

$$F(x) = \int_0^x \{f(x) - \tfrac{1}{2}a_0\}\, dx.$$

$F(x)$ is absolutely continuous and therefore of bounded variation (Appendix, Th. 13 and Def. 11). $F(x)$ is also of period 2π. In fact

$$F(x + 2\pi) = \int_x^{x+2\pi} f(x)dx - a_0\pi + F(x)$$

$$= \int_0^{2\pi} f(x)\, dx - a_0\pi + F(x) = F(x).$$

By the Dirichlet-Jordan theorem (cf. Th. 18 or Th. 41′) we obtain

$$(3)\qquad F(x) = \tfrac{1}{2}\alpha_0 + \sum_{k=1}^{\infty} [\alpha_k \cos kx + \beta_k \sin kx],$$

$$(4)\qquad \alpha_k = \frac{1}{\pi}\int_0^{2\pi} F(x) \cos kx\, dx; \quad \beta_k = \frac{1}{\pi}\int_0^{2\pi} F(x) \sin kx\, dx,$$

and the series on the right of (3) is uniformly convergent in $[-\pi, \pi]$ and by the periodicity, uniformly convergent in every interval.

Integrating by parts, for $k > 0$, we have

$$(5_1)\quad \alpha_k = \frac{-1}{k\pi}\int_0^{2\pi} F'(x) \sin kx\, dx = \frac{-1}{k\pi}\int_0^{2\pi} \{f(x) - \tfrac{1}{2}a_0\} \sin kx dx = -\frac{b_k}{k},$$

and analogously

$$\beta_k = \frac{a_k}{k},$$

and putting $x = 0$ in (3), we obtain

$$(6)\qquad 0 = \tfrac{1}{2}\alpha_0 + \sum_{k=1}^{\infty} \alpha_k = \tfrac{1}{2}\alpha_0 - \sum_{k=1}^{\infty} \frac{b_k}{k},$$

whence it follows that $\sum_{k=1}^{\infty} b_k/k$ is convergent.

It is not generally true, however, that $\sum_{k=1}^{\infty} a_k/k$ is convergent (cf. L. Tonelli: [111b] p. 246).

From (3) with the help of (5_1), (5_2) and (6) follows

$$\int_0^x f(x)dx = \sum_{k=1}^{\infty} \frac{b_k}{k} + \tfrac{1}{2}a_0 x + \sum_{k=1}^{\infty} \left[-\frac{b_k}{k}\cos kx + \frac{a_k}{k}\sin kx \right],$$

which is precisely (2) written in a slightly different form.

2. The preceding theorem can be generalized as follows:

THEOREM 24. If $f(x)$ is periodic with period 2π and integrable in $[0, 2\pi]$; if

$$f(x) \sim \tfrac{1}{2}a_0 + \sum_{k=1}^{\infty} [a_k \cos kx + b_k \sin kx], \tag{1}$$

if $g(x)$ is another function integrable in $[c, d]$ not necessarily belonging to $[0, 2\pi]$, $c < d$; if $f(x)$ and $g(x)$ are square integrable respectively in $[0, 2\pi]$ and $[c, d]$, or $g(x)$ is of bounded variation in $[c, d]$, then for all x in $[c, d]$

$$\int_c^x f(x) g(x)\, dx = \tfrac{1}{2}a_0 \int_c^x g(x)\, dx + \sum_{k=1}^{\infty} \left[a_k \int_c^x g(x) \cos kx\, dx + b_k \int_c^x g(x) \sin kx\, dx\right],$$

and the right side is uniformly convergent for all x in $[c, d]$.

The theorem is also true under the assumption that $f(x)$ is of bounded variation in $[0, 2\pi]$ and $g(x)$ is integrable in $[c, d]$. (Cf. L. Tonelli: [111 b] p. 343).

Proof. If $f(x)$ and $g(x)$ are square integrable the proof is similar to that of Theorem 32 of Chapter I. In fact if $s_n(x)$ denotes the sum of the first $n + 1$ terms of (1), then by the Schwarz inequality we have

$$\begin{aligned} &\Bigg| \int_c^x f(x)g(x)dx - \tfrac{1}{2}a_0 \int_c^x g(x)dx \\ &- \sum_{k=1}^{n} \left[a_k \int_c^x g(x) \cos kx\, dx + b_k \int_c^x g(x) \sin kx\, dx\right] \Bigg| \\ &= \left| \int_c^x f(x)g(x)\, dx - \int_c^x s_n(x)g(x)dx \right| = \left| \int_c^x [f(x) - s_n(x)]g(x)dx \right| \\ &\leqq \left[\int_c^d [f(x) - s_n(x)]^2 dx\right]^{1/2} \left[\int_c^d g^2(x)dx\right]^{1/2}, \end{aligned}$$

whence as $n \to \infty$, the first part of the theorem follows from Theorem 5, which is still valid if the interval $[c, d]$ is substituted for the interval $[-\pi, \pi]$.

If $g(x)$ is of bounded variation in $[0, 2\pi]$ and $g(x) = g_1(x) - g_2(x)$, where $g_1(x)$ and $g_2(x)$ are non-negative and non-increasing in $[c, d]$, then by the second theorem of the mean [Appendix, Th. 18]

we have

$$\left|\int_c^x [f(x) - s_n(x)]g(x)dx\right|$$
$$\leqq \left|\int_c^x [f(x) - s_n(x)]g_1(x)dx\right| + \left|\int_c^x [f(x) - s_n(x)]g_2(x)dx\right|$$
$$\leqq g_1(c)\left|\int_c^{\xi_1} [f(x) - s_n(x)]\,dx\right| + g_2(c)\left|\int_c^{\xi_2} [f(x)-s_n(x)]dx\right|,$$
$$c \leqq \xi_1, \quad \xi_2 \leqq d.$$

But by the preceding theorem we have uniformly for all ξ in $[c, d]$

$$\lim_{n\to\infty} \int_c^{\xi} [f(x) - s_n(x)]dx = 0,$$

and consequently, the theorem is completely proved.

3. The following criterion of Hardy and Littlewood ([44c]; [70]) is of special interest since the convergence at a point x of the Fourier series of $f(x)$, periodic with period 2π and integrable, is deduced from a hypothesis on the behaviour of $f(x)$ in a neighborhood of x and the order of magnitude of its Fourier coefficients.

THEOREM 25. If at a point x

$$\lim_{h\to 0} |f(x + h) - f(x)|\,|\log h| = 0, \tag{7}$$

(the hypothesis implies the continuity of $f(x)$ at the point x and $\log h$ denotes the principal value of the logarithm of h) and if the Fourier coefficients of $f(x)$ satisfy the inequalities

$$|a_k| < \frac{C}{k^\delta}, \; |b_k| < \frac{C}{k^\delta}, \; (k = 1, 2, \ldots; \delta > 0, C = \text{constant}) \tag{8}$$

then at the point x

$$f(x) = \tfrac{1}{2}a_0 + \sum_{k=1}^{\infty} [a_k \cos kx + b_k \sin kx]. \tag{9}$$

To prove the theorem it is sufficient to show

$$0 = \tfrac{1}{2}a_0 - f(x) + \sum_{k=1}^{\infty} [a_k \cos kx + b_k \sin kx].$$

Consider the function (even with respect to t)

$$F(t) = \frac{f(x+t)+f(x-t)}{2} - f(x) = \frac{[f(x+t)-f(x)]+[f(x-t)-f(x)]}{2}. \tag{10}$$

For the coefficients of its Fourier series

$$F(t) \sim \tfrac{1}{2}\alpha_0 + \sum_{k=1}^{\infty} \alpha_k \cos kt \tag{11}$$

we have

$$\tfrac{1}{2}\alpha_0 = \tfrac{1}{2}a_0 - f(x), \quad \alpha_k = a_k \cos kx + b_k \sin kx, \quad (k = 1, 2, \ldots).$$

Indeed

$$\tfrac{1}{2}\alpha_0 = \frac{1}{\pi}\int_0^{\pi} F(t)dt = \frac{1}{2\pi}\left[\int_0^{\pi} f(x+t)dt + \int_0^{\pi} f(x-t)\,dt\right] - f(x)$$

$$= \frac{1}{2\pi}\left[\int_x^{x+\pi} f(u)\,du - \int_x^{x-\pi} f(u)\,du\right] - f(x) = \frac{1}{2\pi}\int_{x-\pi}^{x+\pi} f(u)du - f(x)$$

$$= \frac{1}{2\pi}\int_0^{2\pi} f(u)du - f(x) = \tfrac{1}{2}a_0 - f(x),$$

and analogously

$$\alpha_k = \frac{2}{\pi}\int_0^{\pi} F(t)\cos kt\,dt = \frac{1}{\pi}\left[\int_0^{\pi} f(x+t)\cos kt\,dt + \int_0^{\pi} f(x-t)\cos kt\,dt\right]$$

$$= \frac{1}{\pi}\left[\int_x^{x+\pi} f(u)\cos k(x-u)du - \int_x^{x-\pi} f(u)\cos k(x-u)du\right]$$

$$= \frac{1}{\pi}\int_{x-\pi}^{x+\pi} f(u)\cos k(x-u)du = \frac{1}{\pi}\int_0^{2\pi} f(u)\cos k(x-u)du$$

$$= a_k \cos kx + b_k \sin kx.$$

Now to deduce (9), it suffices to show that for (11) we have

$$F(0) = 0 = \tfrac{1}{2}\alpha_0 + \sum_{k=1}^{\infty} \alpha_k. \tag{12}$$

From (7) and (8) follows

$$\lim_{h\to 0} | F(h) || \log h | = 0, \quad | \alpha_k | < \frac{2C}{k^\delta}, \quad (k = 1, 2, \ldots),$$

and there is no loss of generality in supposing $\alpha_0 = 0$, since otherwise we could consider the function $F_1(t) = F(t) - \frac{1}{2}\alpha_0(1 - \cos t)$, for which all the required conditions are satisfied.

Under this supposition, we observe that for the sum $s_n(0)$ of the first $n+1$ terms of the series (11) at the point 0, we have [Sec. 4.2, (4)]

$$s_n(0) = \sum_{k=1}^{n} \alpha_k = \frac{1}{\pi}\int_0^{2\pi} F(\alpha) \frac{\sin (n+\frac{1}{2})\alpha}{2\sin\frac{\alpha}{2}} d\alpha = \frac{1}{\pi}\int_0^{\pi} F(\alpha) \frac{\sin (n+\frac{1}{2})\alpha}{\sin\frac{\alpha}{2}} d\alpha$$

$$= \frac{1}{\pi}\int_0^{\pi} F(\alpha) \left[\frac{\sin (n+\frac{1}{2})\alpha}{\sin \alpha/2} - \cos n\alpha\right] d\alpha + \tfrac{1}{2}\alpha_n$$

$$= \frac{1}{\pi}\int_0^{\pi} F(\alpha) \frac{\sin n\alpha}{\tan\frac{\alpha}{2}} d\alpha + \tfrac{1}{2}\alpha_n,$$

and since $\lim_{n\to\infty} \alpha_n = 0$ (cf. Th. 13), to prove the theorem it is sufficient to show that if

$$I_n = \frac{1}{\pi}\int_0^{\pi} F(\alpha) \frac{\sin n\alpha}{\tan\frac{\alpha}{2}} d\alpha,$$

then

$$\lim_{n\to\infty} I_n = 0. \tag{13}$$

If we let $r = \delta/2$ (there is no loss of generality in assuming $0 < \delta < 1$) we have

$$I_n = P_n + Q_n + R_n \tag{14}$$

where

$$P_n = \frac{1}{\pi}\int_0^{1/n} F(\alpha)\frac{\sin n\alpha}{\tan\frac{\alpha}{2}}\,d\alpha, \qquad Q_n = \frac{1}{\pi}\int_{1/n}^{1/n^r} F(\alpha)\frac{\sin n\alpha}{\tan\frac{\alpha}{2}}\,d\alpha,$$

$$R_n = \frac{1}{\pi}\int_{1/n^r}^{\pi} F(\alpha)\frac{\sin n\alpha}{\tan\frac{\alpha}{2}}\,d\alpha.$$

Since for $0 < x < \pi/2$, $x/\sin x < \pi/2$, it follows that

$$|P_n| = \left|\frac{2}{\pi}n\int_0^{1/n} F(\alpha)\frac{\sin n\alpha}{n\alpha}\cos\frac{\alpha}{2}\frac{\frac{\alpha}{2}}{\sin\frac{\alpha}{2}}\,d\alpha\right| \leqq \max_{0\leqq\alpha\leqq 1/n}.\ |F(\alpha)|,$$

whence

$$\lim_{n\to\infty} P_n = 0. \tag{15_1}$$

Letting $\varepsilon(\alpha) = \max_{0\leqq h\leqq\alpha}.\ [\,|F(h)|\,|\log h|\,]$ we have

$$\frac{\pi}{2}|Q_n| \leqq \int_{1/n}^{1/n^r}\frac{|F(\alpha)|}{\alpha}\,d\alpha \leqq \varepsilon\left(\frac{1}{n^r}\right)\int_{1/n}^{1/n^r}\frac{d\alpha}{\alpha\,|\log\alpha|}$$

$$= -\varepsilon\left(\frac{1}{n^r}\right)\Big[\log|\log\alpha|\Big]_{1/n}^{1/n^r} = \varepsilon\left(\frac{1}{n^r}\right)\log\frac{1}{r} = \varepsilon\left(\frac{1}{n^r}\right)\log\frac{2}{\delta},$$

and therefore

$$\lim_{n\to\infty} Q_n = 0, \tag{15_2}$$

consequently to prove (13) it suffices to show

$$\lim_{n\to\infty} R_n = 0. \tag{15_3}$$

Since $F(t) \sim \sum_{k=1}^{\infty} \alpha_k \cos kt$, it follows from Theorem 24 that

$$R_n = \frac{1}{\pi}\sum_{k=1}^{\infty}\alpha_k\int_{1/n^r}^{\pi}\frac{\sin n\alpha\cos k\alpha}{\tan\frac{1}{2}\alpha}\,d\alpha.$$

But by the second theorem of the mean (Appendix, Th. 18)

$$\int_{1/n^r}^{\pi} \frac{\sin n\alpha \cos k\alpha}{\tan \frac{1}{2}\alpha}\, d\alpha = \frac{1}{\tan \dfrac{1}{2n^r}} \int_{1/n^r}^{\xi} \sin n\alpha \cos k\alpha \, d\alpha$$

$$< 2n^r \int_{1/n^r}^{\xi} \sin n\alpha \cos k\alpha \, d\alpha = n^r \int_{1/n^r}^{\xi} [\sin (n+k)\alpha + \sin (n-k)\alpha] d\alpha,$$

whence for $k \neq n$ $[n^{-r} \leqq \xi \leqq \pi]$

$$\left| \alpha_k \frac{1}{\pi} \int_{1/n^r}^{\pi} \frac{\sin n\alpha \cos k\alpha}{\tan \frac{1}{2}\alpha}\, d\alpha \right|$$

$$\leqq \frac{n^r}{\pi} \left| \left[-\frac{1}{n+k} \cos (n+k)\alpha - \frac{1}{n-k} \cos (n-k)\alpha \right]_{\alpha=1/n^r}^{\alpha=\xi} \right| |\alpha_k| \leqq \frac{8C}{\pi k^{\delta}} \frac{n^r}{|n-k|}.$$

and since also

$$\left| \alpha_n \int_{1/n^r}^{\pi} \frac{\sin nt \cos nt}{\tan \frac{1}{2}t}\, dt \right| \leqq \frac{4Cn^r}{n^{\delta}} = \frac{4C}{n^{\delta/2}} \to 0 \text{ for } n \to \infty,$$

to prove (15_3) it suffices to show that if

$$R_n' = \sum_{k=1}^{n-1} \frac{n^r}{(n-k)k^{\delta}}, \qquad R_n'' = \sum_{k=n+1}^{\infty} \frac{n^r}{(k-n)k^{\delta}}$$

(16_1) $\lim R'_n = 0,$ (16_2) $\lim R''_n = 0.$

If m denotes the largest integer contained in $n/2$, then

$$R_n' \leqq \frac{n^r}{\dfrac{n}{2}} \sum_{k=1}^{m} \frac{1}{k^{\delta}} + \frac{n^r}{\left(\dfrac{n}{2}\right)^{\delta}} \sum_{k=m+1}^{n-1} \frac{1}{n-k},$$

and since

$$R_n'' < n^{r-\delta} \sum_{k=n+1}^{2n} \frac{1}{k-n} + 2n^r \sum_{k=2n+1}^{\infty} \frac{1}{k^{\delta+1}},$$

and

$$\tfrac{1}{2} + \ldots + \frac{1}{m} < \int_0^{m-1} \frac{1}{x+1}\,dx = \log m,$$

$$\sum_{k=2}^{m} \frac{1}{k^\delta} < \int_0^{m-1} \frac{1}{(x+1)^\delta}\,dx = \frac{m^{1-\delta}-1}{1-\delta}, \qquad (\delta \neq 1),$$

$$\sum_{k=m+1}^{\infty} \frac{1}{k^{1+\delta}} < \frac{1}{\delta m^\delta},$$

(cf. Ch. III, Sec. 11.2, series of Legendre polynomials of functions of bounded variation) (16_1) and (16_2) follow immediately and the theorem is proved.

6. *Fejér* $(C, 1)$ *Summability of Fourier Series*

1. THEOREM 26. If $\{b_n\}$ is an increasing sequence of positive numbers, diverging to $+\infty$, and $\{a_n\}$ is any sequence of real terms, then

$$(1) \qquad \overline{\lim_{n\to\infty}} \frac{a_{n+1}-a_n}{b_{n+1}-b_n} \geqq \overline{\lim_{n\to\infty}} \frac{a_n}{b_n} \geqq \underline{\lim_{n\to\infty}} \frac{a_n}{b_n} \geqq \underline{\lim_{n\to\infty}} \frac{a_{n+1}-a_n}{b_{n+1}-b_n}.$$

In particular if $\lim_{n\to\infty} (a_{n+1}-a_n)/(b_{n+1}-b_n)$ exists and is finite, then $\lim_{n\to\infty} a_n/b_n$ also exists and the two limits are equal; if $\lim_{n\to\infty} (a_{n+1}-a_n)/(b_{n+1}-b_n) = +\infty$ $[-\infty]$ then also $\lim_{n\to\infty} a_n/b_n = +\infty$ $[-\infty]$.

Proof. Suppose first

$$\overline{\lim_{n\to\infty}} \frac{a_{n+1}-a_n}{b_{n+1}-b_n} = M, \qquad \underline{\lim_{n\to\infty}} \frac{a_{n+1}-a_n}{b_{n+1}-b_n} = m.$$

By definition, given $\sigma > 0$, there exists an n_0 such that for $n \geqq n_0$

$$m - \sigma < \frac{a_{n+1}-a_n}{b_{n+1}-b_n} < M + \sigma,$$

$$(m-\sigma)(b_{n+1}-b_n) < a_{n+1}-a_n < (M+\sigma)(b_{n+1}-b_n).$$

Taking $n = n_0, n_0+1, \ldots, n_0+k-1$ and adding, we obtain

$$(m - \sigma)(b_{n_0+k} - b_{n_0}) < a_{n_0+k} - a_{n_0} < (M + \sigma)(b_{n_0+k} - b_{n_0}).$$

Dividing by b_{n_0+k} we have

$$(m - \sigma)\left(1 - \frac{b_{n_0}}{b_{n_0+k}}\right) < \frac{a_{n_0+k}}{b_{n_0+k}} - \frac{a_{n_0}}{b_{n_0+k}} < (M + \sigma)\left(1 - \frac{b_{n_0}}{b_{n_0+k}}\right),$$

but since

$$\lim_{k\to\infty} b_{n_0}/b_{n_0+k} = 0, \ \lim_{k\to\infty} a_{n_0}/b_{n_0+k} = 0,$$

it follows for k sufficiently large that

$$m - 2\sigma \leqq \frac{a_{n_0+k}}{b_{n_0+k}} \leqq M + 2\sigma$$

and since σ is arbitrary, (1) follows at once.

There remains to consider the case

$$\varliminf_{n\to\infty} \frac{a_{n+1} - a_n}{b_{n+1} - b_n} = +\infty$$

and therefore

$$\lim_{n\to\infty} \frac{a_{n+1} - a_n}{b_{n+1} - b_n} = +\infty.$$

and the analog

$$\varlimsup_{n\to\infty} \frac{a_{n+1} - a_n}{b_{n+1} - b_n} = -\infty.$$

By hypothesis, for $L > 0$ there exists an n_0 such that for $n \geqq n_0$

$$\frac{a_{n+1} - a_n}{b_{n+1} - b_n} > L,$$

and consequently

$$a_{n+1} - a_n > L(b_{n+1} - b_n).$$

Taking $n = n_0, n_0 + 1, n_0 + 2, \ldots, n_0 + k - 1$ and adding, we obtain

$$a_{n_0+k} - a_{n_0} > L(b_{n_0+k} - b_{n_0})$$

$$\frac{a_{n_0+k}}{b_{n_0+k}} - \frac{a_{n_0}}{b_{n_0+k}} > L\left(1 - \frac{b_{n_0}}{b_{n_0+k}}\right)$$

and since

$$\lim_{k\to\infty} b_{n_0}/b_{n_0+k} = 0, \quad \lim_{k\to\infty} a_{n_0}/b_{n_0+k} = 0,$$

for k sufficiently large we have

$$a_{n_0+k}/b_{n_0+k} > L - 1$$

and therefore $\lim_{n\to\infty} a_n/b_n = +\infty$. Q.E.D.

THEOREM 27. If a sequence $\{\alpha_n\}$ of real terms has a limit, finite or infinite, then the sequence

$$\left\{\frac{\alpha_1 + \alpha_2 + \ldots + \alpha_n}{n}\right\}$$

of which the nth term is the arithmetic mean of the first n terms of the given sequence, has the same limit.

Proof. Let $\lim_{n\to\infty} \alpha_n = A$, and

$$a_n = \alpha_1 + \alpha_2 + \ldots + \alpha_n, \quad b_n = n,$$

then

$$(a_n - a_{n-1})/(b_n - b_{n-1}) = \alpha_n,$$

whence

$$\lim_{n\to\infty} (a_n - a_{n-1})/(b_n - b_{n-1}) = A,$$

and consequently by the preceding theorem

$$\lim_{n\to\infty} a_n/b_n = \lim_{n\to\infty} \frac{\alpha_1 + \alpha_2 + \ldots + \alpha_n}{n} = A.$$

The proof is similar if $\lim_{n\to\infty} \alpha_n = +\infty$, $\lim_{n\to\infty} \alpha_n = -\infty$.

THEOREM 28. If the two sequences $\{a_n\}$, $\{b_n\}$ have limits A and B respectively, then (Cesaro [18])

$$\lim_{n\to\infty} \frac{a_1b_n + a_2b_{n-1} + \ldots + a_{n-1}b_2 + a_nb_1}{n} = AB. \tag{2}$$

Proof. Clearly

$$\lim_{n\to\infty} | a_n | = | A |$$

whence, by Theorem 27,

$$\lim_{n\to\infty} \frac{| a_1 | + | a_2 | + \ldots + | a_n |}{n} = | A |.$$

Now let $\nu = [n/2]$, that is, ν denotes the largest integer contained in $n/2$. Then $\lim_{n\to\infty} \nu/n = 1/2$, whence

$$\lim_{n\to\infty} \frac{|a_1|+|a_2|+\ldots+|a_\nu|}{n} = \lim_{n\to\infty} \frac{\nu}{n}\,\frac{|a_1|+|a_2|+\ldots+|a_\nu|}{\nu} = \tfrac{1}{2}\,| A |,$$

and consequently if $\alpha > | A |/2$, there exists an n_0 such that for $n > n_0$

$$\frac{1}{n}[| a_1 | + | a_2 | + \ldots + | a_\nu |] < \alpha, \quad n > n_0, \quad \nu = \left[\frac{n}{2}\right].$$

If now the sequence $\{ \sigma_n \}$ is defined by

$$\sigma_n = \frac{1}{n}[a_1b_n + a_2b_{n-1} + \ldots + a_\nu b_{n-\nu+1} - B(a_1 + a_2 + \ldots + a_\nu)] \tag{3}$$

$$\sigma_n = \frac{1}{n}[a_1(b_n - B) + a_2(b_{n-1} - B) + \ldots + a_\nu(b_{n-\nu+1} - B)],$$

then, since $n - \nu \to \infty$ as $n \to \infty$, we may suppose that for any preassigned $\varepsilon > 0$, n_0 is so chosen that $| b_r - B | < \varepsilon/\alpha$ for $r > n - \nu$. It therefore follows that

$$| \sigma_n | < \frac{1}{n}\,\frac{\varepsilon}{\alpha}\,[| a_1 | + | a_2 | + \ldots + | a_\nu |] < \varepsilon$$

whence $\lim_{n\to\infty}\sigma_n = 0$ and from (3) and Theorem 27 we have

$$\lim_{n\to\infty} \frac{a_1b_n+a_2b_{n-1}+\ldots+a_\nu b_{n-\nu+1}}{n} = B\lim\frac{a_1+a_2+\ldots+a_\nu}{n} = \tfrac{1}{2}AB$$

Analogously

$$\lim_{n\to\infty} \frac{a_nb_1 + a_{n-1}b_2 + \ldots + a_{\nu+1}b_{n-\nu}}{n} = \tfrac{1}{2}AB$$

whence, by addition, (2) follows.

2. The usual definition of the sum of a series can be generalized on the basis of the preceding theorems so that a series which converges to S, or diverges to $+\infty$ or $-\infty$, according to the usual definition, has the same property according to the new definition. On the other hand there exist series which are convergent according to the new definition but are not convergent according to the usual definition. The discussion will be limited here to what is necessary to formulate various propositions concerning Cesaro summability. (Cesaro [18].)

Given a sequence $\{a_n\}$, let

$$s_n^{(0)} = a_0 + a_1 + \ldots + a_n$$

$$s_n^{(1)} = s_0^{(0)} + s_1^{(0)} + \ldots + s_n^{(0)}$$

and, in general, for any integer k, let

$$(1) \qquad s_n^{(k)} = s_0^{(k-1)} + s_1^{(k-1)} + \ldots + s_n^{(k-1)}, \quad (k = 1, 2, \ldots).$$

The sums $s_n^{(k)}$ are linearly expressible in terms of the sums $s_n^{(0)}$ and it is easily verified that the number of these sums which occur is $\binom{n+k}{k}$. This property clearly holds for $k = 0$. Continuing by induction it appears that the right side of (1) contains

$$\binom{k-1}{k-1} + \binom{k}{k-1} + \ldots + \binom{n+k-1}{k-1} = \binom{n+k}{k}$$

terms $s_n^{(0)}$, which is precisely the result to be proved.

DEFINITION 5. If

$$C_n^{(k)} = s_n^{(k)} \Big/ \binom{n+k}{k}, \qquad (n = 0, 1, 2, \ldots) \tag{2}$$

we shall call $C_n^{(k)}$ the *Cesaro mean of order* k of the first $n + 1$ terms of the sequence $s_0^{(0)}, s_1^{(0)}, \ldots, s_n^{(0)}, \ldots$ and if $\lim_{n\to\infty} C_n^{(k)} = A$, where A is finite, for some value of k, we shall say that $\sum_{n=0}^{\infty} a_n$ is *Cesaro summable of order k to sum A*, or simply *summable* (C, k) *to sum A*.

THEOREM 29. If the series $\sum_{n=0}^{\infty} a_n$ has a sum A in the ordinary sense, then its (C, k) sum is also equal to A; if the series diverges to $+\infty$ $[-\infty]$ then its (C, k) sum is also $+\infty$ $[-\infty]$.

This state of affairs is expressed by saying that (C, k) summability satisfies the *condition of permanence of the sum* (*first condition of regularity*). (Cf. Hardy, Divergent Series p. 10.)

Proof. Writing $\alpha_n = s_n^{(k)}$, $\beta_n = \binom{n+k}{k}$, we have

$$\alpha_n - \alpha_{n-1} = s_n^{(k-1)}, \quad \beta_n - \beta_{n-1} = \binom{n+k}{k} - \binom{n+k-1}{k} = \binom{n+k-1}{k-1}$$

whence

$$\alpha_n/\beta_n = C_n^{(k)}, \ (\alpha_n - \alpha_{n-1})/(\beta_n - \beta_{n-1}) = C_n^{(k-1)}$$

and by Theorem 26, if $\lim_{n\to\infty} C_n^{(k-1)}$ exists, then $\lim_{n\to\infty} C_n^{(k)}$ also exists and the two limits are equal. But since the ordinary sum of the series $\sum_{n=0}^{\infty} a_n$ coincides with $\lim_{n\to\infty} C_n^{(0)}$ the theorem is proved.

THEOREM 30. If the two series $\sum_{n=0}^{\infty} a_n$, $\sum_{n=0}^{\infty} b_n$ are summable (C, k) to A and B respectively, then the series $\sum_{n=0}^{\infty} (\lambda a_n + \mu b_n)$ has (C, k) sum equal to $\lambda A + \mu B$ (*second condition of regularity*).

Proof. Clearly the following relation exists between the means, $C_n^{(k)}$, $\bar{C}_n^{(k)}$, $\bar{\bar{C}}_n^{(k)}$ of the three series in question

$$C_n^{(k)} = \lambda \bar{C}_n^{(k)} + \mu \bar{\bar{C}}_n^{(k)}.$$

Passing to the limit as $n \to \infty$, the theorem follows.

THEOREM 31. If we write

$$A_n^{(k)} = \binom{n+k}{n} = \binom{n+k}{k} \tag{3}$$

then the following identities hold

$$s_n^{(k)} = \sum_{m=0}^{n} A_{n-m}^{(k)} a_m = \sum_{m=0}^{n} A_{n-m}^{(k-1)} s_m^{(0)} = \sum_{m=0}^{n} A_{n-m}^{(k-l-1)} s_m^{(l)}, \; (k>l\geqq 0). \tag{4}$$

Proof. We first establish the identity

$$(5_1) \quad s_n^{(k)} = \binom{n+k}{n} a_0 + \binom{n+k-1}{n-1} a_1 + \dots$$

$$+ \binom{k+1}{1} a_{n-1} + \binom{k}{0} a_n = \sum_{m=0}^{n} \binom{n-m+k}{n-m} a_m.$$

This is clearly true for $k = 0$. The identity follows by induction, since

$$s_n^{(k+1)} = \sum_{i=0}^{n} s_i^{(k)} = \sum_{i=0}^{n} \sum_{m=0}^{i} \binom{i-m+k}{i-m} a_m$$

$$= \sum_{m=0}^{n} a_m \left[\sum_{i=m}^{n} \binom{i-m+k}{i-m} \right] = \sum_{m=0}^{n} \binom{n-m+k+1}{n-m} a_m.$$

Since, moreover, by virtue of (1) the $s_n^{(k)}$ result from $k+1$ operations on the terms of the sequence $\{a_n\}$, and therefore k operations on the terms of the sequence $\{s_n^0\}$, we have also the identity

$$(5_2) \qquad s_n^{(k)} = \sum_{m=0}^{n} A_{n-m}^{(k-1)} s_m^{(0)}.$$

By similar considerations, we obtain more generally

$$(5_3) \qquad s_n^{(k)} = \sum_{m=0}^{n} A_{n-m}^{(k-l-1)} s_m^{(l)}.$$

THEOREM 32. If the means $C_n^{(k)}$ of order k of the sums $s_0^{(0)}$, $s_1^{(0)}, s_2^{(0)}, \dots, s_n^{(0)}, \dots$ have for limit, as $n \to \infty$, A $[+\infty, -\infty]$, then the means of order k of the sequence $s_p^{(0)}, s_{p+1}^{(0)}, \dots, s_{p+n}^{(0)}, \dots$, obtained from the preceding sequence by suppressing the first p terms, also have for limit, as $n \to \infty$, $A[+\infty, -\infty]$, (*third condition of regularity*).

Proof. Let $T_{n,p}$ denote the mean of order k of the first $n+1$ terms of the sequence $s_p^{(0)}, s_{p+1}^{(0)}, \ldots, s_{p+n}^{(0)}, \ldots$. Then by (2) and (4)

$$T_{n,p} = \frac{1}{A_n^{(k)}} \sum_{i=0}^{n} A_{n-i}^{(k-1)} s_{i+p}^{(0)} = \frac{1}{A_n^{(k)}} \sum_{m=p}^{(n+p)} A_{n+p-m}^{(k-1)} s_m^{(0)}$$

$$= \frac{1}{A_n^{(k)}} \sum_{m=0}^{(n+p)} A_{n+p-m}^{(k-1)} s_m^{(0)} - \frac{1}{A_n^{(k)}} \sum_{m=0}^{(p-1)} A_{n+p-m}^{(k-1)} s_m^{(0)}.$$

Moreover

$$C_{n+p}^{(k)} = \frac{1}{A_{n+p}^{(k)}} \sum_{m=0}^{(n+p)} A_{n+p-m}^{(k-1)} s_m^{(0)}$$

and therefore

$$T_{n,p} = \frac{A_{n+p}^{(k)}}{A_n^{(k)}} C_{n+p}^{(k)} - \frac{1}{A_n^{(k)}} \sum_{m=0}^{(p-1)} A_{n+p-m}^{(k-1)} s_m^{(0)}. \tag{6}$$

Now by (3)

$$\frac{A_{n+p}^{(k)}}{A_n^{(k)}} = \frac{(n+p+k)(n+p+k-1)\ldots(n+p+1)}{(n+k)(n+k-1)\ldots(n+1)};$$

$$\frac{A_{n+p-m}^{(k-1)}}{A_n^{(k)}} = k\,\frac{(n+p-m+k-1)\ldots(n+p-m+1)}{(n+k)\ldots(n+1)},$$

and consequently

$$\lim_{n\to\infty} A_{n+p}^{(k)}/A_n^{(k)} = 1, \lim_{n\to\infty} A_{n+p-m}^{(k-1)}/A_n^{(k)} = 0.$$

Since the sum which occurs on the right side of (6) has a finite number, p, of terms, by passing to the limit in (6), as $n \to \infty$, the theorem follows.

THEOREM 33. If the two series $\sum_{n=0}^{\infty} a_n$, $\sum_{n=0}^{\infty} b_n$ are summable respectively (C, k) and (C, l), then the Cauchy product (see

Hardy's *Divergent Series*, p. 227) $\sum_{n=0}^{\infty} w_n$, where

$$w_n = a_0 b_n + a_1 b_{n-1} + \ldots + a_{n-1} b_1 + a_n b_0,$$

of the two series is summable $(C, k + l + 1)$ and its sum is equal to the product of the sums of the two factors, (*fourth condition of regularity*).

Proof. The proof will be limited to the case where $k = l = 0$. To this end, we write

$$A_n = \sum_{r=0}^{n} a_r, \quad B_n = \sum_{r=0}^{n} b_r, \quad W_n = \sum_{r=0}^{n} w_r,$$

and let

$$\lim_{n\to\infty} A_n = A, \quad \lim_{n\to\infty} B_n = B.$$

Then

$$W_n = a_0 B_n + a_1 B_{n-1} + \ldots + a_{n-1} B_1 + a_n B_0,$$

$$W_0 + W_1 + W_2 + \ldots + W_n = A_0 B_n + A_1 B_{n-1} + \ldots + A_{n-1} B_1 + A_n B_0$$

and therefore by Theorem 28, we obtain

$$(7) \qquad \lim_{n\to\infty} \frac{W_0 + W_1 + W_2 + \ldots + W_n}{n+1} = AB.$$

3. In view of (2), (3) and (4) of the preceding section the Cesaro means $C_n^{(k)}$ can also be defined by the following

$$(1) \quad C_n^{(k)} = \sum_{m=0}^{n} \binom{n - m + k - 1}{n - m} s_m^{(0)} \Big/ \binom{n+k}{n}, \quad \left[s_m = s_m^{(0)} = \sum_{r=0}^{m} a_r \right]$$

which has a meaning not only for non-negative integer values of k, but also for every real number k positive or zero and for every real negative number k which is not an integer. We are therefore led to the following definition:

DEFINITION 6. If $\lim_{n\to\infty} C_n^{(k)} = A$ where A is finite, for any real value of k, other than a negative integral value, we shall say that the series $\sum_{n=0}^{\infty} a_n$ is summable (C, k) to sum A.

Actually the only cases which are considered in analysis are those for which $k > -1$ for reasons which will be explained forthwith.

In order for a method of summability not to lead to paradoxes or contradictions in applications, it will be required to satisfy the following *conditions of regularity.*

i. The method sums a convergent series to its ordinary sum (or a divergent series to $+\infty, -\infty$). In other words, if $\lim_{n\to\infty}$ gen. s_n denotes the value of the series $\sum_{n=0}^{\infty} a_n$ according to a given method of summability and $\lim_{n\to\infty} s_n$ exists, then

$$\lim_{n\to\infty} \text{gen. } s_n = \lim_{n\to\infty} s_n.$$

(Cf. Hardy's definition of *totally regular, Divergent Series* p. 10.)

ii. The method satisfies the *distributive law,* namely, if $\lim_{n\to\infty}$ gen. s_n, $\lim_{n\to\infty}$ gen. s'_n exist, then for any constants a and b,

$$\lim_{n\to\infty} \text{gen. } [as_n + bs'_n] = a \lim_{n\to\infty} \text{gen. } s_n + b \lim_{n\to\infty} \text{gen. } s'_n.$$

iii. The suppression or alteration of the first n terms of the sequence $\{s_n\}$ does not affect the summability of the series nor alter the value of the sum. In other words, if p is any positive integer, then

$$\lim_{n\to\infty} \text{gen. } s_{n+p} = \lim_{n\to\infty} \text{gen. } s_n.$$

iv. If two series and their Cauchy product are summable, then the sum of the product series equals the product of the sums of the two given series.

It can be shown that (C, k) summability is regular if and only if $k \geqq 0$. Specifically the following theorem, which generalizes Theorem 29, can be established.

THEOREM 29′. If a series is summable (C, k) and $k' > k > -1$, then it is summable (C, k'), and the two (C, k) and (C, k') sums coincide. (For a discussion of methods of summability, see Knopp, *Infinite series,* Ch. XIII; Hobson, *Theory of Functions of a Real Variable,* II, pp. 65–98).

The following remarks concerning summability of series are of interest in the applications of the theory.

REMARKS. If $0 < \varrho < 1$, then

$$\frac{1}{(1-\varrho)^{k+1}} = \sum_{n=0}^{\infty} (-1)^n \binom{-k-1}{n} \varrho^n = \sum_{n=0}^{\infty} \binom{n+k}{n} \varrho^n = \sum_{n=0}^{\infty} A_n^{(k)} \varrho^n .$$

Forming the Cauchy product of this series and $\sum_{n=0}^{\infty} a_n \varrho^n$ gives

$$\frac{1}{(1-\varrho)^{k+1}} \sum_{n=0}^{\infty} a_n \varrho^n$$

$$= \sum_{n=0}^{\infty} \left[\binom{n+k}{n} a_0 + \binom{n+k-1}{n-1} a_1 + \ldots + \binom{k}{0} a_n \right] \varrho^n = \sum_{n=0}^{\infty} s_n^{(k)} \varrho^n ,$$

whence it appears that the coefficients of $s_n^{(k)}$ of ϱ^n in the product series divided by $A_n^{(k)}$ are the Cesaro means $C_n^{(k)}$ of order k of the first $n+1$ terms of the series $\sum_{n=0}^{\infty} a_n$.

4a. The following theorem gives sufficient conditions under which $(C, 1)$ summability implies $(C, 0)$ summability.

THEOREM 34. (Hardy-Landau). If the series

$$a_0 + a_1 + a_2 + \ldots + a_n + \ldots \tag{1}$$

is summable $(C, 1)$, and if $na_n > -A$, $n = 0, 1, 2, \ldots$ (or $na_n < A$, $n = 0, 1, 2, \ldots$) then the series (1) is also summable $(C, 0)$. (Hardy [44a], Landau [64c]).

Proof. Let

$$s_n = \sum_{k=0}^{n} a_k = a_0 + a_1 + \ldots + a_n \tag{2}$$

and define the sequence $\{ \sigma_n \}$ by the relations

$$(n+1)\sigma_n = \sum_{k=0}^{n} s_k = s_0 + s_1 + \ldots + s_n , \quad [\sigma_n = C_n^{(1)}]. \tag{3}$$

By hypothesis

$$na_n > -A, \quad (n = 0, 1, 2, \ldots), \quad \lim_{n\to\infty} \sigma_n = S.$$

Therefore we must prove $\lim_{n\to\infty} s_n = S$.

If p is an integer such that $0 < p < n$, then

$$(n+1)\sigma_n = s_0 + s_1 + \ldots + s_n;\ (n-p+1)\sigma_{n-p} = s_0 + s_1 + \ldots + s_{n-p}$$

whence

$$(4)\quad \sum_{k=n-p+1}^{n} s_k = (n+1)\sigma_n - (n-p+1)\sigma_{n-p}$$
$$= (n+1)(\sigma_n - \sigma_{n-p}) + p\sigma_{n-p}.$$

$$(5)\quad \sum_{l=n+1}^{n+p} s_l = (n+p+1)\sigma_{n+p} - (n+1)\sigma_n$$
$$= (n+1)(\sigma_{n+p} - \sigma_n) + p\sigma_{n+p}.$$

In (4) the index k satisfies the relation $n - p + 1 \leqq k \leqq n$, whence if $k < n$, then $s_k = s_n - (a_n + a_{n-1} + \ldots + a_{k+1})$ and since $-a_n < A/n$ it follows that

$$s_k < s_n + A/n + \ldots + A/(k+1),$$

whence

$$s_k < s_n + \frac{(n-k)A}{k+1}.$$

Moroever, since

$$n - k \leqq p - 1 < p,\ n - p < n - p + 1 \leqq k < k + 1,$$

it follows that

$$(6)\qquad s_k < s_n + \frac{pA}{n-p},\quad n - p + 1 \leqq k \leqq n.$$

In (5) the index l satisfies the relation $n + 1 \leqq l \leqq n + p$, whence

$$s_l = s_n + (a_{n+1} + \ldots + a_l)$$
$$> s_n - A/(n+1) - \ldots - A/l > s_n - (l-n)\,A/(n+1)$$

and finally

$$(7)\qquad s_l > s_n - \frac{pA}{n},\quad n + 1 \leqq l \leqq n + p.$$

By (6) and (7), we obtain from (4) and (5)

$$ps_n + \frac{p^2A}{n-p} > (n+1)(\sigma_n - \sigma_{n-p}) + p\sigma_{n-p},$$

$$ps_n - \frac{p^2A}{n} < (n+1)(\sigma_{n+p} - \sigma_n) + p\sigma_{n+p},$$

and dividing by p

$$(8)\qquad \begin{cases} s_n + \dfrac{pA}{n-p} > \dfrac{n+1}{p}(\sigma_n - \sigma_{n-p}) + \sigma_{n-p}, \\ s_n - \dfrac{pA}{n} < \dfrac{n+1}{p}(\sigma_{n+p} - \sigma_n) + \sigma_{n+p}. \end{cases}$$

If $0 < \varepsilon < 1$, so that $\lim_{n\to\infty}(n+1)\varepsilon = \infty$, and p denotes the maximum integer contained in $(n+1)\varepsilon$, then

$$\lim_{n\to\infty} p = \infty,\ p \leqq (n+1)\varepsilon < p+1$$

$$\frac{p}{n+1} \leqq \varepsilon < \frac{p}{n+1} + \frac{1}{n+1},\quad \lim_{n\to\infty}\left|\frac{p}{n+1} - \varepsilon\right| \leqq \lim_{n\to\infty}\frac{1}{n+1} = 0,$$

$$\lim_{n\to\infty}\frac{p}{n+1} = \varepsilon,\ \lim_{n\to\infty}\frac{p}{n} = \lim_{n\to\infty}\frac{p}{n+1}\,\frac{n+1}{n} = \varepsilon,\quad \lim_{n\to\infty}\frac{n+1}{p} = \frac{1}{\varepsilon},$$

$$\lim_{n\to\infty}\frac{p}{n-p} = \lim_{n\to\infty}\frac{p/n}{1-p/n} = \frac{\varepsilon}{1-\varepsilon}.$$

and since by hypothesis $\lim_{n\to\infty}\sigma_n = S$, and therefore $\lim_{n\to\infty}(\sigma_{n\pm p} - \sigma_n) = 0$ it follows from (8) that the superior and inferior limits of the sequence $\{s_n\}$ satisfy the relations

$$\varliminf_{n\to\infty} s_n + \frac{\varepsilon}{1-\varepsilon}A \geqq S,\qquad \varlimsup_{n\to\infty} s_n - \varepsilon A \leqq S,$$

and therefore

$$S - \frac{\varepsilon}{1-\varepsilon}A \leqq \varliminf_{n\to\infty} s_n \leqq \varlimsup_{n\to\infty} s_n \leqq S + \varepsilon A.$$

Since ε is arbitrary, $\varepsilon A/(1-\varepsilon)$ and εA are arbitrarily small in absolute value and consequently

$$\overline{\lim_{n\to\infty}} s_n = \underline{\lim_{n\to\infty}} s_n = S$$

which proves that $\lim_{n\to\infty} s_n$ exists and equals S. (For an extensive bibliography of summability of series by the method of arithmetic means, see E. Kogbetlianz [58a].

4b. In view of (8) we have the following

COROLLARY. If the terms $a_n(x)$ of the series $a_0(x) + a_1(x) + + \ldots + a_n(x) + \ldots$ are functions of x in $[a, b]$, if $na_n(x) > -A$, (or $[na_n(x) < A]$) where A is a constant, and if the $(C, 1)$ sum of the series converges uniformly to $S(x)$ in $[a, b]$, then the $(C, 0)$ sum also converges uniformly to $S(x)$ in $[a, b]$.

5. Let $f(x)$ be periodic with period 2π and integrable in $[-\pi,\pi]$ and suppose

$$f(x) \sim \tfrac{1}{2}a_0 + \sum_{k=1}^{\infty} (a_k \cos kx + b_k \sin kx) \tag{1}$$

$$a_k = \frac{1}{\pi}\int_{-\pi}^{\pi} f(x) \cos kx\, dx, \quad b_k = \frac{1}{\pi}\int_{-\pi}^{\pi} f(x) \sin kx\, dx\,,$$

$$(k = 0, 1, 2, \ldots).$$

Let

$$\alpha_0 = \tfrac{1}{2}a_0, \quad \alpha_k = a_k \cos kx + b_k \sin kx. \tag{2}$$

If we write

$$S_k = \alpha_0 + \alpha_1 + \ldots + \alpha_k; \quad \sigma_n = \frac{1}{n}\sum_{k=0}^{(n-1)} S_k$$

the determination of the $(C, 1)$ sum of $\sum_{k=0}^{\infty} \alpha_k$ is then equivalent to determining the $\lim_{n\to\infty} \sigma_n$.

Since by Sec. 4.2, (4)

$$S_k = \frac{1}{\pi}\int_{-\pi}^{\pi} f(x + \alpha)\, \frac{\sin (k + \frac{1}{2})\alpha}{2 \sin \dfrac{\alpha}{2}}\, d\alpha$$

it follows that

$$\sigma_n = \frac{1}{n\pi}\int_{-\pi}^{\pi} f(x+\alpha)\,\frac{\sum\limits_{k=0}^{(n-1)} \sin\,(k+\tfrac{1}{2})\alpha}{2\sin\dfrac{\alpha}{2}}\,d\alpha$$

and from the identity

$$\sum_{k=0}^{(n-1)} \sin\,(k+\tfrac{1}{2})\alpha = \frac{\sum\limits_{k=0}^{(n-1)} 2\sin\dfrac{\alpha}{2}\sin\,(k+\tfrac{1}{2})\,\alpha}{2\sin\dfrac{\alpha}{2}}$$

$$= \frac{\sum\limits_{k=0}^{(n-1)} \{\cos k\alpha - \cos\,(k+1)\alpha\}}{2\sin\dfrac{\alpha}{2}} = \frac{1-\cos n\alpha}{2\sin\dfrac{\alpha}{2}} = \left(\sin^2\frac{n}{2}\alpha\right)/\sin\frac{\alpha}{2}$$

follows

$$\sigma_n = \frac{1}{n\pi}\int_{-\pi}^{\pi} f(x+\alpha)\,\frac{\sin^2\dfrac{n}{2}\alpha}{2\sin^2\dfrac{\alpha}{2}}\,d\alpha.$$

Substituting 2α for α, we obtain the Fejér integral expression for the mean σ_n

$$\sigma_n = \frac{1}{n\pi}\int_{-\pi/2}^{\pi/2} f(x+2\alpha)\left(\frac{\sin n\alpha}{\sin\alpha}\right)^2 d\alpha, \quad \text{(Fejér [33a]).} \tag{4}$$

Dividing the interval of integration $[-\pi/2, \pi/2]$ into the two subintervals $[-\pi/2, 0]$, $[0, \pi/2]$ and substituting $-\alpha$ for α in the integral over the first of these subintervals, we obtain

$$\alpha_n = \frac{1}{n\pi}\int_0^{\pi/2} [f(x+2\alpha)+f(x-2\alpha)]\left(\frac{\sin n\alpha}{\sin\alpha}\right)^2 d\alpha. \tag{5}$$

For $f(x) = 1$, $\alpha_0 = 1$, $\alpha_k = 0$, $S_k = 1$, $\sigma_n = 1$, whence by (5),

$$1 = \frac{2}{n\pi}\int_0^{\pi/2}\left(\frac{\sin n\alpha}{\sin\alpha}\right)^2 d\alpha. \tag{6}$$

Therefore

$$S(x) = \frac{2}{n\pi}\int_0^{\pi/2} S(x)\left(\frac{\sin n\alpha}{\sin \alpha}\right)^2 d\alpha. \tag{6'}$$

It follows from (5), (6′) that a necessary and sufficient condition for the $(C, 1)$ sum of (1) at a point x to be equal to $S(x)$, that is for $\lim_{n\to\infty} \sigma_n = S(x)$, is

$$0 = \lim_{n\to\infty} \frac{1}{n\pi}\int_0^{\pi/2} [f(x + 2\alpha) + f(x - 2\alpha) - 2S(x)]\left(\frac{\sin n\alpha}{\sin \alpha}\right)^2 d\alpha.$$

Finally, writing (cf. Sec. 4.4, (4))

$$\boxed{\varphi(x, \alpha) = f(x + 2\alpha) + f(x - 2\alpha) - 2S(x)} \tag{7}$$

we obtain

THEOREM 35. A necessary and sufficient condition for the $(C, 1)$ sum of the Fourier series of a function $f(x)$, periodic with period 2π and integrable in $[-\pi, \pi]$, to equal $S(x)$ at a point x is that corresponding to an arbitrary $\omega > 0$ there exists an integer $n_\omega > 0$ such that for every integer $n > n_\omega$

$$\left| \frac{1}{n\pi}\int_0^{\pi/2} \varphi(x, \alpha)\left(\frac{\sin n\alpha}{\sin \alpha}\right)^2 d\alpha \right| < \omega.$$

Now let

$$A = \frac{1}{n\pi}\int_0^{\pi/2} \varphi(x, \alpha)\left(\frac{\sin n\alpha}{\sin \alpha}\right)^2 d\alpha - \frac{1}{n\pi}\int_0^{\pi/2} \varphi(x, \alpha)\left(\frac{\sin n\alpha}{\alpha}\right)^2 d\alpha$$

then

$$| A | = \left| \frac{1}{n\pi}\int_0^{\pi/2} \varphi(x, \alpha) \sin^2 n\alpha\left(\frac{1}{\sin^2 \alpha} - \frac{1}{\alpha^2}\right) d\alpha \right|$$

$$\leqq \frac{1}{n\pi}\int_0^{\pi/2} |\varphi(x, \alpha)| \left| \frac{1}{\sin^2 \alpha} - \frac{1}{\alpha^2} \right| d\alpha,$$

and since $1/\sin^2 \alpha - 1/\alpha^2$ is continuous, and therefore bounded, in $[0, \pi/2]$, we may assume $| 1/\sin^2 \alpha - 1/\alpha^2 | < L$ for $0 \leqq \alpha \leqq \pi/2$ whence

$$(8) \qquad |A| < \frac{L}{n\pi}\int_0^{\pi/2} |\varphi(x, \alpha)|\, d\alpha,$$

and therefore the limits as $n \to \infty$ of the two integrals

$$(9) \quad \frac{1}{n\pi}\int_0^{\pi/2} \varphi(x, \alpha)\left(\frac{\sin n\alpha}{\sin \alpha}\right)^2 d\alpha, \quad \frac{1}{n\pi}\int_0^{\pi/2} \varphi(x, \alpha)\left(\frac{\sin n\alpha}{\alpha}\right)^2 d\alpha$$

have the same behaviour. Consequently

THEOREM 36. A necessary and sufficient condition for the $(C, 1)$ sum of the Fourier series of a function $f(x)$, periodic with period 2π and integrable in $[-\pi, \pi]$, to equal $S(x)$ at a point x is that corresponding to an arbitrary $\omega > 0$ there exists an integer $n_\omega > 0$ such that for every integer $n > n_\omega$

$$(10) \qquad \left|\frac{1}{n\pi}\int_0^{\pi/2} \varphi(x, \alpha)\left(\frac{\sin n\alpha}{\alpha}\right)^2 d\alpha\right| < \omega.$$

Moreover, if $S(x)$ is continuous in $[a, b]$ where $[a, b] \in [-\pi, \pi]$ (We assume here and in the following theorems that $S(-\pi) = S(\pi)$ if $a = -\pi$, and $b = \pi$,), and the number n_ω corresponding to an arbitrary ω is independent of x then the $(C, 1)$ sum of the Fourier series of $f(x)$ converges uniformly to $S(x)$ in $[a, b]$.

Indeed, if $|S(x)| < L'$ in $[a, b]$, then by (7) and (8)

$$|A| < \frac{L}{n\pi} M$$

where

$$M = \int_0^{\pi/2} |f(x + 2\alpha)|\, d\alpha + \int_0^{\pi/2} |f(x - 2\alpha)|\, d\alpha + L'\pi$$

$$= \int_{-\pi/2}^{\pi/2} |f(x + 2\alpha)|\, d\alpha + L'\pi = \tfrac{1}{2}\int_{-\pi}^{\pi} |f(t)|\, dt + L'\pi.$$

Condition (10) can be modified as follows.

If $0 < \varepsilon < \pi/2$ then

$$\Omega = \left|\frac{1}{n\pi}\int_0^{\pi/2} \varphi(x, \alpha)\left(\frac{\sin n\alpha}{\alpha}\right)^2 d\alpha - \frac{1}{n\pi}\int_0^{\varepsilon} \varphi(x, \alpha)\left(\frac{\sin n\alpha}{\alpha}\right)^2 d\alpha\right|$$

$$= \frac{1}{n\pi}\left|\int_\varepsilon^{\pi/2} \varphi(x, \alpha)\left(\frac{\sin n\alpha}{\alpha}\right)^2 d\alpha\right|,$$

but if $0 < \varepsilon \leqq \alpha$, then

$$(\sin n\alpha/\alpha)^2 \leqq 1/\alpha^2 \leqq 1/\varepsilon^2$$

whence

$$0 \leqq \Omega \leqq \frac{1}{n\pi\varepsilon^2}\int_\varepsilon^{\pi/2} |\,\varphi(x, \alpha)\,|\, d\alpha,$$

which implies $\lim_{n\to\infty} \Omega = 0$ and therefore:

THEOREM 36′. A necessary and sufficient condition for the $(C, 1)$ sum of the Fourier series of a function $f(x)$, periodic with period 2π and integrable in $[-\pi, \pi]$, to equal $S(x)$ at a point x is that for an arbitrary $\omega > 0$ there exists an ε satisfying $0 < \varepsilon \leqq \pi/2$ and an integer n_ω such that for every $n > n_\omega$

$$\left|\frac{1}{n\pi}\int_0^\varepsilon \varphi(x, \alpha)\left(\frac{\sin n\alpha}{\alpha}\right)^2 d\alpha\right| < \omega.$$

If $S(x)$ is continuous in $[a, b]$, where $[a, b] \in [-\pi, \pi]$, and the numbers ε and n_ω are independent of x, then the $(C, 1)$ sum of the Fourier series of $f(x)$ converges uniformly to $S(x)$ in $[a, b]$.

6. THEOREM 37. Let $f(x)$ be periodic, with period 2π, and integrable in $[-\pi, \pi]$, and let

$$\varphi(x, \alpha) = f(x + 2\alpha) + f(x - 2\alpha) - 2S(x).$$

Then, if

$$\lim_{\alpha\to 0} \varphi(x, \alpha) = 0$$

the $(C, 1)$ sum of the Fourier series of $f(x)$ at the point x equals $S(x)$.

Proof. Clearly, if $\lim_{\alpha\to 0} \varphi(x, \alpha) = 0$, then for $\omega > 0$ there exists an $\varepsilon > 0$ such that $|\,\varphi(x, \alpha)\,| < \omega$ for $0 \leqq \alpha \leqq \varepsilon$.

Then, with the help of (6) of the preceding section, we have

$$\left|\frac{1}{n\pi}\int_0^\varepsilon \varphi(x, \alpha)\left(\frac{\sin n\alpha}{\alpha}\right)^2 d\alpha\right| \leqq \frac{\omega}{n\pi}\int_0^\varepsilon \left(\frac{\sin n\alpha}{\alpha}\right)^2 d\alpha$$

$$\leqq \frac{\omega}{n\pi}\int_0^{\pi/2} \left(\frac{\sin n\alpha}{\sin \alpha}\right)^2 d\alpha = \frac{\omega}{2} < \omega$$

and therefore the theorem follows from Theorem 36′.

The following important corollaries are immediate consequences of the theorem just proved.

COROLLARY 1. The $(C, 1)$ sum of the Fourier series of $f(x)$ equals $f(x)$ at every point where $f(x)$ is regular (Appendix, Def. 15).

COROLLARY 2. The $(C, 1)$ sum of the Fourier series of $f(x)$ equals $[f(x+) + f(x-)]/2$ at every point x where $f(x)$ has a discontinuity of the first kind.

COROLLARY 3. If the Fourier series of $f(x)$ is convergent [i.e. summable $(C, 0)$] at a point x where $f(x)$ is regular, then its sum equals $f(x)$.

Proof. Clearly the $(C, 0)$ and $(C, 1)$ sums of $f(x)$ at the point x are equal (Th. 29), but by hypothesis the $(C, 1)$ sum equals $f(x)$ by Corollary 1.

COROLLARY 4. If the Fourier series of $f(x)$ is convergent at a point x of discontinuity of the first kind, its sum equals $[(f(x+) + f(x-)]/2$.

From Theorems 36′ and 37, by setting $S(x) = f(x)$, we obtain.

THEOREM 38. If $f(x)$ is periodic, with period 2π, integrable in $[-\pi, \pi]$, and continuous in $[a, b]$, where $[a, b] \epsilon (-\pi, \pi]$, then the $(C, 1)$ sum of the Fourier series of $f(x)$ converges uniformly to $f(x)$ in every interval interior to $[a, b]$.

In particular, if $f(x)$ is continuous in $[-\pi, \pi]$ and $f(-\pi) = f(+\pi)$ then the $(C, 1)$ sum of the Fourier series of $f(x)$ converges uniformly to $f(x)$ in $[-\pi, \pi]$

7. THEOREM 39. (Lebesgue [67a], p. 94). If $f(x)$ is integrable in $[-\pi, \pi]$, the $(C, 1)$ sum of its Fourier series equals $f(x)$ almost everywhere.

Proof. We must show that, with the exception of a set of measure zero, for x in $[-\pi, \pi]$

$$\lim_{n\to\infty} \sigma_n = f(x),$$

or, letting

$$\varphi(x, \alpha) = f(x + 2\alpha) + f(x - 2\alpha) - 2f(x) \tag{1}$$

that

$$\lim_{n\to\infty} \frac{1}{n\pi} \int_0^\varepsilon \varphi(x, \alpha)\left(\frac{\sin n\alpha}{\alpha}\right)^2 d\alpha = 0, \quad 0 < \varepsilon \leqq \pi/2$$

almost everywhere for x in $[-\pi, \pi]$.

Since $\alpha - \sin \alpha \geqq 0$, $\alpha(1 - \sin \alpha) \geqq 0$, for $\alpha > 0$, by addition we have $2\alpha \geqq (1 + \alpha) \sin \alpha$, $\sin \alpha/\alpha \leqq 2/(1 + \alpha)$, and therefore it suffices to prove that

$$\lim_{n\to\infty} \frac{4n}{\pi} \int_0^\varepsilon | \varphi(x, \alpha) | \frac{1}{(1 + n\alpha)^2} d\alpha = 0, \tag{2}$$

almost everywhere in $[-\pi, \pi]$.

Let

$$\Phi(x, \alpha) = \int_0^\alpha | \varphi(x, \beta) | \, d\beta. \tag{3}$$

Then, integrating by parts, we have

$$\frac{4n}{\pi} \int_0^\varepsilon |\varphi(x, \alpha)| \frac{1}{(1+n\alpha)^2} d\alpha = \frac{4}{\pi} \frac{n}{(1+n\varepsilon)^2} \Phi(x, \varepsilon) + \frac{8}{\pi} \int_0^\varepsilon \Phi(x, \alpha) \frac{n^2}{(1+n\alpha)^2} d\alpha$$

$$\frac{4n}{\pi} \int_0^\varepsilon |\varphi(x, \alpha)| \frac{1}{(1+n\alpha)^2} d\alpha \leqq \frac{4}{\pi} \frac{n}{(1+n\varepsilon)^2} \Phi(x, \varepsilon) + \frac{8}{\pi} \int_0^\varepsilon \frac{\Phi(x, \alpha)}{\alpha} \frac{n}{(1+n\alpha)^2} d\alpha.$$

Now if $\varepsilon > 0$, $\lim_{n\to\infty} 4/\pi \cdot n/(1 + n\varepsilon)^2 \cdot \Phi(x, \varepsilon) = 0$, and so the theorem will follow if we show that, almost everywhere in $[-\pi, \pi]$, for an arbitrary $\sigma > 0$, there exists an $\varepsilon > 0$ such that

$$\int_0^\varepsilon \frac{\Phi(x, \alpha)}{\alpha} \frac{n}{(1 + n\alpha)^2} d\alpha < \sigma.$$

Clearly

$$\frac{\Phi(x, \alpha)}{\alpha} = \frac{1}{\alpha} \int_0^\alpha | \varphi(x, \beta) | \, d\beta$$

$$\leqq \frac{1}{\alpha} \int_0^\alpha |f(x + 2\beta) - f(x)| \, d\beta + \frac{1}{\alpha} \int_0^\alpha | f(x - 2\beta) - f(x) | \, d\beta,$$

$$\frac{\Phi(x, \alpha)}{\alpha} \leqq \frac{1}{2\alpha} \int_x^{x+2\alpha} | f(t) - f(x) | \, dt + \frac{1}{2\alpha} \int_x^{x-2\alpha} | f(t) - f(x) | \, dt,$$

and since almost everywhere in $[-\pi, \pi]$ we have (Appendix, Th. 16)

$$\lim_{\alpha\to 0} \frac{1}{2\alpha}\int_x^{x+2\alpha} |f(t) - f(x)|\, dt = 0, \quad \lim_{\alpha\to 0} \frac{1}{2\alpha}\int_x^{x-2\alpha} |f(t) - f(x)|\, dt = 0,$$

it follows that corresponding to $\sigma > 0$, there exists an $\varepsilon > 0$ such that $0 < \Phi(x, \alpha)/\alpha < \sigma$ for $0 < \alpha \leqq \varepsilon$, and, consequently,

$$\int_0^{\varepsilon} \frac{\Phi(x, \alpha)}{\alpha} \frac{n}{(1 + n\alpha)^2} d\alpha \leqq \sigma \int_0^{\varepsilon} \frac{n}{(1 + n\alpha)^2} d\alpha < \sigma.$$

8. THEOREM 40. (Lebesgue [67a], p. 45). If $f(x)$ is of bounded variation in $[-\pi, \pi]$ then its Fourier constants satisfy the relations

$$a_k = O\left(\frac{1}{k}\right), \quad b_k = O\left(\frac{1}{k}\right).$$

Proof. By hypothesis, the following identity holds in $[-\pi, \pi]$

$$f(x) = f_1(x) - f_2(x)$$

where $f_1(x)$ and $f_2(x)$ are two non-negative, non-decreasing functions (Appendix, Th. 12).

Then

$$a_k = \frac{1}{\pi}\int_{-\pi}^{\pi} f_1 \cos kx\, dx - \frac{1}{\pi}\int_{-\pi}^{\pi} f_2 \cos kx\, dx$$

and applying the second theorem of the mean to the two integrals (Appendix, Th. 18) we have

$$|a_k| \leqq \frac{f_1(\pi)}{\pi}\left|\int_{\xi_1}^{\pi} \cos kx\, dx\right| + \frac{f_2(\pi)}{\pi}\left|\int_{\xi_2}^{\pi} \cos kx\, dx\right|$$

$$\leqq \frac{f_1(\pi) + f_2(\pi)}{\pi}\frac{1}{k}[|\sin k\xi_1| + |\sin k\xi_2|] < \frac{2}{\pi}[f_1(\pi) + f_2(\pi)]\frac{1}{k}.$$

A similar argument can be made for the constants b_k.

THEOREM 41. (Dirichlet-Jordan [55b]). If $f(x)$ is of bounded variation in $[-\pi, \pi]$ then at every point of $[-\pi, \pi]$ its

Fourier series converges to $f(x)$ or to $[f(x+)+f(x-)]/2$, depending on whether $f(x)$ is continuous or discontinuous at the point.

Proof. Depending on whether the point x is a point of continuity or finite discontinuity (all the points of discontinuity of a function of bounded variation are points of finite discontinuity), its $(C, 1)$ sum converges to $f(x)$ or $[f(x+)+f(x-)]/2$, (Th. **37**, Cor. 1, 2) and since by the preceding theorem $a_k \cos kx + b_k \sin kx = O(1/k)$, it follows by the Hardy-Landau theorem (Th. **34**) that the $(C, 0)$ sum coincides with the $(C, 1)$ sum and therefore our theorem is established.

THEOREM **41′**. If $f(x)$ is continuous and of bounded variation in $[-\pi, \pi]$, its Fourier series converges uniformly to $f(x)$ in every interval interior to $[-\pi, \pi]$. If $f(-\pi)=f(\pi)$ the series converges uniformly in $[-\pi, \pi]$.

Proof. Using Theorem **38** and the corollary of Theorem **34**, the argument proceeds as in the proof of the preceding theorem.

REMARK. If instead of supposing $f(x)$ is of bounded variation in $[-\pi, \pi]$ we suppose $f(x)$ is of bounded variation in an interval $[a, b]$ belonging to $[-\pi, \pi]$, the conclusions of Theorems **41** and **41′** are still valid for the points or intervals interior to $[a, b]$. Indeed, by Riemann's theorem (Th. **14**) the behaviour of the Fourier series of $f(x)$ at the points interior to $[a, b]$ is the same as the behaviour of the Fourier series of a function of bounded variation in $[-\pi, \pi]$ which coincides with $f(x)$ in $[a, b]$.

Thus we have proved Theorem 18 as announced in Sec. 4.7.

7. (C,k) Summability $(k>0)$ of Fourier Series

1. LEMMA. Let $0 \leqq k \geqq 1$, and consider the development in series

$$\frac{1}{(1-z)^k} = \alpha_0 + \alpha_1 z + \ldots + \alpha_n z^n + \ldots \tag{1}$$

$$\alpha_0 = 1;\ \alpha_n = (-1)^n \binom{-k}{n} = \binom{n+k-1}{n} = \frac{(n+k-1)\ldots(k+1)k}{n!}.$$

If $n \geqq 0$, $|z| \leqq 1, z \neq 1$, we have the following inequality

$$(2) \qquad |\alpha_0 + \alpha_1 z + \ldots + \alpha_n z^n| < \frac{H}{|1-z|^k},$$

where H is a constant independent of n and z but dependent on k. In Ch. III, Sec. 10.1, 2 (Stieltjes' bounds for Legendre polynomials) we shall establish by elementary considerations that if $k = \frac{1}{2}$, we may take $H = \sqrt{2}$.

Proof. If $k = 0$, (2) is obviously true. We shall suppose, therefore, that $k > 0$.

Clearly

$$\alpha_n/\alpha_{n+1} = (n+1)/(n+k) \geqq 1,$$

whence

$$\alpha_0 \geqq \alpha_1 \geqq \alpha_2 \geqq \ldots \geqq \alpha_n \geqq \ldots.$$

Moreover, by the property of the Γ function and Stirling's formula (See Appendix, Ths. 19 and 20)

$$\alpha_n = \frac{\Gamma(n+k)}{\Gamma(k)\Gamma(n+1)}$$

$$= \frac{1}{\Gamma(k)} \frac{\sqrt{2\pi(n+k-1)}}{\sqrt{2\pi n}} \left(\frac{n+k-1}{e}\right)^{n+k-1} \left(\frac{e}{n}\right)^n e^{\frac{\theta}{12(n+k-1)} - \frac{\theta_1}{12n}}$$

$$0 < \theta, \quad \theta_1 < 1,$$

whence

$$\alpha_n = \frac{1}{\Gamma(k)} \left(1 + \frac{k-1}{n}\right)^{k-1/2} \frac{1}{e^{k-1}} \left(1 + \frac{k-1}{n}\right)^n \frac{1}{n^{1-k}} e^{\frac{\theta}{12(n+k-1)} - \frac{\theta_1}{12n}}.$$

Therefore, there exist two positive constants A and B such that

$$(3) \qquad \frac{B}{n^{1-k}} < \alpha_n = \binom{n+k-1}{n} < \frac{A}{n^{1-k}}, \qquad (n = 0, 1, \ldots).$$

Moreover

$$|(1-z)(\alpha_{n+1}z^{n+1} + \alpha_{n+2}z^{n+2} + \ldots + \alpha_{m+1}z^{m+1})|$$
$$= |\alpha_{n+1}z^{n+1} + (\alpha_{n+2} - \alpha_{n+1})z^{n+2} + \ldots + (\alpha_{m+1} - \alpha_m)z^{m+1} - \alpha_{m+1}z^{m+2}|$$
$$\leqq \alpha_{n+1} + (\alpha_{n+1} - \alpha_{n+2}) + (\alpha_{n+3} - \alpha_{n+2}) + \ldots + (\alpha_m - \alpha_{m+1}) + \alpha_{m+1}$$
$$= 2\alpha_{n+1}$$

whence, by (3)

$$|\alpha_{n+1}z^{n+1}+\alpha_{n+2}z^{n+2}+\ldots+\alpha_{m+1}z^{m+1}| \leqq \frac{2\alpha_{n+1}}{|1-z|} < \frac{2A}{n^{1-k}|1-z|}.$$

Consequently if $|1-z| \leqq 1/n$, then

$$|\alpha_0+\alpha_1 z+\ldots+\alpha_n z^n| \leqq \alpha_0+\alpha_1+\ldots+\alpha_n$$

$$< A\left(\frac{1}{1^{1-k}}+\frac{1}{2^{1-k}}+\ldots+\frac{1}{n^{1-k}}\right) < A\int_0^n \frac{1}{x^{1-k}}\,dx = \frac{A}{k}n^k \leqq \frac{A}{k|1-z|}.$$

If, on the other hand, $|1-z| > 1/n$, then

$$|\alpha_0+\alpha_1 z+\ldots+\alpha_n z^n| = \left|\frac{1}{(1-z)^k}-(\alpha_{n+1}z^{n+1}+\alpha_{n+2}z^{n+2}+\ldots)\right|$$

$$\leqq \frac{1}{|1-z|^k}+|\alpha_{n+1}z^{n+1}+\alpha_{n+2}z^{n+2}+\ldots|$$

$$\leqq \frac{1}{|1-z|^k}+\frac{2A}{|1-z|\,n^{1-k}} \leqq \frac{1}{|1-z|^k}+\frac{2A|1-z|^{1-k}}{|1-z|} = \frac{2A+1}{|1-z|^k}.$$

This proves the lemma.

2. Let $f(x)$ be periodic, with period 2π, integrable in $[-\pi, \pi]$ and let

$$(4_1) \qquad f(x) \sim \tfrac{1}{2}a_0+\sum_{n=1}^{\infty}[a_n \cos nx + b_n \sin nx],$$

$$(4_2) \qquad a_n = \frac{1}{\pi}\int_{-\pi}^{\pi} f(x)\cos nx\,dx, \quad b_n = \frac{1}{\pi}\int_{-\pi}^{\pi} f(x)\sin nx\,dx.$$

In the preceding section we studied the $(C, 1)$ summability of (4_1), so that by Theorem 29′ (Sec. 6.3, d), it will suffice to consider (C, k) summability where $0 \leqq k < 1$.

To express the $C_n^{(k)}$ mean of the first $n+1$ terms of (4_1) we denote by

$$(5) \qquad s_n^{(k)}(\alpha) = \mathfrak{S}_n^{(k)}(\alpha)/A_n^{(k)},$$

$$A_n^{(k)} = 1; \quad A_n^{(k)} = \binom{n+k}{n}, \quad (n = 1, 2, \ldots)$$

the mean of order k of the first $n+1$ terms of the series

$$(6) \qquad 1 + 2\cos\alpha + 2\cos 2\alpha + \ldots + 2\cos n\alpha + \ldots,$$

where [Sec. 4.3, (3)]

$$(7) \quad s_n^{(0)}(\alpha) = 1 + 2\cos\alpha + \ldots + 2\cos n\alpha = \frac{\sin(n+\frac{1}{2})\alpha}{\sin\frac{1}{2}\alpha},$$

and [Sec. 6.2, (4)]

$$(8) \qquad \mathfrak{S}_n^{(k)}(\alpha) = A_n^{(k-1)}s_0^{(0)} + A_{n-1}^{(k-1)}s_1^{(0)} + \ldots + A_0^{(k-1)}s_n^{(0)}$$
$$= \binom{k+n}{n} + \binom{k+n-1}{n-1} 2\cos\alpha + \ldots + \binom{k}{0} 2\cos n\alpha.$$

In view of (4_1) and (4_2) and (2) of Sec. 4.2, we have

$$f(x) \sim \frac{1}{2\pi}\int_{-\pi}^{\pi} f(x+\alpha)[1 + 2\cos\alpha + 2\cos 2\alpha + \ldots]d\alpha,$$

and consequently

$$(9) \qquad C_n^{(k)}(x) = \frac{1}{2\pi}\int_{-\pi}^{\pi} f(x+\alpha)s_n^{(k)}(\alpha)d\alpha.$$

We shall call $s_n^{(k)}(\alpha)$ the *kernel of order k* of (C, k) summability.

THEOREM 42. The kernel $s_n^{(k)}(\alpha)$ has the following properties:

1) (10_1) $\int_{-\pi}^{\pi} | s_n^{(k)}(\alpha) | d\alpha < C$, where C is a constant.

2) (10_2)
$$\frac{1}{2\pi}\int_{-\pi}^{\pi} s_n^{(k)}(\alpha)d\alpha = 1.$$

3) If
$$M_n(\delta) = \max_{0<\delta\leq\alpha\leq\pi} | s_n^{(k)}(\alpha) |, \text{ then}$$

$$(10_3) \qquad \lim_{n\to\infty} M_n(\delta) = 0.$$

Proof. To prove the theorem we first establish two auxiliary bounds (12) and (13) below.

From (6), (7), (8) follows

$$\mathfrak{S}_n^{(k)}(\alpha) = \frac{1}{\sin\frac{\alpha}{2}} \sum_{m=0}^{n} A_m^{(k-1)} \sin\left(n - m + \tfrac{1}{2}\right)\alpha$$

$$= \frac{1}{\sin\frac{\alpha}{2}} Im\, e^{i(n+\frac{1}{2})\alpha} \sum_{m=0}^{n} A_m^{(k-1)} e^{-im\alpha}$$

and by the lemma of Sec. 7.1 [for the case $A_m^{(k-1)} = \binom{m+k-1}{m} = \alpha_m$, $z = e^{-i\alpha}$] we have

$$|\,\mathfrak{S}_n^{(k)}(\alpha)\,| \leqq \frac{1}{\sin\frac{\alpha}{2}} \frac{H}{|\,1-e^{-i\alpha}\,|^k}.$$

But

$$|\,1-e^{-i\alpha}\,| = |\,1-\cos\alpha + i\sin\alpha\,| = \sqrt{2-2\cos\alpha} = 2\sin\frac{\alpha}{2}.$$

Therefore, by Jordan's inequality, namely, $(\sin\alpha/2)/(\alpha/2) \geqq 1/(\pi/2) = 2/\pi$, $1/\sin\alpha/2 < \pi/\alpha$ for $0 < \alpha \leqq \pi$, (See Copson, *Theory of Functions of a Complex Variable*, 1935, p. 136) we have

$$(11) \qquad |\,\mathfrak{S}_n^{(k)}(\alpha)\,| \leqq \frac{2H}{\left(2\sin\frac{\alpha}{2}\right)^{k+1}} \leqq \frac{H'}{\alpha^{k+1}},$$

where H' is a constant independent of α and n, and $0 < \alpha \leqq \pi$.

Moreover, (cf. (3) with $k + 1$ replacing k)

$$A_n^{(k)} > Bn^k,$$

whence by (5)

$$|\,s_n^{(k)}(\alpha)\,| < \frac{H'}{\alpha^{k+1}} \frac{1}{Bn^k},$$

and, consequently,

$$(12) \qquad |\,s_n^{(k)}(\alpha)\,| < \frac{L}{n^k\,\alpha^{1+k}}, \qquad (0 \leqq k \leqq 1,\ 0 < \alpha \leqq \pi)$$

where L is a constant. This is the first bound we wished to establish.

The second inequality for $s_n^{(k)}(\alpha)$ is easily derived as follows. For m and n integers satisfying $0 \leqq m \leqq n$, $n > 0$, we have

$$\left|\frac{\sin(m+\frac{1}{2})\alpha}{2\sin\frac{1}{2}\alpha}\right| = \left|\tfrac{1}{2}+\sum_{l=1}^{m}\cos l\alpha\right| \leqq \tfrac{1}{2}+1+\ldots+1 = m+\tfrac{1}{2} < n+1,$$

and by (5), (7), (8), in view of $\sum_{m=0}^{n} A_{n-m}^{(k-1)} = A_n^{(k)}$

$$|\,s_n^{(k)}(\alpha)\,| \leqq (n+1)\sum_{m=0}^{n} A_{n-m}^{(k-1)}/A_n^{(k)} = n+1 < 2n,$$

which establishes the second bound, namely,

$$|\,s_n^{(k)}(\alpha)\,| \leqq 2n, \quad (n > 0). \tag{13}$$

By (5), (12), (13), if $n > 0$

$$\int_{-\pi}^{\pi} |\,s_n^{(k)}(\alpha)\,|\,d\alpha = 2\int_0^{\pi} |\,s_n^{(k)}(\alpha)\,|\,d\alpha = 2\left[\int_0^{1/n} |s_n^{(k)}(\alpha)|\,d\alpha + \int_{1/n}^{\pi} |s_n^{(k)}(\alpha)\,|\,d\alpha\right]$$

$$< 2\left[2+\frac{L}{kn^k}\left(n^k-\frac{1}{\pi^k}\right)\right] < C,$$

where C is a constant, which establishes (10_1).

By (7)

$$\frac{1}{\pi}\int_{-\pi}^{\pi}\frac{\sin(n+\frac{1}{2})\alpha}{2\sin\dfrac{\alpha}{2}}\,d\alpha = 1.$$

Also by (5) and (8)

$$\frac{1}{2\pi}\int_{-\pi}^{\pi} s_n^{(k)}(\alpha)\,d\alpha = \frac{\sum_{m=0}^{n} A_{n-m}^{(k-1)}}{A_n^{(k)}} = 1$$

which establishes (10_2).

Now, by (12) for an arbitrary δ, such that $0 < \delta < \pi$,

$$M_n(\delta) = \max_{\delta \leqq \alpha \leqq \pi} |\, s_n^{(k)}(\alpha) \,| < \frac{L}{n^k \delta^{k+1}}$$

whence $\lim_{n\to\infty} M_n(\delta) = 0$, which is (10_3).

3. Theorems 37 and 38 were extended by M. Riesz [94a, b] and S. Chapmann [19] to (C, k) sums, for $k > 0$, of the Fourier series of an integrable function $f(x)$ in accordance with the following theorem.

THEOREM 43. Let $f(x)$ be a periodic function, with period 2π, integrable in $[-\pi, \pi]$, and let

$$\varphi(x, \alpha) = f(x + 2\alpha) + f(x - 2\alpha) - 2S(x).$$

Then if

$$\lim_{\alpha\to 0} \varphi(x, \alpha) = 0,$$

the (C, k) sum, for $k > 0$, of the Fourier series of $f(x)$ at the point x, equals $S(x)$.

In particular, at the points where $f(x)$ is regular (Appendix, Def. 15) the sum equals $f(x)$.

Further, if $f(x)$ is continuous in $[a, b] \in [-\pi, \pi]$, the sum $C_n^{(k)}(x)$ of the Fourier series of $f(x)$ converges uniformly to $f(x)$ as $n \to \infty$, in every interval interior to $[a, b]$; if $a = -\pi$, $b = \pi$, $f(-\pi) = f(\pi)$ then the sum $C_n^{(k)}(x)$ converges uniformly to $f(x)$ as $n \to \infty$ in every interval $[a, b]$.

Proof. By (9)

$$\begin{aligned} C_n^{(k)}(x) &= \frac{1}{2\pi}\int_0^{\pi} f(x+\alpha)s_n^{(k)}(\alpha)\,d\alpha + \frac{1}{2\pi}\int_{-\pi}^{0} f(x+\alpha)s_n^{(k)}(\alpha)d\alpha \\ &= \frac{1}{2\pi}\int_0^{\pi} [f(x+\alpha)+f(x-\alpha)]s_n^{(k)}(\alpha)d\alpha \\ &= \frac{1}{\pi}\int_0^{\pi/2} [f(x+2\alpha)+f(x-2\alpha)]s_n^{(k)}(2\alpha)d\alpha, \end{aligned}$$

and by (10_2)

$$C_n^{(k)}(x) - S(x) = \frac{1}{\pi}\int_0^{\pi/2} \varphi(x, \alpha)s_n^{(k)}(2\alpha)d\alpha.$$

But for an arbitrary $\varepsilon > 0$, there exists a $\delta > 0$ and less than $\pi/2$ such that $|\varphi(x, \alpha)| < \varepsilon$ for $0 < \alpha < \delta$, whence

$$|C_n^{(k)}(x) - S(x)| \leqq \frac{1}{\pi}\int_0^\delta |\varphi(x,\alpha)||s_n^{(k)}(2\alpha)|\,d\alpha + \frac{1}{\pi}M_n(\delta/2)\int_\delta^{\pi/2}|\varphi(x,\alpha)|\,d\alpha$$

and, therefore, by (10_1)

$$|C_n^{(k)}(x) - S(x)| \leqq \frac{\varepsilon C}{\pi} + \frac{1}{\pi}M_n(\delta/2)\int_0^{\pi/2}|\varphi(x,\alpha)|\,d\alpha.$$

Finally, by virtue of (10_3), the first part of the theorem follows.

If $f(x)$ is continuous in $[a, b]$, the uniform convergence of $C_n^{(k)}(x)$ to $f(x)$ as $n \to \infty$ in every interval interior to $[a, b]$ follows at once by noting, precisely as in the proof of Th. 36, that

$$\left|\int_0^{\pi/2}\varphi(x,\alpha)\,d\alpha\right| \leqq \tfrac{1}{2}\int_{-\pi}^{\pi}|f(t)|\,dt + \pi \max_{a \leqq x \leqq b}|f(x)|.$$

4. The following theorem, due to Hardy [44b], extends Theorem 39 of Lebesgue. It will be stated without proof.

THEOREM 44. If $f(x)$ is integrable in $[-\pi, \pi]$, the (C, k) sum, for $k > 0$, of its Fourier series is equal to $f(x)$ almost everywhere.

8. Poisson's Method of Summing Fourier Series

1. We shall give a brief account of Poisson's method of summing series with a view to considering its application to summing Fourier series.

DEFINITION 7. Given a numerical series $\sum_{n=0}^\infty a_n$, consider the associated series $\sum_{n=0}^\infty a_n \varrho^n$ and suppose that this series is convergent for $0 \leqq \varrho < 1$. Then, if we let

$$S(\varrho) = \sum_{n=0}^{\infty} a_n \varrho^n$$

and

$$\lim_{\rho \to 1-} S(\varrho)$$

exists and equals S, we shall say that S is the generalized Poisson sum (or Abel sum) of the series $\sum_{n=0}^\infty a_n$. (For more details, see

E. Borel, *Leçons sur les séries divergentes*, Paris, 1928, Ch. VI, p. 216).

It follows at once from Abel's theorem on power series (Appendix, Th. 21) that if the series $\sum_{n=0}^{\infty} a_n$ converges to S, then its Poisson sum coincides with S, (cf. G. Sansone: *Lezioni di Analisi Matematica*, Vol. I, 9th ed., Padova 1946, p. 428). It is then easy to prove that Poisson's method of summation is regular.

1a. The following theorem shows the connection between the $(C, 1)$ sum and the Poisson sum of the series $\sum_{n=0}^{\infty} a_n$.

THEOREM 45. (Frobenius [39]). If the series $\sum_{n=0}^{\infty} a_n$ is $(C, 1)$ summable to S, then its generalized Poisson sum is also S, namely,

$$\lim_{\rho \to 1-} \sum_{n=0}^{\infty} a_n \varrho^n = S.$$

Proof. Let $\{ s_n \}$ and $\{\sigma_n \}$ be defined as usual by the relations

$$s_n = a_0 + a_1 + \ldots + a_n; \qquad (n + 1)\sigma_n = s_0 + s_1 + \ldots + s_n.$$

By hypothesis

$$\lim_{n \to \infty} \sigma_n = S. \tag{1}$$

Clearly, if $0 \leqq \varrho < 1$

$$\frac{1}{1 - \varrho} = \sum_{n=0}^{\infty} \varrho^n, \quad \frac{1}{(1 - \varrho)^2} = \sum_{n=0}^{\infty} (n + 1)\varrho^n$$

and since the sequence $\{ \sigma_n \}$ is bounded, it follows that the series $\sum_{n=0}^{\infty} (n + 1)\sigma_n \varrho^n$ is convergent for $0 \leqq \varrho < 1$, and, consequently, for the same values of ϱ

$$\lim_{m \to \infty} m\sigma_{m-1} \varrho^{m-1} = 0. \tag{2}$$

From the relation

$$\sum_{n=0}^{(m-1)} s_n \varrho^n = \sigma_0 + \sum_{n=1}^{(m-1)} [(n + 1)\sigma_n - n\sigma_{n-1}] \varrho^n$$

$$= (1 - \varrho) \sum_{n=0}^{(m-2)} (n + 1) \sigma_n \varrho^n + m\sigma_{m-1} \varrho^{m-1}$$

and (2), it follows that the series $\sum_{n=0}^{\infty} s_n \varrho^n$ is convergent and that

$$\sum_{n=0}^{\infty} s_n \varrho^n = (1-\varrho) \sum_{n=0}^{\infty} (n+1)\, \sigma_n \varrho^n. \tag{3}$$

Moreover

$$\sum_{n=0}^{m} a_n \varrho^n = s_0 + \sum_{n=1}^{m} (s_n - s_{n-1})\, \varrho^n = (1-\varrho) \sum_{n=0}^{(m-1)} s_n \varrho^n + s_m \varrho^m \tag{4}$$

and since from the convergence of the series in (3) it follows that $\lim_{m\to\infty} s_m \varrho^m = 0$, by passing to the limit in (4), we have

$$\sum_{n=0}^{\infty} a_n \varrho^n = (1-\varrho) \sum_{n=0}^{\infty} s_n \varrho^n = (1-\varrho)^2 \sum_{n=0}^{\infty} (n+1)\, \sigma_n \varrho^n.$$

Given $\varepsilon > 0$, there exists a positive integer m_0 such that for $n > m_0$ we have $| S - \sigma_n | < \varepsilon$, and since

$$\left| \sum_{n=0}^{\infty} a_n \varrho^n - S \right| = \left| (1-\varrho)^2 \sum_{n=0}^{\infty} (n+1) \sigma_n \varrho^n - (1-\varrho)^2 S \sum_{n=0}^{\infty} (n+1)\, \varrho^n \right|$$

$$< (1-\varrho)^2 \left| \sum_{n=0}^{m_0} (n+1)(\sigma_n - S)\, \varrho^n \right| + \varepsilon (1-\varrho)^2 \sum_{n=0}^{\infty} (n+1)\, \varrho^n$$

$$\leq (1-\varrho)^2 \left| \sum_{n=0}^{m_0} (n+1)(\sigma_n - S)\, \varrho^n \right| + \varepsilon,$$

ε is arbitrary, and $\lim_{\rho\to 1-} (1-\varrho)^2 \left| \sum_{n=0}^{m_0} (n+1)(\sigma_n - S)\, \varrho^n \right| = 0$, the theorem follows.

Notice finally that if the a_n are functions of θ in $[\theta_1, \theta_2]$ and if $\sigma_n(\theta)$ tends uniformly in $[\theta_1, \theta_2]$ to a bounded function $S(\theta)$, then the convergence of $\sum_{n=0}^{\infty} a_n \varrho^n$ to $S(\theta)$ as $\varrho \to 1-$, is uniform in $[\theta_1, \theta_2]$.

1b. Frobenius theorem is a particular case of the following theorem which will be stated without proof.

THEOREM 46. If the series $\sum_{n=0}^{\infty} a_n$ is (C, k) summable to S for $k > -1$, then its generalized Poisson sum is S. (Cf. K. Knopp, *Theory of Infinite Series*, p. 490).

Accordingly we say that Poisson's method of summation is stronger than any (C, k) method of summation for $k > -1$.

2. Let $f(\theta)$ be integrable in $(0, 2\pi)$, $f(0) = f(2\pi)$, and consider the Fourier series

$$f(\theta) \sim \tfrac{1}{2}a_0 + \sum_{n=1}^{\infty} [a_n \cos n\theta + b_n \sin n\theta], \tag{5}$$

where

$$a_n = \frac{1}{\pi}\int_0^{2\pi} f(\alpha) \cos n\alpha \, d\alpha, \quad b_n = \frac{1}{\pi}\int_0^{2\pi} f(\alpha) \sin n\alpha \, da. \tag{6}$$

To obtain the Poisson sum of the Fourier series of $f(\theta)$, we form the series

$$\tfrac{1}{2} a_0 + \sum_{n=1}^{\infty} \varrho^n [a_n \cos n\theta + b_n \sin n\theta]. \tag{7}$$

If we interpret ϱ, θ as the polar coordinates of a point, it is immediately clear that the series (7) converges uniformly in every circle with center at the origin and radius $\varrho_1 < 1$. Indeed, by Theorem 13 (Lebesgue) $\lim_{n\to\infty} a_n = 0$, $\lim_{n\to\infty} b_n = 0$, whence there exists a constant L such that

$$|\tfrac{1}{2} a_0| < L; \quad |a_n| < L, \quad |b_n| < L, \quad (n = 1, 2, \ldots) \tag{8}$$

and the series of the absolute values of the terms of (7) for $0 \leqq \varrho \leqq \varrho_1$, $0 \leqq \theta \leqq 2\pi$, is clearly convergent by comparison with the convergent series of positive terms

$$2L \sum_{n=0}^{\infty} \varrho_1^n.$$

Let

$$u(\varrho, \theta) = \tfrac{1}{2} a_0 + \sum_{n=1}^{\infty} \varrho^n [a_n \cos n\theta + b_n \sin n\theta]. \tag{7'}$$

Then by (6)

$$u(\varrho, \theta) = \frac{1}{2\pi}\int_0^{2\pi} f(\alpha)\, d\alpha + \sum_{n=1}^{\infty} \frac{\varrho^n}{\pi}\int_0^{2\pi} f(\alpha) \cos n(\alpha - \theta)\, d\alpha. \tag{7''}$$

Now, if $0 \leqq \varrho < 1$

$$\left| \frac{1}{\pi}\int_0^{2\pi} f(\alpha)[\tfrac{1}{2} + \sum_{n=1}^{\infty} \varrho^n \cos n(\alpha - \theta)]d\alpha - \frac{1}{2\pi}\int_0^{2\pi} f(\alpha)d\alpha - \right.$$

$$- \sum_{n=1}^{m} \frac{\varrho^n}{\pi} \int_0^{2\pi} f(\alpha) \cos n(\alpha - \theta) d\alpha \Bigg| \leqq \frac{1}{\pi} \int_0^{2\pi} |f(\alpha)| \, d\alpha \sum_{n=m+1}^{\infty} \varrho^n$$

and since

$$\lim_{m\to\infty} \frac{1}{\pi} \int_0^{2\pi} |f(\alpha)| \, d\alpha \sum_{n=m+1}^{\infty} \varrho^n = 0$$

it follows that

$$\frac{1}{\pi} \int_0^{2\pi} f(\alpha) \, [\tfrac{1}{2} + \sum_{n=1}^{\infty} \varrho^n \cos n(\alpha - \theta)] \, d\alpha$$

$$= \frac{1}{2\pi} \int_0^{2\pi} f(\alpha) d\alpha + \sum_{n=1}^{\infty} \frac{\varrho^n}{\pi} \int_0^{2\pi} f(\alpha) \cos n(\alpha - \theta) d\alpha,$$

and, therefore, by (7″)

$$u(\varrho, \theta) = \frac{1}{\pi} \int_0^{2\pi} f(\alpha) \, [\tfrac{1}{2} + \sum_{n=1}^{\infty} \varrho^n \cos n(\alpha - \theta)] d\alpha.$$

The sum in square brackets under the integral sign can be easily determined as follows. If $|z| < 1$

$$\tfrac{1}{2} + z + z^2 + \ldots + z^n + \ldots = - \tfrac{1}{2} + \frac{1}{1 - z}$$

and if $z = \varrho e^{i(\alpha-\theta)}$, where $0 < \varrho < 1$, we have

$$\tfrac{1}{2} + \varrho \cos (\alpha - \theta) + \varrho^2 \cos 2(\alpha - \theta) + \ldots + \varrho^n \cos n(\alpha - \theta) + \ldots$$

$$= Re\left(- \tfrac{1}{2} + \frac{1}{1 - \varrho e^{i(\alpha-\theta)}}\right)$$

and since

$$Re\left(- \tfrac{1}{2} + \frac{1}{1 - \varrho e^{i(\alpha-\theta)}}\right) = Re\left(- \tfrac{1}{2} + \frac{1 - \varrho e^{-i(\alpha-\theta)}}{1 - 2\varrho \cos (\alpha - \theta) + \varrho^2}\right)$$

$$= - \tfrac{1}{2} + \frac{1 - \varrho \cos (\alpha - \theta)}{1 - 2\varrho \cos (\alpha - \theta) + \varrho^2} = \frac{1 - \varrho^2}{2(1 - 2\varrho \cos (\alpha - \theta) + \varrho^2)}$$

we have finally for $0 \leqq \varrho < 1$ the well-known Poisson integral (Poisson [90a, b])

$$(9) \qquad u(\varrho, \theta) = \frac{1}{2\pi}\int_0^{2\pi} f(\alpha)\frac{1-\varrho^2}{1-2\varrho\cos(\alpha-\theta)+\varrho^2}d\alpha.$$

We are now in a position to express the Poisson sum of the Fourier series of $f(\theta)$ at a point θ, as the limit of an integral, namely

$$\lim_{\rho\to 1-} u(\varrho, \theta) = \frac{1}{2\pi}\lim_{\rho\to 1-}\int_0^{2\pi} f(\alpha)\frac{1-\varrho^2}{1-2\varrho\cos(\alpha-\theta)+\varrho^2}d\alpha.$$

Now, by Corollaries 1, 2, and 3 of Theorem 37, Theorem 39 and Theorem 45, we have

THEOREM 47. If $f(\theta)$ is integrable in $[0, 2\pi]$, $f(0) = f(2\pi)$ then

a) if $f(\theta)$ is continuous at θ_0, $\lim_{\rho\to 1-} u(\varrho, \theta_0) = f(\theta_0)$, that is, the Poisson sum of the Fourier series of $f(\theta)$ is $f(\theta_0)$ (Th. **37**. Cor. 1);

b) If $f(\theta)$ has a discontinuity of the first kind at θ_0,

$$\lim_{\rho\to 1-} u(\varrho, \theta_0) = \frac{f(\theta_0+)+f(\theta_0-)}{2} \qquad \text{(Th. 37, Cor. 2)}$$

c) $\lim_{\rho\to 1-} u(\varrho, \theta) = f(\theta)$ almost everywhere for $0 \leqq \theta \leqq 2\pi$ (Th. 39);

d) if $f(\theta)$ is continuous in $[\theta_1, \theta_2]$, the Poisson sum of the Fourier series of $f(\theta)$ converges uniformly to $f(\theta)$ in every interval interior to $[\theta_1, \theta_2]$ (Th. 38).

3. z^n where n is a positive integer is an analytic function of z. Therefore, if we let $z = \varrho e^{i\theta}$, then $z^n = \varrho^n(\cos n\theta + i \sin n\theta)$, whence the real part $u = \varrho^n \cos n\theta$ and the imaginary part $v = \varrho^n \sin n\theta$ are harmonic functions, that is they are solutions of the equation

$$\Delta_2 u = \frac{\partial^2 u}{\partial x^2} + \frac{\partial^2 u}{\partial y^2} = 0.$$

The function $\varrho^n(a_n \cos n\theta + b_n \sin n\theta)$, where a_n and b_n are constants, is also a harmonic function.

In polar coordinates

$$\Delta_2 u = \frac{\partial^2 u}{\partial \varrho^2} + \frac{1}{\varrho^2}\frac{\partial^2 u}{\partial \theta^2} + \frac{1}{\varrho}\frac{\partial u}{\partial \varrho},$$

and since by the inequalities (8) and in view of the convergence for $|\varrho| < 1$ of the series $2L \sum_{n=1}^{\infty} n\varrho^{n-1}$, $2L \sum_{n=2}^{\infty} n(n-1)\varrho^{n-2}$, it follows from (7') that

$$\frac{\partial^2 u}{\partial \varrho^2} = \sum_{n=2}^{\infty} n(n-1)\varrho^{n-2}[a_n \cos n\theta + b_n \sin n\theta],$$

$$\frac{1}{\varrho^2}\frac{\partial^2 u}{\partial \theta^2} = -\frac{1}{\varrho}(a_1 \cos\theta + b_1 \sin\theta) - \sum_{n=2}^{\infty} n^2\varrho^{n-2}[a_n \cos n\theta + b_n \sin n\theta],$$

$$\frac{1}{\varrho}\frac{\partial u}{\partial \varrho} = \frac{1}{\varrho}(a_1 \cos\theta + b_1 \sin\theta) + \sum_{n=2}^{\infty} n\varrho^{n-2}[a_n \cos n\theta + b_n \sin n\theta].$$

Therefore the $u(\varrho, \theta)$ given by (7') or by (9), is a harmonic function of (ϱ, θ) in the interior of the unit circle and, consequently, under the assumption that $f(\theta)$ is continuous in $[0, 2\pi]$, is a solution of the Dirichlet problem of constructing a function harmonic in the interior of the unit circle, continuous in the closed circular region, and which assumes prescribed values $f(\theta)$ on the boundary, where $f(\theta)$ is continuous.

Finally, if $f(\theta)$ is integrable in $(0, 2\pi)$ by Theorem 39 of Sec. 6. 7 and Theorem 47 above, (7') is a solution of the problem of constructing a function harmonic in the interior of the unit circle which also has $f(\theta)$ for a limit almost everywhere as (ϱ, θ) approaches a point on the circumference along a radius. Cf. L. Tonelli [111b]: *Serie Trigonometriche*, Bologna, 1928, pp. 375—402.

9. *The Fourier Integral* [1]

1a. DEFINITION 8. Let I denote an infinite interval in one or both directions, and let $f(x)$ be defined for every finite value of x in I. Assume that $f(x)$ is of bounded variation in every finite interval $[\alpha, \beta]$ interior to I and let $V(\alpha, \beta)$ denote the corresponding total variation (See Appendix, Def. 17) of $f(x)$. If the totality of values of $V(\alpha, \beta)$ corresponding to all the finite intervals $[\alpha, \beta]$ of I is bounded and if $V(a, b)$ denotes the least upper bound of the

[1] For a complete treatment of the Fourier integral see S. Bochner [9], pp. 1—227.

$V(\alpha, \beta)$ then we shall say that $f(x)$ *is of bounded variation in* I *and that* $V(a, b)$ *is its total variation.*

THEOREM 48. If $f(x)$ is of bounded variation in the infinite interval I then $f(x)$ is the difference of two bounded, nonnegative and nondecreasing functions. (Therefore, also the two limits $\lim_{x\to a+} f(x) = f(a+)$, $\lim_{x\to b-} f(x) = f(b-)$ exist and conversely.)

Proof. Let α be a point interior to I and let $P(\alpha, x), N(\alpha, x)$ denote the total positive variation and the total negative variation respectively (see Appendix, Def. 17) of $f(x)$ in $[\alpha, x]$. Then, clearly

$$f(x) = f(\alpha) + P(\alpha, x) - N(\alpha, x), \quad x \geqq \alpha \tag{1}$$

and since

$$0 \leqq P(\alpha, x) \leqq V(\alpha, x) \leqq V(a, b);\ 0 \leqq N(\alpha, x) \leqq V(\alpha, x) \leqq V(a,b)$$

the functions $P(\alpha, x), N(\alpha, x)$ are bounded in I and therefore the two limits

$$\lim_{x\to b-} P(\alpha, x), \quad \lim_{x\to b-} N(\alpha, x)$$

exist and, consequently, the limit $\lim_{x\to b-} f(x)$ also exists.

Since, moreover

$$\begin{aligned} f(\alpha) &= f(x) + P(x, \alpha) - N(x, \alpha) \\ f(x) &= f(\alpha) - P(x, \alpha) + N(x, \alpha), \qquad x < \alpha \end{aligned} \tag{2}$$

it follows also that the limit $\lim_{x\to a+} f(x)$ exists.

If now we let

$$\lim_{x\to a+} P(x, \alpha) = P(a, \alpha), \quad \lim_{x\to a+} N(x, \alpha) = N(a, \alpha) \tag{3}$$

$$f(a+) = f(\alpha) - P(a, \alpha) + N(a, \alpha)$$

then from (1), (2) and (3) we obtain the required decomposition of $f(x)$:

$$\begin{cases} f(x) = f(a+) + [P(a, \alpha) - P(x, \alpha)] \\ \qquad\qquad - [N(a, \alpha) - N(x, \alpha)] \text{ for } x < \alpha, \\ f(x) = f(a+) + [P(a, \alpha) + P(\alpha, x)] \\ \qquad\qquad - [N(a, \alpha) + N(\alpha, x)] \text{ for } x \geqq \alpha. \end{cases} \tag{4}$$

1b. If $f(x)$ is of bounded variation in I where $I = (-\infty, b]$ or $[a, \infty)$ or $(-\infty, \infty)$ then we shall call $\lim_{x\to-\infty} f(x)$, or $\lim_{x\to+\infty} f(x)$, the value of $f(x)$ at $-\infty$, or $+\infty$, respectively and denote them by $f(-\infty)$ or $f(+\infty)$ respectively.

This convention implies that $f(x)$ is continuous at $-\infty$ or $+\infty$ respectively.

1c. If $f(x)$ is of bounded variation in $[a, +\infty)$ and $f(+\infty)=0$, that is $\lim_{x\to+\infty} f(x) = 0$, then $f(x)$ can be expressed as the difference of two nondecreasing functions $f_1(x)$, $f_2(x)$ each tending to zero as $x \to +\infty$.

Indeed by the preceding theorem, $f(x) = \varphi_1(x) - \varphi_2(x)$ where $\varphi_1(x)$, $\varphi_2(x)$ are bounded, nondecreasing functions, and since by hypothesis $0 = \lim_{x\to+\infty} \varphi_1(x) - \lim_{x\to+\infty} \varphi_2(x)$, if we let $\lim_{x\to+\infty} \varphi_1(x) = l$, we have the required decomposition:

$$f(x) = [\varphi_1(x) - l] - [\varphi_2(x) - l].$$

Writing $f(x) = [l - \varphi_2(x)] - [l - \varphi_1(x)]$ we have the decomposition in terms of two nonincreasing functions.

1d. As in the case of functions of bounded variation in finite intervals (See Hobson, *Theory of Functions of a Real Variable* I, 3rd Edn. 1950, p. 325) we have the following theorem:

THEOREM 49. The points of discontinuity of a function of bounded variation in I are either finite or denumerably infinite.

Proof. Consider the denumerably infinite number of intervals $[-n-1, -n]$ $[n, n+1]$, $(n = 0, 1, \ldots)$. Then the theorem follows by observing that in each of these finite intervals $f(x)$ has the required property.

1e. As in the case of finite intervals, if $f(x)$ is of bounded variation in I and a_n is a point of discontinuity of $f(x)$ interior to I, the differences $f(a_n) - f(a_n -)$, $f(a_n +) - f(a_n)$, $f(a_n +) - f(a_n -)$ are called respectively the left saltus, the right saltus, and the saltus of $f(x)$ at a_n.

If a is finite we consider the right saltus and the saltus at a to be both equal to $f(a+) - f(a)$; if $a = -\infty$, in accordance with the convention of 1b., the right saltus and the saltus are both zero. Similar conventions will be made for b.

1f. If $f(x)$ is of bounded variation in I and $\{a_n\}$ is the sequence of points of discontinuity of $f(x)$, the series

$$\sum_{n=1}^{\infty}[f(a_n)-f(a_n-)],\ \sum_{n=1}^{\infty}[f(a_n+)-f(a_n)],\ \sum_{n=1}^{\infty}[f(a_n+)-f(a_n-)]$$

are absolutely convergent.

This property follows immediately from Theorem 48.

The function $s(x)$ defined at every point x of I so that its value is equal to the sum of the series of the saltuses of $f(x)$ belonging to $[a, x]$, excluding the right saltus at x if one occurs there, is called the saltus function of $f(x)$ as in the case of finite intervals.

The saltus function $s(x)$ is a function of bounded variation, as indeed are the saltus functions of bounded, monotone functions. Moreover, the difference $f(x) - s(x)$ is a continuous function of bounded variation.

Therefore we have for infinite intervals, the same theorem as for finite intervals: Every function of bounded variation is the sum of a continuous function of bounded variation and its saltus function.

2. DEFINITION 9. Let a be finite and let $f(x)$ be defined for every finite value of the interval $[a, +\infty)$ or $(-\infty, a]$; if for every positive number k, $f(x)$ is integrable in $[a, k]$ or $[-k, a]$, respectively, and if $\lim_{k\to+\infty}\int_a^k f(x)dx$ or $\lim_{k\to+\infty}\int_{-k}^a f(x)dx$, respectively, exists and is finite, we shall say that the integral of $f(x)$ converges in $[a, +\infty)$ or $(-\infty, a]$, respectively, and denote these limits by

$$\int_a^{+\infty} f(x)\,dx \quad\text{and}\quad \int_{-\infty}^{a} f(x)\,dx \text{ respectively.}$$

Thus we have by definition

$$\int_a^{+\infty} f(x)\,dx = \lim_{k\to+\infty}\int_a^k f(x)\,dx;\quad \int_{-\infty}^{a} f(x)\,dx = \lim_{k\to+\infty}\int_{-k}^{a} f(x)\,dx.$$

If the integral of $f(x)$ converges in $(-\infty, a]$ and $[a, +\infty)$, then we shall say that the integral of $f(x)$ converges in $(-\infty, \infty)$ and define

$$\int_{-\infty}^{+\infty} f(x)\,dx = \int_{-\infty}^{a} f(x)\,dx + \int_a^{+\infty} f(x)\,dx = \Big[\lim_{\substack{k_1\to+\infty\\ k_2\to+\infty}}\int_{-k_2}^{k_1} f(x)\,dx\Big].$$

DEFINITION 10. Let $f(x)$ be defined at all finite points of $[a, +\infty)$ and integrable in every finite subinterval. If the integral $\int_a^{+\infty} |f(x)|\, dx$ exists, then the integral of $f(x)$ is said to *converge absolutely* in $[a, +\infty)$.

If the integral of $f(x)$ converges absolutely in $[a, +\infty)$, then it converges in $[a, +\infty)$, but the converse is not always true. In general, if $f(x)$ and $g(x)$ are defined in $[a, +\infty)$ and integrable in every finite subinterval, if the integral of $f(x)$ converges absolutely in $[a, +\infty)$ and, finally, if $|\varphi(x)| \leqq |f(x)|$, almost everywhere in $(a, +\infty)$, then the integral of $\varphi(x)$ converges absolutely and

$$\left|\int_a^{+\infty} \varphi(x)\, dx\right| \leqq \int_a^{+\infty} |f(x)|\, dx.$$

It is easily verified that if the integral of $f(x)$ converges absolutely in $[a, +\infty)$, the two integrals $\int_a^{+\infty} |f(x)|\, dx$, $\int_a^{+\infty} f(x)\, dx$ coincide with the corresponding Lebesgue integrals in the same interval (Appendix, Th. 2a), or in other words, $f(x)$ is Lebesgue integrable in the same interval and conversely (Appendix, Th. 2).

DEFINITION 11. Let $f(x)$ be defined in $(-\infty, +\infty)$ and integrable in every finite subinterval. Then if the limit $\lim_{k\to+\infty} \int_{-k}^{k} f(x)\,dx$ exists, it will be called the *Cauchy principal value* of $\int_{-\infty}^{+\infty} f(x)\, dx$.

The principal value may exist without the integral itself being convergent.

3. THEOREM 50. Let $f(x)$ be defined at all finite points of $[A, +\infty)$ where A is finite and satisfy one of the following conditions:

(a) The integral of $f(x)$ is absolutely convergent in $[A, +\infty)$.
(b) $f(x)$ is of bounded variation in $[A, +\infty)$ and $\lim_{x\to+\infty} f(x) = 0$.

Then, for any a such that $A \leqq a$,

$$(5_1) \qquad \lim_{\mu\to\infty} \int_a^{+\infty} f(x) \cos \mu x\, dx = 0,$$

$$(5_2) \qquad \lim_{\mu\to\infty} \int_a^{+\infty} f(x) \sin \mu x\, dx = 0$$

and the convergence of the integrals is uniform with respect to a (cf. Th. 12)

Proof. We shall prove 5_1. The proof of 5_2 is similar.

If $f(x)$ satisfies condition (a), then for $\varepsilon > 0$, by Cauchy's theorem there exists a $k_0 > A$ such that for $k \geqq \bar{k}_0 \geqq k_0$ we have

$$\int_{\bar{k}_0}^{k} |f(x)| \, dx < \varepsilon/2,$$

and therefore also

$$\left| \int_{\bar{k}_0}^{k} f(x) \cos \mu x \, dx \right| \leqq \int_{\bar{k}_0}^{k} |f(x)| \, dx < \varepsilon/2, \tag{6}$$

whence, by Cauchy's theorem, the integral of $f(x) \cos \mu x$ is convergent in $[A, +\infty]$. Indeed from (6) follows for any μ and $\bar{k}_0 \geqq k_0$

$$\left| \int_{\bar{k}_0}^{+\infty} f(x) \cos \mu x \, dx \right| \leqq \varepsilon/2 \tag{7}$$

By Theorem 12, there exists a μ_0 such that for $\mu > \mu_0$, and for any a in $[A, k_0]$ follows

$$\left| \int_{a}^{k_0} f(x) \cos \mu x \, dx \right| < \varepsilon/2 \quad \text{for} \quad A \leqq a, \, \mu > \mu_0 \tag{8}$$

whence by (7) and (8)

$$\left| \int_{a}^{+\infty} f(x) \cos \mu x \, dx \right| < \varepsilon \quad \text{for} \quad A \leqq a, \, \mu > \mu_0.$$

In order to prove the theorem if $f(x)$ satisfies condition (b) it is sufficient to consider the case where $f(x)$ is nondecreasing and $\lim_{x \to +\infty} f(x) = 0$ in view of the remark of Sec. 9. 1c.

For $\varepsilon > 0$, there exists a k_0 such that $|f(x)| < \varepsilon/2$ for $x \geqq k_0$. But $f(k_0) \leqq f(x) \leqq 0$, for x in $[k_0, k]$ where $k_0 \leqq \bar{k}_0 \leqq k$, and by the second theorem of the mean we obtain

$$\left| \int_{\bar{k}_0}^{k} f(x) \cos \mu x \, dx \right| = \left| f(\bar{k}_0) \int_{\bar{k}_0}^{\xi} \cos \mu x \, dx \right| \leqq \frac{2}{\mu} \, |f(\bar{k}_0)| < \frac{\varepsilon}{\mu}$$

which proves that the integral of $f(x) \cos \mu x$ is convergent in $[A, +\infty)$. Moreover

$$\left| \int_{\bar{k}_0}^{+\infty} f(x) \cos \mu x \, dx \right| \leqq \frac{\varepsilon}{\mu} < \varepsilon/2 \quad \text{for } \mu > 2 \text{ and } \bar{k}_0 \geqq k_0.$$

whence by (8) the theorem also follows for $f(x)$ satisfying (b).

4. We now establish the formula

$$\lim_{m\to+\infty}\int_0^{\pi}\frac{\sin m\alpha/2}{\alpha}\,d\alpha=\frac{\pi}{2}, \tag{9}$$

or, the equivalent formula

$$\lim_{m\to+\infty}\int_0^{\pi/2}\frac{\sin m\alpha}{\alpha}\,d\alpha=\frac{\pi}{2}. \tag{9'}$$

From the formula [Sec. 4. 2, (3)]

$$\tfrac{1}{2}+\sum_{k=1}^{n}\cos k\alpha=\frac{\sin\left(n+\frac{1}{2}\right)\alpha}{2\sin\dfrac{\alpha}{2}}$$

by integrating between 0 and π, we have

$$\frac{\pi}{2}=\int_0^{\pi}\frac{\sin\left(n+\frac{1}{2}\right)\alpha}{2\sin\dfrac{\alpha}{2}}\,d\alpha=\int_0^{\pi/2}\frac{\sin(2n+1)\alpha}{\sin\alpha}\,d\alpha$$

$$=\int_0^{\pi/2}\frac{\sin(2n+1)\alpha}{\alpha}\,d\alpha+\int_0^{\pi/2}\sin(2n+1)\alpha\left[\frac{1}{\sin\alpha}-\frac{1}{\alpha}\right]d\alpha,$$

and since by Theorem 12

$$\lim_{n\to+\infty}\int_0^{\pi/2}\left[\frac{1}{\sin\alpha}-\frac{1}{\alpha}\right]\sin(2n+1)\alpha\,d\alpha=0,$$

it follows that

$$\lim_{n\to+\infty}\int_0^{\pi/2}\frac{\sin(2n+1)\alpha}{\alpha}\,d\alpha=\frac{\pi}{2},\quad\left[\lim_{n\to+\infty}\int_0^{(2n+1)\pi/2}\frac{\sin\alpha}{\alpha}\,d\alpha=\frac{\pi}{2}\right] \tag{10}$$

which establishes (9′) for m an odd integer.

In general, if $2n+1$ is the largest odd integer contained in m we have

$$\int_0^{\pi/2}\frac{\sin m\alpha}{\alpha}\,d\alpha=\int_0^{m(\pi/2)}\frac{\sin\alpha}{\alpha}\,d\alpha=\int_0^{(2n+1)\pi/2}\frac{\sin\alpha}{\alpha}\,d\alpha+\int_{(2n+1)\pi/2}^{m(\pi/2)}\frac{\sin\alpha}{\alpha}\,d\alpha. \tag{11}$$

But

$$\left|\int_{(2n+1)\pi/2}^{m(\pi/2)} \frac{\sin\alpha}{\alpha}\,d\alpha\right| \leqq \int_{(2n+1)\pi/2}^{m(\pi/2)} \frac{1}{\alpha}\,d\alpha$$

$$= \log\frac{m}{2n+1} < \log\frac{2n+3}{2n+1} = \log\left(1+\frac{2}{2n+1}\right) \to 0$$

as $n \to \infty$, and (9′) follows at once from (10) and (11).

Substituting $m\alpha$ for α in (9′) we obtain the well-known formula

$$\int_0^{+\infty} \frac{\sin x}{x}\,dx = \frac{\pi}{2}. \tag{9''}$$

From this we now deduce an inequality which will be needed presently.

If we let

$$\varphi(x) = \int_0^x \frac{\sin x}{x}\,dx$$

then

$$\varphi(n\pi) = \int_0^{n\pi} \frac{\sin x}{x}\,dx = \sum_{k=0}^{n-1}\int_{k\pi}^{(k+1)\pi} \frac{\sin x}{x}\,dx = \sum_{k=0}^{n-1}(-1)^k\int_0^{\pi}\frac{\sin t}{t+k\pi}\,dt$$

and since the terms of the last sum have alternating signs and are decreasing in absolute value and $|(\sin x)/x| \leqq 1$, it follows that

$$\pi > \varphi(\pi) > \varphi(3\pi) > \ldots > \varphi(2n-1)\pi > \ldots > \frac{\pi}{2};$$

$$0 < \varphi(2\pi) < \varphi(4\pi) < \ldots < \varphi(2n\pi) < \ldots < \frac{\pi}{2}.$$

But since the points $n\pi$ are precisely the maximum and minimum points of $\varphi(x)$, it follows that for $a > 0$, $b > 0$

$$\left|\int_a^b \frac{\sin x}{x}\,dx\right| < \pi. \tag{9'''}$$

5. THEOREM 51. Let $f(\alpha)$ be integrable in the interval $[x-\pi, x+\pi]$ and let x be a point of continuity or of discontinuity of the first kind of $f(\alpha)$. Then, if the Fourier series of $f(\alpha)$ is convergent at x we have

$$\lim_{m\to+\infty}\frac{1}{\pi}\int_{x-\pi}^{x+\pi} f(\alpha)\,\frac{\sin m(\alpha-x)}{\alpha-x}\,d\alpha=\tfrac{1}{2}[f(x+)+f(x-)] \tag{12}$$

Proof. We recall that if $f(\alpha)$ is integrable and periodic with period 2π, then if the Fourier series of $f(\alpha)$ converges at a point of continuity x or a point of discontinuity of the first kind of $f(\alpha)$, then the sum of its Fourier series at x equals $[f(x+)+f(x-)]/2$ (Th. 37, Corollaries 3 and 4) and therefore (Th. 15^3)

$$\lim_{n\to\infty}\frac{1}{\pi}\int_0^{\varepsilon}\varphi(x,\alpha)\,\frac{\sin n\alpha}{\alpha}\,d\alpha=0, \qquad 0<\varepsilon\leqq\pi/2 \tag{13}$$

where

$$\varphi(x,\alpha)=f(x+2\alpha)+f(x-2\alpha)-[f(x+)+f(x-)].$$

Since (Th. 12)

$$\lim_{n\to\infty}\int_{\varepsilon}^{\pi/2}\varphi(x,\alpha)\,\frac{\sin n\alpha}{\alpha}\,d\alpha=0,$$

(13) is equivalent to

$$\lim_{n\to\infty}\int_0^{\pi/2}\varphi(x,\alpha)\,\frac{\sin n\alpha}{\alpha}\,d\alpha=0,$$

or

$$\lim_{n\to\infty}\left[\frac{1}{\pi}\int_0^{\pi/2}[f(x+2\alpha)+f(x-2\alpha)]\,\frac{\sin n\alpha}{\alpha}\,d\alpha - [f(x+)+f(x-)]\,\frac{1}{\pi}\int_0^{\pi/2}\frac{\sin n\alpha}{\alpha}\,d\alpha\right]=0$$

and by (9′)

$$\lim_{n\to+\infty}\frac{1}{\pi}\int_0^{\pi/2}[f(x+2\alpha)+f(x-2\alpha)]\,\frac{\sin n\alpha}{\alpha}\,d\alpha=\frac{f(x+)+f(x-)}{2}. \tag{14}$$

But

$$\frac{1}{\pi}\int_0^{\pi/2}[f(x+2\alpha)+f(x-2\alpha)]\,\frac{\sin n\alpha}{\alpha}\,d\alpha$$
$$=\frac{1}{\pi}\int_{-\pi/2}^{\pi/2} f(x+2\alpha)\,\frac{\sin n\alpha}{\alpha}\,d\alpha=\frac{1}{\pi}\int_{-\pi}^{\pi} f(x+\alpha)\,\frac{\sin 2^{-1}n\alpha}{\alpha}\,d\alpha$$

whence, by a change of variable ($\alpha' = x + \alpha$) and writing m in place of $2^{-1}n$, we obtain (12).

6. THEOREM 52. Let $f(x)$ be defined for every finite value of x and integrable in every finite interval $[a, b]$ and satisfy one of the following hypotheses:

(a) The integral of $f(x)/x$ is absolutely convergent in $[a, +\infty)$, $a > 0$.

(b) $f(x)/x$ is of bounded variation in $[a, +\infty)$, $a > 0$, and $\lim_{x\to+\infty} f(x)/x = 0$.

If we suppose further that for $b < 0$ an hypothesis analogous to (a) or (b) holds in $(-\infty, b]$, that $f(x)$ is continuous, or has a jump discontinuity at x, and finally, that the Fourier series of $f(x)$, relative to an interval which contains x in its interior, is convergent for x; then the following formula due to Fourier [37a] (first form) holds:

$$\lim_{n\to+\infty} \frac{1}{\pi}\int_{-\infty}^{+\infty} f(\alpha)\frac{\sin n(\alpha - x)}{\alpha - x}\,d\alpha = \tfrac{1}{2}[f(x+) + f(x-)] \tag{15}$$

If, further, $f(x)$ is continuous in $[A, B]$ and the Fourier series of $f(x)$ converges uniformly to $f(x)$ in $[A, B]$, then the convergence to the limit of the left side of (15) is uniform in $[A, B]$.

Proof. By hypothesis, (12) of Theorem 51 holds; consequently (15) will follow if we show that

$$\lim_{n\to\infty} \frac{1}{\pi}\int_{x+\pi}^{+\infty} f(\alpha)\frac{\sin n(\alpha - x)}{\alpha - x}\,d\alpha = 0; \tag{16_1}$$

$$\lim_{n\to\infty} \frac{1}{\pi}\int_{-\infty}^{x-\pi} f(\alpha)\frac{\sin n(\alpha - x)}{\alpha - x}\,d\alpha = 0. \tag{16_2}$$

If we let $\varphi(\alpha) = f(\alpha)/(\alpha - x)$, then for $\alpha \geqq x + \pi$, we have

$$|\,\varphi(\alpha)\,| = \left|\frac{f(\alpha)}{\alpha}\right|\left|\frac{\alpha}{\alpha - x}\right| = \left|\frac{f(\alpha)}{\alpha}\right|\left|1 + \frac{x}{\alpha - x}\right| < \left|\frac{f(\alpha)}{\alpha}\right|\left(1 + \frac{|x|}{\pi}\right),$$

and consequently under the assumption of hypothesis (a), the integral of $\varphi(\alpha)$ is absolutely convergent in $(x+\pi, +\infty)$ and hence by Theorem 50, hypothesis (a), we have

$$(17_1) \qquad \lim_{n\to\infty} \int_{x+\pi}^{+\infty} f(\alpha) \frac{\sin n\alpha}{\alpha - x}\, d\alpha = 0,$$

$$(17_2) \qquad \lim_{n\to\infty} \int_{x+\pi}^{+\infty} f(\alpha) \frac{\cos n\alpha}{\alpha - x}\, d\alpha = 0.$$

Now

$$\int_{x+\pi}^{b} f(\alpha) \frac{\sin n(\alpha - x)}{\alpha - x}\, d\alpha$$

$$= \cos nx \int_{x+\pi}^{b} f(\alpha) \frac{\sin n\alpha}{\alpha - x}\, d\alpha - \sin nx \int_{x+\pi}^{b} f(\alpha) \frac{\cos n\alpha}{\alpha - x}\, d\alpha$$

whence

$$\left| \int_{x+\pi}^{b} f(\alpha) \frac{\sin n(\alpha - x)}{\alpha - x} d\alpha \right| \leqq \left| \int_{x+\pi}^{b} f(\alpha) \frac{\sin n\alpha}{\alpha - x} d\alpha \right| + \left| \int_{x+\pi}^{b} f(\alpha) \frac{\cos n\alpha}{\alpha - x} d\alpha \right|$$

and as b tends to $+\infty$, (16_1) follows from (17_1) and $17_2)$.

Under hypothesis (b), $f(\alpha)/(\alpha - x) = [f(\alpha)/\alpha]\,[1 + x/(\alpha - x)]$ is of bounded variation in $[x+\pi, +\infty)$ and $\lim_{a\to+\infty} f(\alpha)/(\alpha - x) = 0$; consequently (16_1) follows by Theorem 50.

By a similar argument (16_2) can be shown to follow.

The uniform convergence alluded to in the last part of the theorem follows easily in view of Theorem 50 and the fact that under the hypotheses, the left members of (12), (16_1), and (16_2) tend to their limits uniformly for all x in $[A, B]$.

In particular, if $f(x) = |x|^\mu$ where $0 < \mu < 1$; then $|f(x)/x| = |x|^{\mu-1}$ for $x \neq 0$. Consequently, condition (b) of the theorem is satisfied and therefore

$$\lim_{n\to+\infty} \frac{1}{\pi} \int_{-\infty}^{+\infty} |\alpha|^\mu \frac{\sin n(\alpha - x)}{\alpha - x}\, d\alpha = |x|^\mu, \quad (0 < \mu < 1).$$

7. THEOREM 53. Suppose $f(x)$ is defined for every finite value of x, is integrable in every finite interval $[a, b]$ and satisfies one of the following hypotheses:

(a) The integral of $f(x)$ is absolutely convergent in $[a, +\infty)$, $a > 0$;

(b) $f(x)$ is of bounded variation in $[a + \infty)$, $a > 0$ and $\lim_{x\to+\infty} f(x) = 0$.

Suppose further that for $b < 0$, an hypothesis analogous to (a) or (b) holds in $(-\infty, b]$, that $f(x)$ is continuous, or has a jump discontinuity at x, and, finally that the Fourier series of $f(x)$, relative to an interval containing x in its interior, is convergent at x; then the following formula due to Fourier [37b] (second form) holds:

$$\boxed{\frac{1}{\pi}\int_0^{+\infty} dv \int_{-\infty}^{+\infty} f(\alpha)\cos v(\alpha - x)d\alpha = \tfrac{1}{2}[f(x+) + f(x-)]} \tag{18}$$

where, if hypothesis (b) is satisfied, it is understood that

$$\int_0^{+\infty} \ldots dv = \lim_{\sigma\to+0}\int_\sigma^{+\infty} \ldots dv.$$

Proof. Under the assumptions of the hypothesis formula (12) of Theorem 51 holds:

$$\lim_{n\to+\infty} \frac{1}{\pi}\int_{x-\pi}^{x+\pi} f(\alpha)\frac{\sin n(\alpha - x)}{\alpha - x}\, d\alpha = \tfrac{1}{2}[f(x+) + f(x-)]. \tag{19}$$

Now

$$\begin{aligned}\int_{x-\pi}^{x+\pi} f(\alpha)\frac{\sin n(\alpha - x)}{\alpha - x}\, d\alpha &= \int_{x-\pi}^{x+\pi} f(\alpha)\, d\alpha \int_0^n \cos v(\alpha - x)\, dv \\ &= \int_0^n dv \int_{x-\pi}^{x+\pi} f(\alpha)\cos v(\alpha - x)d\alpha,\ ^1\end{aligned}$$

---

[1] The inversion of the order of integration can be justified as follows: If $\varepsilon > 0$, there exists a function $P(\alpha)$ continuous in $(x - \pi,\ x + \pi)$ such that

$$\int_{x-\pi}^{x+\pi} |f(\alpha) - P(\alpha)|\, d\alpha < \varepsilon;$$

whence

whence, by (19)

$$\frac{1}{\pi}\int_0^{+\infty} dv \int_{x-\pi}^{x+\pi} f(\alpha)\cos v(\alpha - x)\,d\alpha = \tfrac{1}{2}[f(x+) + f(x-)].$$

Thus, to prove (18) it will suffice to show that

$$\int_0^{+\infty} dv \int_{x+\pi}^{+\infty} f(\alpha)\cos v(\alpha - x)\,d\alpha = 0,$$

$$\int_0^{+\infty} dv \int_{-\infty}^{x-\pi} f(\alpha)\cos v(\alpha - x)\,d\alpha = 0,$$

or, substituting α for $\alpha - x$, and letting $f(\alpha + x) = \varphi(\alpha)$,

$$(20_1) \qquad \int_0^{+\infty} dv \int_{\pi}^{+\infty} \varphi(\alpha)\cos v\alpha\,d\alpha = 0,$$

$$(20_2) \qquad \int_0^{+\infty} dv \int_{-\infty}^{-\pi} \varphi(\alpha)\cos v\alpha\,d\alpha = 0.$$

We shall give the proof of (20_1). The proof of (20_2) is similar.

Suppose hypothesis (a) holds. For a given positive integer n, there exists a $k_0 > 0$ such that for $k \geqq k_0$

$$\int_k^{+\infty} |\varphi(\alpha)|\,d\alpha < \frac{1}{n^2}, \qquad k \geqq k_0.$$

Then

$$\left|\int_{\pi}^{+\infty} \varphi(\alpha)\cos v\alpha\,d\alpha - \int_{\pi}^{k} \varphi(\alpha)\cos v\alpha\,d\alpha\right|$$

$$= \left|\int_k^{+\infty} \varphi(\alpha)\cos v\alpha\,d\alpha\right| \leqq \int_k^{+\infty} |\varphi(\alpha)|\,d\alpha < \frac{1}{n^2}$$

$$\int_{x-\pi}^{x+\pi} f(\alpha)d\alpha \int_0^n \cos v(\alpha - x)dv = \int_{x-\pi}^{x+\pi} [f(\alpha) - P(\alpha)]d\alpha \int_0^n \cos v(\alpha - x)dv$$

$$+ \int_{x-\pi}^{x+\pi} P(\alpha)d\alpha \int_0^n \cos v(\alpha - x)dv = \theta_1 n\varepsilon + \int_0^n dv \int_{x-\pi}^{x+\pi} P(\alpha)\cos v(\alpha - x)d\alpha$$

$$= \theta_1 n\varepsilon + \int_0^n dv \int_{x-\pi}^{x+\pi} f(\alpha)\cos v(\alpha - x)d\alpha + \int_0^n dv \int_{x-\pi}^{x+\pi} [P(\alpha) - f(\alpha)]\cos v(\alpha - x)d\alpha$$

$$= \theta_1 n\varepsilon + \theta_2 n\varepsilon + \int_0^n dv \int_{x-\pi}^{x+\pi} f(\alpha)\cos v(\alpha - x)\,d\alpha,$$

where $|\theta_1| < 1$, $|\theta_2| < 1$, and since ε is arbitrary, the inversion of order of integration is established.

whence

$$\int_{\pi}^{+\infty} \varphi(\alpha) \cos v\alpha \, d\alpha = \int_{\pi}^{k} \varphi(\alpha) \cos v\alpha \, d\alpha + \frac{\theta_1(k, v)}{n^2},$$

$$| \theta_1(k, v) | < 1, \qquad k \geqq k_0;$$

Therefore

$$\int_0^n dv \int_{\pi}^{+\infty} \varphi(\alpha) \cos v\alpha \, d\alpha = \frac{\theta_1(k)}{n} + \int_0^n dv \int_{\pi}^{k} \varphi(\alpha) \cos v\alpha \, d\alpha$$

$$= \frac{\theta_1(k)}{n} + \int_{\pi}^{k} \varphi(\alpha) \frac{\sin n\alpha}{\alpha} \, d\alpha, \quad | \theta_1(k) | < 1.$$

But the integral $\int_{\pi}^{+\infty} \varphi(\alpha) \sin n\alpha/\alpha \, d\alpha$ is (absolutely) convergent; consequently, letting $k \to +\infty$, we obtain

$$\int_0^n dv \int_{\pi}^{+\infty} \varphi(\alpha) \cos v\alpha \, d\alpha = \frac{\mu(n)}{n} + \int_{\pi}^{+\infty} \frac{\varphi(a)}{\alpha} \sin n\alpha \, d\alpha, \quad | \mu(n) | \leqq 1,$$

whence, letting $n \to \infty$, and noting that the right side approaches zero [Th. 50 (a)], (20_1) follows.

Suppose, now, that hypothesis (b) holds. Then we may assume that $\varphi(\alpha)$ is nondecreasing and that $\varphi(\alpha) \to 0$ as $\alpha \to +\infty$. Take $\sigma > 0$ and consider values of $v \geqq \sigma$. For $\varepsilon > 0$, there exists a k_0 such that $| 2\varphi(k_0)/\sigma | < \varepsilon$. Since, moreover, $\varphi(k_0) \leqq \varphi(\alpha) \leqq 0$, for $\alpha \geqq k_0$, by the second theorem of the mean we have

$$\left| \int_{k_0}^{k} \varphi(\alpha) \cos v\alpha \, d\alpha \right| = \left| \varphi(k_0) \int_{k_0}^{\xi} \cos v\alpha \, d\alpha \right| \leqq \left| \frac{2}{\sigma} \varphi(k_0) \right| < \varepsilon,$$

whence the integral $\int_{\pi}^{+\infty} \varphi(\alpha) \cos v\alpha \, d\alpha$ converges uniformly with respect to v. Consequently

$$\int_{\sigma}^{n} dv \int_{\pi}^{+\infty} \varphi(\alpha) \cos v\alpha \, d\alpha = \lim_{k \to +\infty} \int_{\sigma}^{n} dv \int_{\pi}^{k} \varphi(\alpha) \cos v\alpha \, d\alpha$$

$$= \lim_{k \to +\infty} \int_{\pi}^{k} \varphi(\alpha) \, d\alpha \int_{\sigma}^{n} \cos v\alpha \, dv$$

$$= \lim_{k \to +\infty} \left[\int_{\pi}^{k} \varphi(\alpha) \frac{\sin n\alpha}{\alpha} \, d\alpha - \int_{\pi}^{k} \varphi(\alpha) \frac{\sin \sigma\alpha}{\alpha} \, d\alpha \right].$$

But $\varphi(\alpha)/\alpha$ tends monotonically to zero as $\alpha \to +\infty$. Therefore the following two integrals are convergent

$$\int_{\pi}^{+\infty} \varphi(\alpha) \frac{\sin n\alpha}{\alpha}\, d\alpha, \qquad \int_{\pi}^{+\infty} \varphi(\alpha) \frac{\sin \sigma\alpha}{\alpha}\, d\alpha,$$

whence

$$(21) \quad \int_{\sigma}^{n} dv \int_{\pi}^{+\infty} \varphi(\alpha) \cos v\alpha\, d\alpha = \int_{\pi}^{+\infty} \varphi(\alpha) \frac{\sin n\alpha}{\alpha}\, d\alpha - \int_{\pi}^{+\infty} \varphi(\alpha) \frac{\sin \sigma\alpha}{\alpha}\, d\alpha.$$

By virtue of Theorem 50 (b) we have

$$\lim_{n\to\infty} \int_{\pi}^{+\infty} \varphi(\alpha) \frac{\sin n\alpha}{\alpha}\, d\alpha = 0,$$

and therefore (20_1) will follow from (21) if we show that

$$(22) \qquad \lim_{\sigma\to 0} \int_{\pi}^{+\infty} \varphi(\alpha) \frac{\sin \sigma\alpha}{\alpha}\, d\alpha = 0.$$

For $\varepsilon > 0$, determine k_0 such that $|\varphi(k_0)| < \varepsilon/2\pi$, and consider positive values of $\sigma \leqq \sigma_0$ where $\sigma_0 \int_{\pi}^{k_0} |\varphi(\alpha)|\, d\alpha < \varepsilon/2$.

If $k > k_0$, we have by the second theorem of the mean

$$(23) \quad \left| \int_{\pi}^{k} \varphi(\alpha) \frac{\sin \sigma\alpha}{\alpha}\, d\alpha \right| = \left| \int_{\pi}^{k_0} \varphi(\alpha) \frac{\sin \sigma\alpha}{\alpha}\, d\alpha + \varphi(k_0) \int_{k_0}^{\xi} \frac{\sin \sigma\alpha}{\alpha}\, d\alpha \right|,$$
$$\xi \geqq k_0.$$

However,

$$\left| \int_{\pi}^{k_0} \varphi(\alpha) \frac{\sin \sigma\alpha}{\alpha}\, d\alpha \right| = \left| \sigma \int_{\pi}^{k_0} \varphi(\alpha) \frac{\sin \sigma\alpha}{\sigma\alpha}\, d\alpha \right| \leqq \sigma \int_{\pi}^{k_0} |\varphi(\alpha)|\, d\alpha < \varepsilon/2,$$

$$\left| \varphi(k_0) \int_{k_0}^{\xi} \frac{\sin \sigma\alpha}{\alpha}\, d\alpha \right| < \frac{\varepsilon}{2\pi} \left| \int_{k_0\sigma}^{\xi\sigma} \frac{\sin \alpha}{\alpha}\, d\alpha \right| < \varepsilon/2 \qquad [\text{cf. } (9''')]$$

whence by (23)

$$\left| \int_{\pi}^{k} \varphi(\alpha) \frac{\sin \sigma\alpha}{\alpha}\, d\alpha \right| < \varepsilon, \qquad \left| \int_{\pi}^{+\infty} \varphi(\alpha) \frac{\sin \sigma\alpha}{\alpha}\, d\alpha \right| \leqq \varepsilon \text{ for } 0 < \sigma \leqq \sigma_0$$

which implies (22).

8. We now give a brief procedure for deriving the two integral formulas of Fourier.

Let the integral of $f(x)$ be convergent in $(-\infty, +\infty)$, and consider $f(x)$ in the interval $(-l, l)$. By the transformation $x = lx'/\pi$, we obtain from $f(x)$ a function $\varphi(x')$ defined in $(-\pi, \pi)$ by letting $f(x) = f(lx'/\pi) = \varphi(x')$. If the conditions for convergence of the Fourier series of $\varphi(x')$ are satisfied we can write

$$\tfrac{1}{2}[\varphi(x'+) + \varphi(x'-)] = \tfrac{1}{2}a'_0 + \sum_{k=1}^{\infty}(a'_k \cos kx' + b'_k \sin kx')$$

where

$$a'_k = \frac{1}{\pi}\int_{-\pi}^{\pi}\varphi(\alpha')\cos k\alpha'\,d\alpha' = \frac{1}{l}\int_{-l}^{l} f(\alpha)\cos\frac{k\pi\alpha}{l}\,d\alpha,$$

$$b'_k = \frac{1}{l}\int_{-l}^{l} f(\alpha)\sin\frac{k\pi\alpha}{l}\,d\alpha,$$

whence

$$\tfrac{1}{2}[f(x+) + f(x-)] = \frac{1}{2l}\int_{-l}^{l} f(\alpha)d\alpha + \frac{1}{l}\sum_{k=1}^{\infty}\int_{-l}^{l} f(\alpha)\cos\frac{k\pi}{l}(a - x)d\alpha.$$

Now we take l sufficiently large to make $(1/2l)\int_{-l}^{l} f(\alpha)d\alpha$, negligible, and let $\pi/l = \Delta v$. Then the preceding formula can be written

$$\tfrac{1}{2}[f(x+) + f(x-)] = \frac{1}{\pi}\left[\Delta v\int_{-l}^{l} f(\alpha)\cos\Delta v(\alpha - x)\,d\alpha\right.$$
$$\left. + \Delta v\int_{-l}^{l} f(\alpha)\cos 2\Delta v(\alpha - x)\,d\alpha + \dots\right]$$

Then as $l \to +\infty$, assuming all the limits in question exist, we have

$$(18') \quad \tfrac{1}{2}[f(x+) + f(x-)] = \frac{1}{\pi}\int_{0}^{+\infty} dv\int_{-\infty}^{+\infty} f(\alpha)\cos v(\alpha - x)d\alpha,$$

whence, assuming that reversing the order of integration is justified, we have

$$\tfrac{1}{2}[f(x+)+f(x-)] = \frac{1}{\pi}\lim_{n\to\infty}\int_0^n dv\int_{-\infty}^{+\infty} f(\alpha)\cos v(\alpha - x)\,d\alpha$$

$$= \frac{1}{\pi}\lim_{n\to\infty}\int_{-\infty}^{+\infty} f(\alpha)\,d\alpha\int_0^n \cos v(\alpha - x)\,dv,$$

$$(15') \quad \tfrac{1}{2}[f(x+)+f(x-)] = \frac{1}{\pi}\lim_{n\to\infty}\int_{-\infty}^{+\infty} f(\alpha)\,\frac{\sin n(\alpha - x)}{\alpha - x}\,d\alpha.$$

The procedure just described is not intended to be rigorous. However, the reasoning of Secs. 6 and 7 insures that formulas (15′) and (18′) are applicable whenever $f(x)$ satisfy the conditions prescribed.

9. Suppose $f(x)$ is defined in $[0, +\infty)$ and integrable in every finite interval, satisfies the conditions prescribed in Theorem 53, which assures the validity of (18), and, finally, that $f(x) = f(-x)$ for $x < 0$ (i.e. $f(x)$ is an even function).

If we break the integral in (18) into the two integrals over $(-\infty, 0]$ and $[0, +\infty)$, and take into account that $f(x)$ is even, we obtain

$$\frac{1}{\pi}\int_0^{+\infty} dv\int_0^{+\infty} f(\alpha)\cos v(\alpha - x)d\alpha + \frac{1}{\pi}\int_0^{+\infty} dv\int_0^{+\infty} f(\alpha)\cos v(\alpha + x)\,d\alpha$$

$$= \frac{1}{\pi}\int_0^{+\infty} dv\int_0^{+\infty} f(\alpha)\,[\cos v(\alpha - x) + \cos v(\alpha + x)]\,d\alpha$$

$$= \frac{2}{\pi}\int_0^{+\infty} dv\int_0^{+\infty} f(\alpha)\cos v\alpha\cos vx\,d\alpha,$$

whence

$$(24) \quad \boxed{\frac{2}{\pi}\int_0^{+\infty} dv\int_0^{+\infty} f(\alpha)\cos v\alpha\cos vx\,d\alpha = \tfrac{1}{2}[f(x+)+f(x-)]}$$

$$[f(0+) = f(0-)]$$

which is valid if $f(x)$ satisfies in $(0, +\infty)$ the conditions prescribed in Theorem 53. It should be observed that for the validity at $x = 0$ of the Fourier series of $f(x)$, where $f(-x) = f(x)$, relative to an interval which contains 0 in its interior, it is also necessary that the series converge to $[f(0+)+f(0-)]/2$.

Similarly, suppose $f(x)$ is defined in $[0, +\infty)$ and integrable in every finite interval, satisfies the conditions prescribed in Theorem 53, and finally that $f(x) = -f(-x)$ for $x < 0$ (i.e., $f(x)$ is an odd function). Then we have

$$\frac{1}{\pi}\int_0^{+\infty} dv \int_{-\infty}^{+\infty} f(\alpha) \cos v(\alpha - x) d\alpha$$

$$= \frac{1}{\pi}\int_0^{+\infty} dv \int_0^{+\infty} f(\alpha)[\cos v(\alpha - x) - \cos v(\alpha + x)]\, d\alpha$$

whence

$$\boxed{\frac{2}{\pi}\int_0^{+\infty} dv \int_0^{+\infty} f(\alpha) \sin v\alpha \sin vx\, d\alpha = \tfrac{1}{2}[f(x+) + (f(x-)]} \tag{25}$$

$$[f(0+) = -f(0-)].$$

(24) and (25) are known as Fourier cosine and sine integrals for representation of a function.

Again, if (b) of Theorem 53 is satisfied we agree that $\int_0^{+\infty} \ldots dv$ means $\lim_{\sigma\to+0} \int_\sigma^{+\infty} \ldots dv$.

10. The preceding results will now be illustrated by some examples:

1°) Let $f(x) = x^{-\gamma}$, $x > 0$, $0 < \gamma < 1$. Since $x^{-\gamma}$ is decreasing with $\lim_{x\to+\infty} x^{-\gamma} = 0$, (24) and (25) give for every $x > 0$:

$$\frac{2}{\pi}\int_0^{+\infty} dv \int_{-\infty}^{+\infty} \alpha^{-\gamma} \cos v\alpha \cos vx\, d\alpha = x^{-\gamma}; \tag{26_1}$$

$$\frac{2}{\pi}\int_0^{+\infty} dv \int_0^{+\infty} \alpha^{-\gamma} \sin v\alpha \sin vx\, d\alpha = x^{-\gamma}. \tag{26_2}$$

From (26_1) we can deduce two well-known results. Clearly

$$\frac{2}{\pi}\int_0^{+\infty} dv \int_0^{+\infty} \alpha^{-\gamma} \cos v\alpha \cos vx \, d\alpha$$

$$= \frac{2x^{-\gamma}}{\pi}\int_0^{+\infty} (vx)^{-(1-\gamma)} \cos vx \, d\,(vx) \int_0^{+\infty} (v\alpha)^{-\gamma} \cos v\alpha \, d(v\alpha)$$

and therefore, writing $vx = t$, $v\alpha = t$, we have

$$(27_1) \qquad \int_0^{+\infty} t^{-(1-\gamma)} \cos t \, dt \int_0^{+\infty} t^{-\gamma} \cos t \, dt = \frac{\pi}{2}.$$

Similarly from (26_2), we obtain

$$(27_2) \qquad \int_0^{+\infty} t^{-(1-\gamma)} \sin t \, dt \int_0^{+\infty} t^{-\gamma} \sin t \, dt = \frac{\pi}{2},$$

whence for $\gamma = \frac{1}{2}$ we have

$$(28) \qquad \boxed{\int_0^{+\infty} \frac{\cos t}{\sqrt{t}}\, dt = \int_0^{+\infty} \frac{\sin t}{\sqrt{t}}\, dt = \sqrt{\frac{\pi}{2}}}$$

The convergence of the integrals to a positive value is proved as follows:

$$\int_0^{+\infty} \frac{\sin t}{\sqrt{t}}\, dt = \int_0^{\pi} \frac{\sin t}{\sqrt{t}}\, dt + \int_{\pi}^{2\pi} \frac{\sin t}{\sqrt{t}}\, dt + \int_{2\pi}^{3\pi} \frac{\sin t}{\sqrt{t}}\, dt + \dots$$

$$= \int_0^{\pi} \frac{\sin t}{\sqrt{t}}\, dt - \int_0^{\pi} \frac{\sin t}{\sqrt{t+\pi}}\, dt + \int_0^{\pi} \frac{\sin t}{\sqrt{t+2\pi}}\, dt - \dots$$

and thus the series on the right is an alternating series of decreasing terms tending to zero and the sign is that of the first term of the series. As regards the other integral, we have

$$(29) \qquad \begin{aligned} I = \int_0^{+\infty} \frac{\cos x}{\sqrt{x}}\, dx &= \int_0^{\pi/2} \frac{\cos x}{\sqrt{x}}\, dx - \int_{\pi/2}^{3\pi/2} \frac{|\cos x|}{\sqrt{x}}\, dx \\ &\quad + \int_{3\pi/2}^{5\pi/2} \frac{|\cos x|}{\sqrt{x}}\, dx - \int_{5\pi/2}^{7\pi/2} \frac{|\cos x|}{\sqrt{x}}\, dx + \dots, \end{aligned}$$

$$I = \int_0^{+\infty} \frac{\cos x}{\sqrt{x}}\, dx = \int_0^{\pi/2} \frac{\cos x}{\sqrt{x}}\, dx - \int_{\pi/2}^{3\pi/2} \frac{|\cos x|}{\sqrt{x}}\, dx$$

$$+ \int_{\pi/2}^{3\pi/2} \frac{|\cos x|}{\sqrt{x+\pi}}\, dx - \int_{\pi/2}^{3\pi/2} \frac{|\cos x|}{\sqrt{x+2\pi}}\, dx + \ldots$$

For $0 < x \leqq \pi/2$ we have

$$\cos x = 1 - \frac{x^2}{2!} + \frac{x^2}{4!}\cos \theta x > 1 - \frac{x^2}{2}$$

whence

$$\int_{\pi/2}^{\pi/2} \frac{\cos x}{\sqrt{x}}\, dx > \int_0^{\pi/2} (x^{-1/2} - \tfrac{1}{2} x^{3/2})\, dx = \sqrt{\frac{\pi}{2}}\left[2 - \frac{\pi^2}{20}\right];$$

moreover,

$$\int_{\pi/2}^{3\pi/2} \frac{|\cos x|}{\sqrt{x}}\, dx < \int_{\pi/2}^{3\pi/2} x^{-1/2}\, dx = 2\sqrt{\frac{\pi}{2}}\,[\sqrt{3} - 1],$$

But

$$2 - \frac{\pi^2}{20} > 2(\sqrt{3} - 1), \left[4 - 2\sqrt{3} > \frac{\pi^2}{20}, 20(4 - 2 \cdot 1.74) > (3.15)^2\right],$$

whence from (29) follows $I > 0$.

Substituting x^2 for t in (28) we obtain

$$\int_0^{+\infty} \cos x^2\, dx = \int_0^{+\infty} \sin x^2\, dx = \tfrac{1}{2}\sqrt{\frac{\pi}{2}}. \tag{30}$$

2°) Let $f(x) = 1$ for $0 \leqq x \leqq 1$, $f(x) = 0$ for $x > 1$.
From (24), we have for $0 \leqq x < 1$

$$(31_1)\quad 1 = \frac{2}{\pi}\int_0^{+\infty} dv \int_0^1 \cos v\alpha \cos vx\, d\alpha = \frac{2}{\pi}\int_0^{+\infty} \frac{\sin v \cos vx}{v}\, dv,\ 0 \leqq x < 1.$$

For $x = 1$, since $f(1-) + f(1+) = 1$, we have

$$(31_2)\quad \tfrac{1}{2} = \frac{2}{\pi}\int_0^{+\infty} \frac{\sin v \cos v}{v}\, dv.$$

Finally, for $x > 1$,

$$0 = \frac{2}{\pi}\int_0^{+\infty} \frac{\sin v \cos vx}{v}\, dv. \tag{31_3}$$

The integral $2/\pi \int_0^{+\infty} \sin v \cos vx/v\, dv$ is known as the Dirichlet discontinuous factor (Dirichlet [29b]).

From (31_1), (31_2), (31_3) follows immediately for $\lambda > 0$, $\mu > 0$

$$\frac{2}{\pi}\int_0^{+\infty} \frac{\sin \lambda x \cos \mu x}{x}\, dx = 1, \quad \tfrac{1}{2}, \quad 0 \tag{32}$$

according as

$$\mu < \lambda, \quad \mu = \lambda, \quad \mu > \lambda.$$

These results can also be deduced directly from (9″) of Sec. 9.4.

3^0) Let $f(x) = 0$ for $0 \leqq x < a$ and $x > b$, $a < b$, and $f(x) = x$ for $a < x < b$.

Then

$$\int_a^b \alpha \sin v\alpha \sin vx\, d\alpha = \sin vx \int_a^b \alpha \sin v\alpha\, d\alpha$$

$$= \sin vx \left[\frac{a \cos av - b \cos bv}{v} + \frac{\sin bv - \sin av}{v^2}\right]$$

whence by (25) we obtain

$$\frac{2}{\pi}\int_0^{\infty} \sin vx \left[\frac{a \cos av - b \cos bv}{v} + \frac{\sin bv - \sin av}{v^2}\right] dv$$

$$= 0, \ \tfrac{1}{2}a, \ x, \ \tfrac{1}{2}b, \ 0$$

according as x satisfies the relations

$$x < a, \ x = a, \ a < x < b, \ x = b, \ x > b.$$

4°) Let $f(x) = e^{-ax}$ where $a > 0$.

Since

$$\int_0^{+\infty} e^{-a\alpha} \cos v\alpha\, d\alpha = \frac{a}{a^2 + v^2}, \quad \int_0^{+\infty} e^{-a\alpha} \sin v\alpha\, d\alpha = \frac{v}{a^2 + v^2}.$$

From (24) and (25) follows

$$\frac{2a}{\pi}\int_0^{+\infty}\frac{\cos vx}{a^2+v^2}dv = e^{-ax}, \quad \frac{2}{\pi}\int_0^{+\infty}\frac{v\sin vx}{a^2+v^2}dv = e^{-ax}$$

or the formulas due to Laplace

$$(32_1) \qquad \int_0^{+\infty}\frac{\cos vx}{a^2+v^2}dv = \frac{\pi e^{-ax}}{2a}, \quad \text{where } a>0,\ x\geqq 0,$$

$$(32_2) \qquad \int_0^{+\infty}\frac{v\sin vx}{a^2+v^2}dv = \frac{\pi}{2}e^{-ax}, \quad \text{where } a>0,\ x>0.$$

10. Gibbs' Phenomenon

1. We have shown (Sec. 6, Th. 41) that if $f(x)$ is periodic with period 2π, integrable, of bounded variation in an interval with center at x_0, then its Fourier series converges at x_0 to $[f(x_0+)+f(x_0-)]/2$, It is of interest to examine the behaviour of the partial sums of the series in the neighborhood of x_0. It will be sufficient to examine first the simple case of the function $f(x)$ defined as follows:

$$(1) \qquad \begin{cases} f(x) = c & \text{for } 0<x<\pi, \quad (c=\text{const},\ c>0),\\ f(x) = -c & \text{for } -\pi<x<0,\end{cases}$$

$$f(-\pi) = f(0) = f(\pi) = 0.$$

Since as we shall see in Sec. 10. 2, the general case reduces to this case, $f(x)$ admits of the following expansion [Sec. 4. 6, (4)]

$$f(x) = \frac{4}{\pi}c\left[\sin x+\frac{\sin 3x}{3}+\frac{\sin 5x}{5}+\ldots\right]$$

and the sum $S_n(x)$ of the first n terms is

$$S_n(x) = \frac{1}{\pi}\int_{-\pi}^{\pi}f(x+\alpha)\frac{\sin(n+\frac{1}{2})\alpha}{2\sin(\alpha/2)}d\alpha,$$

Following the argument of Sec. 4. 4, we have for $0<\varepsilon<\pi/4$,

$$S_n(x) = \frac{1}{\pi}\int_0^{\varepsilon} [f(x+2\alpha)+f(x-2\alpha)]\frac{\sin(2n+1)\alpha}{\alpha}\,d\alpha + R_n(x),$$

where $\lim_{n\to\infty} R_n(x) = 0$, uniformly for all x in $[-\pi, \pi]$.

Now consider the function

$$\Phi_n(x) = \frac{1}{\pi}\int_0^{\varepsilon} [f(x+2\alpha)+f(x-2\alpha)]\,\frac{\sin(2n+1)\alpha}{\alpha}\,d\alpha,$$

for x in $[-\varepsilon, \varepsilon]$.

Then

$$S_n(x) = \Phi_n(x) + R_n(x)$$

and the behaviour of $S_n(x)$ is seen to depend essentially on that of $\Phi_n(x)$.

It will be sufficient to consider the values of x in $[0, \varepsilon]$. Clearly

$$\Phi_n(0) = 0.$$

Since $f(x-2\alpha) + f(x+2\alpha) = 0$ for $x - 2\alpha < 0$, we have for $0 < x \leqq \varepsilon$,

$$\Phi_n(x) = \frac{2c}{\pi}\int_0^{x/2}\frac{\sin(2n+1)\alpha}{\alpha}\,d\alpha = \frac{2c}{\pi}\int_0^{(n+\frac{1}{2})x}\frac{\sin\alpha}{\alpha}\,d\alpha,$$

and therefore (Sec. 9. 4) the relative maxima of $\Phi_n(x)$ occur at

$$\pi/(n+\tfrac{1}{2}), \qquad 3\pi/(n+\tfrac{1}{2}), \ldots$$

and the relative minima occur at

$$2\pi/(n+\tfrac{1}{2}), \qquad 4\pi/(n+\tfrac{1}{2}), \ldots$$

Since, moreover,

$$\Phi\left(\frac{\pi}{n+\frac{1}{2}}\right) > \Phi\left(\frac{3\pi}{n+\frac{1}{2}}\right) > \ldots; \quad \Phi\left(\frac{2\pi}{n+\frac{1}{2}}\right) < \Phi\left(\frac{4\pi}{n+\frac{1}{2}}\right) < \ldots,$$

it follows, by taking into account (9″) of Sec. 9. 4, that the absolute maximum of $\Phi_n(x)$ is

$$\Phi\left(\frac{\pi}{n+\frac{1}{2}}\right) = \frac{2c}{\pi}\int_0^{\pi}\frac{\sin\alpha}{\alpha}\,d\alpha = \frac{2c}{\pi}\left[\int_0^{+\infty}\frac{\sin\alpha}{\alpha}\,d\alpha - \int_{\pi}^{+\infty}\frac{\sin\alpha}{\alpha}\,d\alpha\right],$$

$$\Phi\left(\frac{\pi}{n+\frac{1}{2}}\right) = c - \frac{2c}{\pi}\int_{\pi}^{+\infty}\frac{\sin\alpha}{\alpha}\,d\alpha,$$

which is clearly independent of n. Similarly the absolute minimum is

$$\Phi\left(\frac{2\pi}{n+\frac{1}{2}}\right) = c - \frac{2c}{\pi}\int_{2\pi}^{+\infty} \frac{\sin\alpha}{\alpha}\,d\alpha,$$

which is also independent of n.

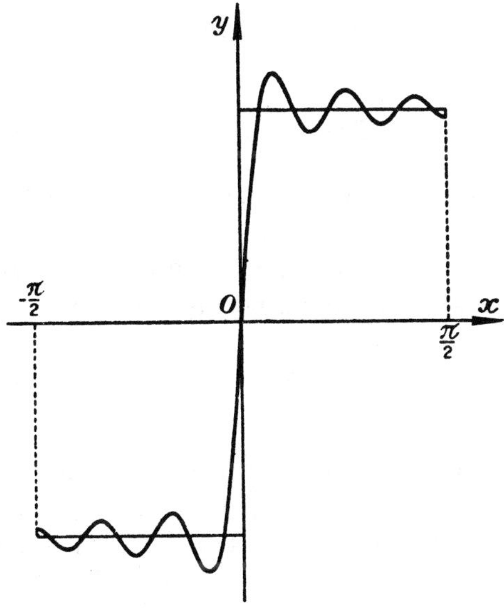

Fig. 10.

From the tables of Jahnke-Emde [[30], [54]] we find that

$$-\int_{\pi}^{+\infty} \frac{\sin\alpha}{\alpha}\,d\alpha = 0.\ 28114\ldots, \quad \int_{2\pi}^{+\infty} \frac{\sin\alpha}{\alpha}\,d\alpha = 0.\ 15264\ldots$$

whence the difference between the absolute maximum and the absolute minimum is

$$-\frac{2c}{\pi}\int_{\pi}^{2\pi} \frac{\sin\alpha}{\alpha}\,d\alpha = \frac{2c}{\pi}\,0.\ 43378\ldots$$

and since $\lim_{n\to\infty} \pi/(n+\frac{1}{2}) = 0$, $\lim_{n\to\infty} 2\pi/(n+\frac{1}{2}) = 0$, we conclude:

The curve $y = S_n(x)$ is symmetrical with respect to the origin and passes through the origin; if $c > 0$ $(c < 0)$, the absolute maximum [minimum] of $S_n(x)$ for $0 \leqq x \leqq \pi/4$ tends to

$$c + \frac{2c}{\pi} 0.28114\ldots, \quad \left[c - \frac{2c}{\pi} 0.15264\ldots\right]$$

as $n \to \infty$ and the absolute minimum [maximum] for the same values of x tends to

$$c - \frac{2c}{\pi} 0.15264\ldots, \quad \left[c + \frac{2c}{\pi} 0.28114\ldots\right]$$

as $n \to \infty$. Moreover, these values are assumed for values of x which approach zero as $n \to \infty$.

For $x > 0$, the curve $y = S_n(x)$ performs oscillations about the line $y = c$ of maximum amplitude tending to

$$-\frac{2\,|\,c\,|}{\pi}\int_{\pi}^{2\pi} \frac{\sin\alpha}{\alpha}\,d\alpha = \frac{2\,|\,c\,|}{\pi} 0.43378\ldots$$

as $n \to \infty$.

The so-called Gibbs phenomenon of the Fourier series at the origin consists of the fact that the projection on the y-axis of the curve $y = S_n(x)$ for x in $(-\pi/4, \pi/4)$, does not tend to the segment whose extremities are the points with ordinates $-c$ and c, but rather to the segment whose extremities are the points with ordinates

$$\pm c\left[1 + \frac{2}{\pi}\left|\int_{\pi}^{+\infty} \frac{\sin\alpha}{\alpha}\,d\alpha\right|\right].$$

This phenomenon was noted by J. W. Gibbs [41] in a letter to *Nature*. For the general case of Sec. 10.2 of the text, see M. Bôcher [8a, b].

2. Let $f(x)$ be periodic of period 2π, integrable, of bounded variation and continuous in $[x_0 - \delta, x_0 + \delta]$, except at x_0 where it has a finite discontinuity

$$f(x_0 -) \neq f(x_0 +).$$

Define $\varphi(x)$ as follows:

$$\varphi(x) = \frac{f(x_0-) - f(x_0+)}{2} \quad \text{for} \quad x_0 - \pi \leqq x \leqq x_0,$$

$$\varphi(x) = \frac{f(x_0+) - f(x_0-)}{2} \quad \text{for} \quad x_0 < x < x_0 + \pi,$$

$$\varphi(x + 2k\pi) = \varphi(x) \quad \text{for} \quad k = 0, \pm 1, \pm 2, \ldots,$$

and define $\psi(x)$ as follows:

$$\psi(x) = f(x) - \varphi(x) \quad \text{for} \quad x \neq x_0 + 2k\pi, \quad (k = 0, \pm 1, \pm 2, \ldots)$$

$$\psi(x) = \frac{f(x_0+) + f(x_0-)}{2} \quad \text{for} \quad x = x_0 + 2k\pi, \quad (k = 0, \pm 1, \pm 2, \ldots).$$

Then

$$\lim_{x \to x_0+} \psi(x) = f(x_0+) - \frac{f(x_0+) - f(x_0-)}{2} = \frac{f(x_0+) + f(x_0-)}{2},$$

$$\lim_{x \to x_0-} \psi(x) = f(x_0-) - \frac{f(x_0-) - f(x_0+)}{2} = \frac{f(x_0+) + f(x_0-)}{2},$$

and consequently $\psi(x)$ is periodic with period 2π, integrable, continuous and of bounded variation in $(x_0 - \delta, x_0 + \delta)$.

$f(x) = \varphi(x) + \psi(x)$, for every $x \neq x_0 + 2k\pi$ and if $S_n(x)$, $S_n^{(1)}(x)$, $S_n^{(2)}(x)$ denote the partial sums of the first $n + 1$ terms of the Fourier series of $f(x)$, $\varphi(x)$, $\psi(x)$ respectively, then

$$S_n(x) = S_n^{(1)}(x) + S_n^{(2)}(x).$$

The sum $S_n^{(2)}(x)$ converges uniformly in $[x_0 - \delta/4, x_0 + \delta/4]$ to the continuous function $\psi(x)$ [Sec. 6, Th. 41′] and in particular

$$\lim_{n \to \infty} S_n^{(2)}(x_0) = \frac{f(x_0+) + f(x_0-)}{2}.$$

The sum $S_n^{(1)}(x)$ corresponding to $\varphi(x)$ has the behavior described in Sec. 10.1. If

$$f(x_0+) - f(x_0-) > 0, \quad [f(x_0+) - f(x_0-) < 0]$$

the absolute maximum [minimum] of $S_n^{(1)}(x)$ in $[x_0 - \delta/4, x_0+\delta/4]$ is

$$\frac{f(x_0+)-f(x_0-)}{2} - \frac{f(x_0+)-f(x_0-)}{\pi}\int_{\pi}^{+\infty}\frac{\sin\alpha}{\alpha}\,d\alpha,$$

and the absolute minimum [maximum] of $S_n^{(1)}(x)$ in $[x_0 - \delta/4, x_0 + \delta/4]$ is

$$-\frac{f(x_0+)-f(x_0-)}{2} + \frac{f(x_0+)-f(x_0-)}{\pi}\int_{\pi}^{+\infty}\frac{\sin\alpha}{\alpha}\,dv.$$

Consequently, if $f(x_0+) > f(x_0-)$, $[f(x_0+) < f(x_0-)]$, the maximum [minimum] of $S_n(x)$, in an interval sufficiently small to the right of x_0, tends to

$$\text{(1)} \qquad f(x_0+) - \frac{f(x_0+)-f(x_0-)}{\pi}\int_{\pi}^{+\infty}\frac{\sin\alpha}{\alpha}\,d\alpha,$$

as $n \to \infty$, and the minimum [maximum] in an interval sufficiently small to the left of x_0 tends to

$$\text{(2)} \qquad f(x_0-) + \frac{f(x_0+)-f(x_0-)}{\pi}\int_{\pi}^{+\infty}\frac{\sin\alpha}{\alpha}\,d\alpha.$$

The curve $y = S_n(x)$ in an interval sufficiently small to the left and right of x_0 oscillates about the curve $y = f(x)$. The arc of the curve $y = S_n(x)$ whose extremities are the absolute maximum and absolute minimum tends as $n \to \infty$ to the segment of the line $x = x_0$ whose extremities are the points with ordinates (1) and (2) and of length

$$|f(x_0+) - f(x_0-)|\left[1 + \left|\frac{2}{\pi}\int_{\pi}^{+\infty}\frac{\sin\alpha}{\alpha}\,d\alpha\right|\right].$$

This is the Gibbs phenomenon for a function $f(x)$ which has a finite discontinuity and satisfies the hypothesis stated at the point x_0.

3a. DEFINITION 12. Let $f(x)$ be periodic with period 2π, integrable, of bounded variation and continuous in $(x_0 - \delta, x_0 + \delta)$ except at x_0 where it has a finite discontinuity. Let $C_n^{(k)}(x)$ denote the mean of order k of the first $n+1$ terms of the Fourier series of $f(x)$ at the point x, and project orthogonally on the line $x = x_0$ the arc of the curve $y = C_n^{(k)}(x)$ corresponding to the values of a sufficiently small neighborhood of x_0. If the segment of the line $x = x_0$ so obtained tends as $n \to \infty$ to a segment which does not coincide with the segment whose extremities are the points $[x_0, f(x_0 -)]$, $[x_0, f(x_0 +)]$, then we shall say that the (C, k) sum of the Fourier series of $f(x)$ presents the Gibbs phenomenon at x_0.

In view of the results of Sections 1 and 2, we have

THEOREM 55. If $f(x)$ is periodic with period 2π, integrable, of bounded variation and continuous in $(x_0 - \delta, x_0 + \delta)$ except at x_0, where it has a finite discontinuity, then the $(C, 0)$ sum of its Fourier series presents the Gibbs phenomenon at x_0.

3b. We shall prove

THEOREM 56. If $f(x)$ satisfies the hypotheses of Theorem 55, the $(C, 1)$ sum of its Fourier series does not present the Gibbs phenomenon at x_0.

Proof. Proceeding as in Sections 10.1 and 10.2 it suffices to consider the function $f(x)$ defined by (1) of Sec. 10.1.

For $0 < \varepsilon < \pi/4$, by the results of Sec. 6.5, we have for the mean $\sigma_n(x)$ of the first order

$$\sigma_n(x) = \frac{1}{n\pi}\int_0^{\varepsilon} [f(x+2\alpha)+f(x-2\alpha)] \left(\frac{\sin n\alpha}{\alpha}\right)^2 d\alpha + R_n(x)$$

where $\lim_{n\to\infty} R_n(x) = 0$ uniformly for all x in $[-\pi, \pi]$.

For $0 < x \leqq \varepsilon$, we have

$$\sigma_n(x) = \frac{2c}{n\pi}\int_0^{x/2} \left(\frac{\sin n\alpha}{\alpha}\right)^2 d\alpha + R_n(x),$$

$$\sigma_n(x) = \frac{2c}{\pi}\int_0^{nx/2} \left(\frac{\sin \alpha}{\alpha}\right)^2 d\alpha + R_n(x).$$

Since $\lim_{n\to\infty} \sigma_n(x) = c$ it follows that

$$\int_0^{+\infty} \left(\frac{\sin \alpha}{\alpha}\right)^2 d\alpha = \frac{\pi}{2}. \tag{3}$$

whence

$$\sigma_n(x) = c\left[1 - \frac{2}{\pi}\int_{nx/2}^{+\infty}\left(\frac{\sin\alpha}{\alpha}\right)^2 d\alpha\right],$$

and since an analogous formula holds for $-\varepsilon \leqq x < 0$, it follows that in the interval $(-\varepsilon, \varepsilon)$, the limit as $n \to \infty$ of the orthogonal projection on the y-axis of the curve $y = \sigma_n(x)$ is the segment whose extremities are $(0, -c)$, $(0, c)$ and therefore the Gibbs phenomenon does not occur.

3c. We state without proof the following theorem due to H. Cramer (cf. also A. Zygmund [121]).

THEOREM 57. There exists a positive constant k_0 such that if $f(x)$ satisfies the hypotheses stated in 3a., then the (C, k) sum of the Fourier series of $f(x)$ presents the Gibbs phenomenon for $0 \leqq k < k_0$, but not for $k \geqq k_0$.

This value of k_0 is precisely the value of k for which the maximum of the absolute value of the integral

$$\int_0^1 (1-t)^k\left(\frac{\sin xt}{t}\right) dt$$

considered as a function of x, is $\pi/2$. The value of k_0 computed by T. H. Gronwall [43] to 6 decimal places is

$$k_0 = 0.439551 \ldots .$$

11. Inequalities for the Partial Sums of Fourier Series of a Function of Bounded Variation

1a. In the applications we shall make use of the following theorem (cf. e.g. Sec. 12.1b).

THEOREM 58. If $f(x)$ is periodic with period 2π, integrable, of bounded variation in an interval $[a, b]$ belonging to $[-\pi, \pi]$, then for a fixed interval $[c, d]$ interior to $[a, b]$, there exists a constant M such that for the sum $S_n(x)$ of the first $n + 1$ terms of the Fourier series the following inequality holds

$$(1) \qquad | S_n(x) | \leqq M, \quad (n = 0, 1, \ldots, c \leqq x \leqq d),$$

where the constant M depends on c and d, but is independent of n.

Proof. Take ε so that $0 < \varepsilon < \text{Min}\,[(c-a)/2,\ (b-d)/2]$; then, by referring to the derivation of (5) of Sec. 4.4, it is clear that

$$S_n(x) = \frac{1}{\pi}\int_0^{\pi/2} [f(x+2\alpha)+f(x-2\alpha)]\,\frac{\sin\,(2n+1)\alpha}{\sin\alpha}\,d\alpha = I + J \tag{2}$$

where

$$I = \frac{1}{\pi}\int_0^{\varepsilon} [f(x+2\alpha)+f(x-2\alpha)]\,\frac{\sin\,(2n+1)\alpha}{\sin\alpha}\,d\alpha,$$

$$J = \frac{1}{\pi}\int_{\varepsilon}^{\pi/2} [f(x+2\alpha)+f(x-2\alpha)]\,\frac{\sin\,(2n+1)\alpha}{\sin\alpha}\,d\alpha.$$

For the integral J we have $0 < 1/\sin\alpha \leqq 1/\sin\varepsilon$, $|\sin\,(2n+1)\alpha| \leqq 1$, whence

$$|\,J\,| \leqq \frac{1}{\pi\sin\varepsilon}\int_{\varepsilon}^{\pi/2} |\,f(x+2\alpha)+f(x-2\alpha)\,|\,d\alpha \leqq \frac{1}{\pi\sin\varepsilon}\int_{-\pi/2}^{\pi/2} |\,f(x+2\alpha)|\,d\alpha,$$

or

$$|\,J\,| \leqq \frac{1}{2\pi\sin\varepsilon}\int_{-\pi}^{\pi} |\,f(\alpha)\,|\,d\alpha. \tag{3}$$

Proceeding now to the consideration of the integral I, we can assume that $f(x) = f_1(x) - f_2(x)$ in $[a, b]$ where $f_1(x)$ and $f_2(x)$ are nonnegative nondecreasing functions since $f(\alpha)$ is of bounded variation in $[a, b]$ (Appendix, Th. 12), and we can write

$$I = I_{1,1} - I_{1,2} - I_{2,1} + I_{2,2} \tag{4}$$

where

$$I_{l,1} = \frac{1}{\pi}\int_0^{\varepsilon} \frac{\alpha}{\sin\alpha}\,f_l(x+2\alpha)\,\frac{\sin\,(2n+1)\alpha}{\alpha}\,d\alpha, \quad (l = 1, 2),$$

$$-\,I_{l,2} = \frac{1}{\pi}\int_0^{\varepsilon} \frac{\alpha}{\sin\alpha}\,f_l(x-2\alpha)\,\frac{\sin\,(2n+1)\alpha}{\alpha}\,d\alpha, \quad (l = 1, 2).$$

In the interval $[0, \varepsilon]$ the function $\alpha/\sin\alpha$ is positive and increases monotonically from 1 to $\varepsilon/\sin\varepsilon$. Consequently, by the second theorem of the mean (Appendix, Th. 18), we have

$$I_{l,1} = \frac{\varepsilon}{\pi \sin \varepsilon} \int_{\varepsilon_1}^{\varepsilon} f_l(x + 2\alpha) \frac{\sin (2n+1)\alpha}{\alpha} d\alpha, \quad (0 \leqq \varepsilon_1 \leqq \varepsilon),$$

and since $f_l(x + 2\alpha)$ is positive and non decreasing in $[\varepsilon_1, \varepsilon]$ and not greater than $f_l(b)$, we have

$$I_{l,1} = \frac{\varepsilon}{\pi \sin \varepsilon} f_l(b) \int_{\varepsilon_2}^{\varepsilon} \frac{\sin (2n+1)\alpha}{\alpha} d\alpha = \frac{\varepsilon f_l(b)}{\pi \sin \varepsilon} \int_{(2n+1)\varepsilon_2}^{2(n+1)\varepsilon} \frac{\sin \alpha}{\alpha} d\alpha,$$

$$(\varepsilon_1 \leqq \varepsilon_2 \leqq \varepsilon),$$

and by (9‴) of Sec. 9.4

$$(5_1) \qquad |I_{l,1}| \leqq \frac{\varepsilon}{\sin \varepsilon} f_l(b).$$

Similarly

$$(5_2) \qquad |I_{l,2}| \leqq \frac{\varepsilon}{\sin \varepsilon} f_l(b)$$

Finally (1) follows from (2), (3), (4), (5_1) and (5_2).

1b. REMARK. If $f(x)$ is of bounded variation in $[-\pi, \pi]$, then (1) holds for all x in $[-\pi, \pi]$.

12. Applications of Fourier Series

1a. Consider a strip (infinite thin plate) of the xy plane whose base is the segment AB of the x-axis, where $A \equiv (-\pi/2, 0)$, $B \equiv (\pi/2, 0)$, and whose sides are the half lines AC and BD with equations $x = -\pi/2$, $x = \pi/2$, in the half plane $y \geqq 0$, and assume that it is isotropic with respect to the conduction of heat, that the sides AC and BD are held at constant zero temperature, and that at every point x of the base AB, $-\pi/2 < x < \pi/2$ the temperature of the plate is independent of the time and is a continuous function $f(x)$ of x.

Suppose that there is a flow of heat through the strip or plate such that at every point (x, y), the temperature is given as a function $T(x, y)$ continuous at every point (x, y) except possibly at A or B. Thus $T(x, y)$ is harmonic, namely, at every interior point of the strip:

$$\frac{\partial^2 T}{\partial x^2} + \frac{\partial^2 T}{\partial y^2} = 0; \tag{1}$$

$T(x, y)$ also satisfies the following boundary conditions

$$T(-\pi/2, y) = 0 \quad \text{for} \quad y \geqq 0; \tag{2_1}$$

$$T(\pi/2, y) = 0 \quad \text{for} \quad y \geqq 0; \tag{2_2}$$

$$T(x, 0) = f(x) \quad \text{for} \quad -\pi/2 < x < \pi/2. \tag{3}$$

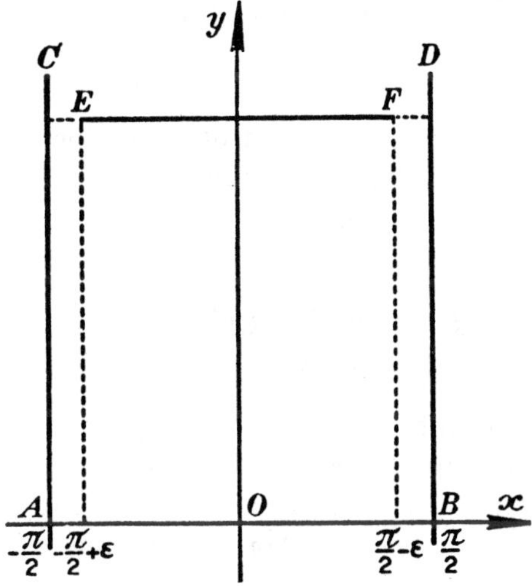

Fig. 11.

If we suppose $f(x)$ is continuous and of bounded variation in the interval $(-\pi/2, \pi/2)$ (we allow for the possibility that $f(x)$ may not vanish for $x = -\pi/2$ or $x = \pi/2$), then the problem of Fourier consists in determining a function $T(x, y)$ continuous in the whole strip, except possibly at the points $(-\pi/2, 0)$, $(\pi/2, 0)$, which satisfies (1) at every interior point and conditions (2_1), (2_2) and (3) on the boundary, and, finally, the condition

$$|T(x, y)| < Me^{-y}, \qquad M, \text{ const.} \tag{4}$$

This problem, for the case $f(x) = 1$, was the first to be considered by Fourier in his celebrated *Theorie Analytique de la Chaleur* Paris, 1822; *Oeuvres*, I, Paris, 1888, p. 144. For the case considered in the text, see H. Poincaré [89], a).

In 1b we shall prove the existence of $T(x, y)$ and give its analytic expression; in 2c we shall prove the uniqueness of $T(x, y)$ under the assumption that (4) is replaced by the conditions that there exist two positive numbers y_1 and y_2 such that

$$(5_1) \qquad |T(x, y)| \leqq \Phi(x), \quad \text{for} \quad y \leqq y_2$$

and $\Phi(x)$ integrable in the sense of Lebesgue in $(-\pi/2, \pi/2)$.

$$(5_2) \qquad |T(x, y)| < Ae^{\lambda y}, \quad A > 0, \quad \lambda < 1, \quad \text{and} \quad y \geqq y_1.$$

1b. Suppose there is a solution of (1) of the form

$$T(x, y) = g(x)h(y)$$

which satisfies (2_1), (2_2) and (4).

Then from (1) follows

$$g''(x)h(y) + g(x)h''(y) = 0,$$
$$-g''(x)/g(x) = h''(y)/h(y) = c,$$

where c is a constant.

It is not possible to have $c = -l^2$, for then $g''(x) - l^2g(x) = 0$, $g(x) = c_1 \sinh l(x - c_2)$ where c_1 and c_2 are constants and $g(x)$ could not vanish at either $-\pi/2$ or $\pi/2$. Therefore $c = l^2$, whence

$$(6_1) \qquad g''(x) + l^2g(x) = 0,$$

$$(6_2) \qquad h''(y) - l^2h(y) = 0.$$

From (6_2) and (4) follows $h(y) = e^{-ly}$ where l is constant, $l > 0$, and from (6_1) follows

$$g(x) = a \cos lx + b \sin lx,$$

where a and b are constants.

Since $g(x)$ vanishes at $-\pi/2$ and $\pi/2$ it follows that $a \cos (l\pi/2) = 0$, $b \sin (l\pi/2) = 0$, which imply that either $b = 0$ and $l = 2m - 1$, or $a = 0$, $l = 2m$, $(m = 1, 2, \ldots)$ and consequently a formal solution of (1) satisfing (2_1) and (2_2) is given by the series

$$(7) \qquad T(x, y) = a_1 e^{-y} \cos x + b_2 e^{-2y} \sin 2x + a_3 e^{-3y} \cos 3x + b_4 e^{-4y} \sin 4x + \dots$$

where the a_{2n-1} and the b_{2n} are two arbitrary sequences of real constants.

In order for (3) to be satisfied, we must have

$$(8) \quad f(x) = \alpha_1 \cos x + b_2 \sin 2x + a_3 \cos 3x + b_4 \sin 4x + \dots;$$

for $-\pi/2 < x < \pi/2$. Substituting $\pi - x$ or $-\pi - x$ for x changes the sign of each term in (8) so that if we define $f(x)$ in $[-\pi, -\pi/2)$ and $(\pi/2, \pi]$ as follows

$$f(x) = -f[-\pi - x] \quad \text{for} \quad -\pi \leqq x < -\pi/2,$$
$$f(x) = -f(\pi - x) \quad \text{for} \quad \pi/2 < x \leqq \pi,$$

then $f(x)$ has precisely the expansion (8), and the coefficients a_{2n-1}, b_{2n} are given by the formulas

$$(9_1) \quad a_{2n-1} = \frac{1}{\pi}\int_{-\pi}^{\pi} f(x) \cos (2n-1)x \, dx = \frac{2}{\pi}\int_{-\pi/2}^{\pi/2} f(x) \cos (2n-1)x \, dx,$$

$$(9_2) \qquad b_{2n} = \frac{2}{\pi}\int_{-\pi/2}^{\pi/2} f(x) \sin 2\,nx \, dx.$$

By the remark of Sec. 6.8 the series on the right side of (8) converges uniformly to $f(x)$ in every interval interior to $(-\pi/2, \pi/2)$ and therefore (3) follows.

Since $f(x)$ is of bounded variation, we have by a theorem of Lebesgue (Sec. 6, Th. 40)

$$|a_{2n-1}| < L/(2n-1), \quad |b_{2n}| < L/2n, \quad (n = 1, 2, \dots)$$

where L is a constant, and therefore for every set of points (x, y) for which $y \geqq d > 0$, in view of the convergence of the numerical series

$$L \sum_{m=1}^{\infty} \frac{1}{m} e^{-md}, \quad L \sum_{m=1}^{\infty} e^{-md}, \quad L \sum_{m=1}^{\infty} m e^{-md},$$

the series (7) and also the series obtained from (7) by term by differentiation with respect to x or with respect to y are absolutely and uniformly convergent for $y \geqq d > 0$. It follows that

$$\lim_{(x, y)\to(-\pi/2, y)} T(x, y) = 0, \qquad \lim_{(x, y)\to(\pi/2, y)} T(x, y) = 0$$

and also that $T(x, y)$ satisfies (1).

By the remark of Sec. 11.1b, if $S_n(x)$ denotes the sum of the first n terms of (8), there exists a constant M such that

$$| S_n(x) | \leqq M, \quad (n = 1, 2, \ldots, -\pi/2 \leqq x \leqq \pi/2);$$

whence, if $T_n(x, y)$ denotes the sum of the first n terms of (7), we have, by applying the Brunacci-Abel transformation (cf. Ch. III, Sec. 10.1),

$$T_n(x, y) = S_1(x)e^{-y} + [S_2(x) - S_1(x)]e^{-2y} + \ldots + [S_n(x) - S_{n-1}(x)]e^{-ny},$$

$$T_n(x, y) = S_1(x)[e^{-y} - e^{-2y}] + S_2(x)[e^{-2y} - e^{-3y}) + \ldots + S_{n-1}(x)[e^{-(n-1)y} - e^{-ny}] + S_n(x)e^{-ny},$$

$$| T_n(x, y) | \leqq M[e^{-y} - e^{-2y}) + (e^{-2y} - e^{-3y}) + \ldots + (e^{-(n-1)y} - e^{-ny}) + e^{-ny}],$$

$$| T_n(x, y) \leqq Me^{-y}$$

and (4) follows immediately if we pass to the limit as $n \to \infty$.

Let $R_n(x, y)$ denote the remainder after n terms of (7). Then since (8) converges uniformly in every interval (α, β) interior to $(-\pi/2, \pi/2)$, we have by Th. 41′ of Sec. 6, that for $\sigma > 0$, there exists an n_0 such that if $n \geqq n_0$, the remainder $R_n(x, 0)$ satisfies the inequality

$$| R_n(x, 0) | < \sigma, \quad (n \geqq n_0, \alpha \leqq x \leqq \beta),$$

and by the previous arguments

$$| R_n(x, y) | \leqq \sigma e^{-(n+1)y} < \sigma.$$

Therefore (7) is uniformly convergent in a sufficiently small neighborhood of every interior point of $(-\pi/2, \pi/2)$, and consequently

$$\lim_{(x,\,y)\to(x,\,0)} T(x, y) = T(x, 0) = f(x), \quad (-\pi/2 < x < \pi/2),$$

whence $T(x, y)$ satisfies all the conditions of the problem.

1c. We shall show now that the function $T(x, y)$, which is continuous at every finite point of the strip, except possibly at $(-\pi/2, 0)$, $(\pi/2, 0)$, and which satisfies conditions (1), (2_1), (2_2), (3), (4), is unique.

We shall replace (4) by the less restrictive conditions
(4′) $|T(x, y)| < Ae^{\lambda y}$, where A and λ are constants, $\lambda < 1$, and $y \geqq y_1 > 0$;
(4″) $|T(x, y)| \leqq \Phi(x)$, for $y \leqq y_2$, $(y_2 > 0)$, $-\pi/2 < x < \pi/2$ where $\Phi(x)$ is Lebesgue integrable in $(-\pi/2, \pi/2)$.

Clearly if the two functions $T(x, y)$ and $\bar{T}(x, y)$ both satisfy conditions (4′) and (4″), then $u(x, y) = T(x, y) - \bar{T}(x, y)$ satisfies the conditions:

i. $u(x, y)$ is continuous at every finite point of the strip, except possibly at $(-\pi/2, 0)$, $(\pi/2, 0)$;

ii. At every interior point (x, y) of the strip $u(x, y)$ is harmonic, that is

$$\frac{\partial^2 u}{\partial x^2} + \frac{\partial^2 u}{\partial y^2} = 0, \quad [-\pi/2 < x < \pi/2,\ y > 0]. \tag{10}$$

iii. $u(x, y)$ satisfies on the boundary the conditions

$$(11_1) \qquad u(-\pi/2, y) = 0 \quad \text{for} \quad y \geqq 0;$$

$$(11_2) \qquad u(\pi/2, y) = 0 \quad \text{for} \quad y \geqq 0,$$

$$(11_3) \qquad u(x, 0) = 0 \quad \text{for} \quad -\pi/2 < x < \pi/2;$$

iv. There exist three positive constants A, y_1, y_2, a number $\lambda < 1$, and a function $\Phi(x)$ defined and integrable in $(-\pi/2, \pi/2)$ such that

$$(12_1) \qquad |u(x, y)| < Ae^{\lambda y} \quad \text{for} \quad y \geqq y_1, \ (\lambda < 1),$$

$$(12_2) \qquad |u(x, y)| < \Phi(x) \quad \text{for} \quad y \leqq y_2, \ -\pi/2 < x < \pi/2.$$

The uniqueness of $T(x, y)$ will be established by proving that a function $u(x, y)$ satisfying conditions i, ii, iii, iv vanishes identically.

The proof can be completed by the so-called method of the partial Laplace transform on a finite interval of integration of M. Picone (cf. M. Picone [88b] pp. 775—793).

Suppose $0 < \varepsilon < \pi/2$, and for every

$$\lambda_n = n\pi/(\pi - 2\varepsilon), \quad (n = 1, 2, \ldots) \tag{13}$$

consider the function

$$v_n(x, \varepsilon) = \sin \lambda_n(x - \varepsilon + \pi/2), \tag{14}$$

which vanishes for $x = -\pi/2 + \varepsilon$, $x = \pi/2 - \varepsilon$.

From (10), integrating along the segment EF with extremities $E \equiv (-\pi/2 + \varepsilon, y)$, $F \equiv (\pi/2 - \varepsilon, y)$, follows:

$$\int_{-\pi/2+\varepsilon}^{\pi/2-\varepsilon} \frac{\partial^2 u}{\partial x^2} v_n(x, \varepsilon)dx + \int_{-\pi/2+\varepsilon}^{\pi/2-\varepsilon} \frac{\partial^2 u}{\partial y^2} v_n(x, \varepsilon)dx = 0, \tag{15}$$

and defining $u_n(\varepsilon, y)$, by

$$\int_{-\pi/2+\varepsilon}^{\pi/2-\varepsilon} u(x, y)v_n(x, \varepsilon)dx = u_n(\varepsilon, y), \quad (y > 0), \tag{16}$$

and integrating by parts twice on the left side of (15) we obtain

$$\frac{d^2u_n(\varepsilon, y)}{dy^2} - \lambda_n^2 u_n(\varepsilon, y) = \lambda_n[u(\pi/2-\varepsilon, y)(-1)^n - u(-\pi/2+\varepsilon, y)].$$

This linear differential equation of the second order has for solution, $(y > 0)$,

$$(17_1) \quad u_n(\varepsilon, y) = A_n(\varepsilon) \sinh \lambda_n(y - y_0) + u_n(\varepsilon, y_0) \cosh \lambda_n(y - y_0)$$
$$+ \int_{y_0}^{y} [u(\pi/2 - \varepsilon, s)(-1)^n - u(-\pi/2 + \varepsilon, s)] \sinh \lambda_n(y - s)ds,$$

where y_0 is a preassigned positive number and the constant $A_n(\varepsilon)$ is determined, for example, by the condition

$$(17_2) \quad u_n(\varepsilon, 2y_0) = A_n(\varepsilon) \sinh ny_0 + u_n(\varepsilon, y_0) \cosh \lambda_n y_0$$
$$+ \int_{y_0}^{2y_0} [u(\pi/2 - \varepsilon, s)(-1)^n - u(-\pi/2 + \varepsilon, s)] \sinh \lambda_n(2y_0 - s)ds.$$

(Cf. for example, Sansone: *Lezioni di Analisi Matematica*, II, (7th Ed.) p. 455).
From (13), (14), (15) follows

$$\lim_{\varepsilon\to 0} \lambda_n = n, \quad \lim_{\varepsilon\to 0} v_n(x, \varepsilon) = \sin n(x + \pi/2)$$

$$\lim_{\varepsilon\to 0} u_n(\varepsilon, y) = u_n(y)$$

where

$$\boxed{\int_{-\pi/2}^{\pi/2} u(x, y) \sin n(x + \pi/2)dx = u_n(y)} \tag{18}$$

and from (12_1) follows

$$| u_n(y) | \leqq \pi A e^{\lambda y} \quad (\text{for } y \geqq y_1, \; n = 1, 2, \ldots). \tag{12'}$$

Passing to the limit as $\varepsilon \to 0$ in (17_2) and taking into account (11_1) and (11_2) we obtain

$$u_n(2y_0) = [\lim_{\varepsilon\to 0} A_n(\varepsilon)] \sinh ny_0 + u_n(y_0) \cosh ny_0.$$

Thus the limit as $\varepsilon \to 0$ of $A_n(\varepsilon)$ exists; we shall denote it by A_n. Passing to the limit as $\varepsilon \to 0$ in (17_1), we obtain

$$u_n(y) = A_n \sinh n(y - y_0) + u_n(y_0) \cosh n(y - y_0),$$

$$2u_n(y) = [A_n + u_n(y_0)]e^{n(y-y_0)} + [A_n - u_n(y_0)]e^{-n(y-y_0)} \tag{19}$$

and by virtue of (12_1) it follows that

$$A_n + u_n(y_0) = 0 \tag{20_1}$$

and therefore

$$2u_n(y) = [A_n - u_n(y_0)]e^{-n(y-y_0)}. \tag{21}$$

Now suppose $\{ \bar{y}_\nu \}$ is a sequence of positive terms converging to zero and set $y = \bar{y}_\nu$ in (18). By (12_2), passing to the limit under the integral sign is permissible. Moreover, if B is a set of nonoverlapping subintervals of $(-\pi/2, \pi/2)$ then

$$\left| \int_B u(x, \bar{y}_\nu) \sin n(x + \pi/2)dx \right| \leqq \int_B \Phi(x)dx,$$

which establishes the equi-absolute convergence (Appendix, Def. 13) of the integrals of the sequence $\{u(x, \bar{y}_\nu) \sin n(x+\pi/2)\}$ in the interval $(-\pi/2, \pi/2)$. Then by Theorem 8 of the Appendix, we have $\lim_{\nu\to\infty} u_n(\bar{y}_\nu) = 0$ and consequently, by (21),

$$A_n - u_n(y_0) = 0. \tag{20$_2$}$$

Finally, from (19), (20_1), and (20_2) follows $u_n(y) \equiv 0$ for $y \geqq 0$. Moreover, the system $\{\sin n(x + \pi/2)\}$ is orthogonal and complete in $(-\pi/2, \pi/2)$ since this system transforms by $x + \pi/2 = t$ into the system $\{\sin nt\}$ for t in $(0, \pi)$, which is complete (cf. Footnote to Th. 4, Sec. 2.3) whence by (18), for a fixed $y>0$, $u(x, y)$, considered as a function of x in $(-\pi/2, \pi/2)$, vanishes almost everywhere and since it is continuous it vanishes identically.

13. The Fourier Transform

1a. Let $h(x)$ and $l(x)$ be two real functions of the real variable x, defined for every x, integrable in every finite interval interior to $(-\infty, 0)$ or $(0, \infty)$ and suppose that for every finite value of the real variable v, the following limit exists and is finite:

$$\lim_{\substack{\sigma\to+0\\ n\to+\infty}} \Big[\int_\sigma^n [h(\alpha) + il(\alpha)][\cos v\alpha - i \sin v\alpha]d\alpha$$

$$+ \int_{-n}^{-\sigma} [h(\alpha) + il(\alpha)][\cos v\alpha - i \sin v\alpha]d\alpha\Big]$$

or, if we write $f(\alpha) = h(\alpha) + il(\alpha)$, we suppose

$$\lim_{\substack{\sigma\to+0\\ n\to+\infty}} \Big[\int_\sigma^n f(\alpha)e^{-iv\alpha}\, d\alpha + \int_{-n}^{-\sigma} f(\alpha)e^{-iv\alpha}d\alpha\Big]$$

exists and is finite. Such a limit will be denoted by the symbol

$$\underset{(0,\,\infty)}{P.V.} \int_{-\infty}^{\infty} f(\alpha)e^{-iv\alpha}\, d\alpha$$

and will be called the principal value of the integral relative to the points 0 and ∞.

REMARK. This definition is an extension of definition 11 in Sec. 9.2, of the Cauchy principal value of an integral between infinite limits. The existence of the Cauchy principal value of the integral $\int_{-\infty}^{\infty} f(\alpha)e^{-iv\alpha}\,d\alpha$ according to Def. 11 of Sec. 9.2, implies the existence of $P.V._{(0,\infty)} \int_{-\infty}^{\infty} f(\alpha)e^{-iv\alpha}\,d\alpha$ but not conversely.

The function $F(v)$ defined by the relation

$$F(v) = \underset{(0,\infty)}{P.V.} \frac{1}{\sqrt{2\pi}} \int_{-\infty}^{\infty} f(\alpha)e^{-iv\alpha}\,d\alpha \tag{I}$$

will be called the Fourier transform of $f(x)$ and we shall write

$$F(v) = J\{ f(x); v \}. \tag{I'}$$

Clearly if $f_1(x)$ and $f_2(x)$ both admit a Fourier transform and c_1 and c_2 are constants, then

$$J\{ c_1 f_1(x) + c_2 f_2(x); v \} = c_1 J\{ f_1(x); v \} + c_2 J\{ f_2(x); v \}. \tag{1}$$

1b. Let $h(x)$ and $l(x)$ be finite and regular (Appendix, Def. 5) for every finite value of x, integrable in every finite interval, and each satisfying at least one of the following conditions:

(i) The integral of $h(x)$ $[l(x)]$ is absolutely convergent in (a, ∞), where $a > 0$.

(ii) $h(x)$ $[l(x)]$ is of bounded variation in (a, ∞), $a > 0$,

$$\text{and } \lim_{x\to+\infty} h(x) = 0, \quad [\lim_{x\to+\infty} l(x) = 0].$$

(iii) For some $b > 0$, both functions $h(x)$ and $l(x)$ satisfy in $(-\infty, b)$ a condition analogous to (i) or (ii).

Under these conditions, by the argument in Theorem 50, it follows that for any real v the following integrals are convergent (Def. 9, Sec. 8.2):

$$\frac{1}{\sqrt{2\pi}} \int_{-\infty}^{\infty} h(\alpha) \cos v\alpha\,d\alpha, \quad \frac{1}{\sqrt{2\pi}} \int_{-\infty}^{\infty} h(\alpha) \sin v\alpha\,d\alpha$$

$$\frac{1}{\sqrt{2\pi}} \int_{-\infty}^{\infty} l(\alpha) \cos v\alpha\,d\alpha, \quad \frac{1}{\sqrt{2\pi}} \int_{-\infty}^{\infty} l(\alpha) \sin v\alpha\,d\alpha$$

and if

$$f(x) = h(x) + il(x), \tag{2}$$

then the Fourier transform $F(v)$ of $f(x)$ exists. If, further,

$$H(v) = \frac{1}{\sqrt{2\pi}}\int_{-\infty}^{\infty} h(\alpha)e^{-iv\alpha}\, d\alpha, \tag{3_1}$$

$$L(v) = \frac{1}{\sqrt{2\pi}}\int_{-\infty}^{\infty} l(\alpha)e^{-iv\alpha}\, d\alpha, \tag{3_2}$$

then

$$F(v) = H(v) + iL(v). \tag{4}$$

Again, suppose that $h(x)$ and $l(x)$, satisfy the condition:

(iv) The Fourier series of $h(x)$ and $l(x)$, relative to a finite interval with x as an interior point, converge respectively to $h(x)$ and $l(x)$ [This condition is clearly satisfied if $h(x)$ and $l(x)$ are of bounded variation in $(-\infty, \infty)$ (Th. 41)].

Then by (3_1)

$$\begin{aligned}
&\int_{\sigma}^{n} e^{ivx} H(v)\, dv + \int_{-n}^{-\sigma} e^{-ivx} H(v) dv \\
&= \int_{\sigma}^{n} [e^{ivx} H(v) + e^{-ivx} H(-v)] dv \\
&= \frac{2}{\sqrt{2\pi}}\int_{\sigma}^{n}\int_{-\infty}^{\infty} h(\alpha) \cos v(\alpha - x) d\alpha
\end{aligned}$$

whence, passing to the limit as $\sigma \to +0$ and $n \to +\infty$, and taking into account the second form of Fourier's formula [Sec. 9.7, (18)]

$$h(x) = \underset{(0,\,\infty)}{P.V.} \frac{1}{\sqrt{2\pi}}\int_{-\infty}^{\infty} H(v)e^{ivx}\, dv,$$

and similarly by (3_2)

$$l(x) = \underset{(0,\,\infty)}{P.V.} \frac{1}{\sqrt{2\pi}}\int_{-\infty}^{\infty} L(v)e^{ivx}\, dv,$$

and therefore by (2)

$$\text{(II)} \qquad f(v) = \underset{(0,\,\infty)}{P.V.} \frac{1}{\sqrt{2\pi}} \int_{-\infty}^{\infty} F(\alpha) e^{iv\alpha} d\alpha.$$

Thus we have proved the following:

THEOREM 59. If $f(x) = h(x) + il(x)$ is defined in $(-\infty, \infty)$ and $h(x)$ and $l(x)$ satisfy conditions i), ii), iii), iv), then (I) can be inverted by (II).

(II) is called the Fourier formula of inversion or reciprocity, and with the notation (I′), it can be written

$$\text{(II}'\text{)} \qquad f(v) = J\{ F(x); - v \}.$$

2. For the convenience of the reader we give next the Fourier transforms of several elementary functions.

i) Let

$$\text{(5)} \quad f(x) = \sqrt{\pi/2} \quad \text{if} \quad |x| \leqq a, \; f(x) = 0 \quad \text{if} \quad |x| > a, \; a > 0.$$

Clearly, $F(0) = a$, and if $v \neq 0$

$$F(v) = \tfrac{1}{2} \int_{-a}^{a} e^{-iv\alpha}\, d\alpha = \frac{\sin av}{v}.$$

Consequently

$$\text{(5}'\text{)} \qquad J\{ f(x); v \} = \frac{\sin av}{v} \text{ if } v \neq 0, \; J\{ f(x); 0 \} = a.$$

ii) If c and w are constants, $c > 0$, let

$$\text{(6)} \quad f(x) = e^{-cx+iwx} \quad \text{if} \quad x \geqq 0; \; f(x) = 0, \text{ if } x < 0.$$

Then

$$\frac{1}{\sqrt{2\pi}} \int_0^{\infty} e^{-c\alpha + i(w-v)\alpha}\, d\alpha$$

$$= \frac{1}{\sqrt{2\pi}} \int_0^{\infty} e^{-c\alpha} \cos (w - v)\alpha \, d\alpha + \frac{i}{\sqrt{2\pi}} \int_0^{\infty} e^{-c\alpha} \sin (w - v)\alpha \, d\alpha$$

$$= \frac{1}{\sqrt{2\pi}} \frac{c}{c^2 + (w - v)^2} + \frac{i}{\sqrt{2\pi}} \frac{w - v}{c^2 + (w - v)^2} = \frac{1}{\sqrt{2\pi}} \frac{i}{w - v + ic}$$

whence

$$J\{f(x); v\} = \frac{1}{\sqrt{2\pi}} \frac{i}{w - v + ic}. \tag{6'}$$

iii) Let

$$f(x) = 1/\sqrt{|x|}. \tag{7}$$

Then

$$F(v) = \frac{1}{\sqrt{2\pi}} \int_{-\infty}^{\infty} \frac{\cos \alpha v}{\sqrt{|\alpha|}} d\alpha - \frac{i}{\sqrt{2\pi}} \int_{-\infty}^{\infty} \frac{\sin \alpha v}{\sqrt{|\alpha|}} d\alpha$$

$$= \frac{2}{\sqrt{2\pi}} \int_{0}^{\infty} \frac{\cos \alpha v}{\sqrt{|\alpha|}} d\alpha = \frac{1}{\sqrt{|v|}} \sqrt{\frac{2}{\pi}} \int_{0}^{\infty} \frac{\cos t}{\sqrt{t}} dt$$

and by (28) of Sec. 9.10, $F(v) = 1/\sqrt{|v|}$, or

$$\frac{1}{\sqrt{|v|}} = J\left\{\frac{1}{\sqrt{|x|}}; v\right\}. \tag{7'}$$

3. As we shall see in the next section, the following theorem concerning the Fourier transform is important for applications.

THEOREM 60. *Suppose $f(x)$ is defined in $(-\infty, \infty)$ and that its Fourier transform exists. Suppose further that all derivatives of $f(x)$ through order $r (r \geqq 1)$ exist in $(-\infty, \infty)$ and that those through order $r - 1$ are absolutely continuous in every finite interval and, finally, that*

$$0 = \lim_{|x|\to\infty} f(x) = \lim_{|x|\to\infty} f'(x) = \ldots = \lim_{|x|\to\infty} f^{(r-1)}(x). \tag{8}$$

Then

$$J\{f^{(r)}(x); v\} = (iv)^r J\{f(x); v\}. \tag{9}$$

Indeed by the theorem on integration by parts

$$\frac{1}{\sqrt{2\pi}} \int_{-n}^{n} f'(\alpha) e^{-iv\alpha} d\alpha = \frac{1}{\sqrt{2\pi}} \Big[f(\alpha) e^{-iv\alpha} \Big]_{\alpha=-n}^{\alpha=n} + iv \frac{1}{\sqrt{2\pi}} \int_{-n}^{n} f(\alpha) e^{-iv\alpha} d\alpha.$$

Now since $|e^{-iv\alpha}| = 1$, the first term on the right approaches zero as a limit by virtue of (8) and the limit of the second term is

$$iv\left[P.V.\frac{1}{\sqrt{2\pi}}\int_{-\infty}^{\infty} f(\alpha)e^{-iv\alpha}\,d\alpha\right],$$

whence

$$J\{f'(x); v\} = iv[J\{f(x); v\}]$$

Therefore (9) follows by induction.

4a. The Fourier transform provides a powerful method of finding an analytical expression for the solutions of some problems of mathematical physics. If we can operate formally on the two sides of the equations for the given problem with the Fourier transform and take into account the transform of the initial conditions and the limits, we may occasionally obtain the explicit expression for the Fourier transform of the solution. Then the formula of inversion (II) will give the explicit expression for the solution of the problem. The procedure itself, where it is legitimate in every respect, provides the proof of existence and uniqueness.

4b. The statements in 4a will be illustrated by an example. Consider the classical problem of small oscillations of a stretched, flexible string of infinite length which is oscillating about the initial rectilinear position.

If (x, y) denotes the coordinates of a point of the string, then $y = y(x, t)$ will denote the displacement of the point of the string from the x-axis at time t. Suppose

$$y(x, 0) = f(x), \quad (-\infty < x < \infty) \tag{10}$$

is the equation of the string at time $t = 0$,

$$\left.\frac{\partial y\,(x, t)}{\partial t}\right|_{t=0} = y(x, 0) = \varphi(x), \quad (-\infty < x < \infty) \tag{11}$$

is the transverse velocity of the point x of the string at time $t = 0$; further that $f(x)$, $f'(x)$, $\varphi(x)$, $\varphi'(x)$, $\varphi''(x)$ are continuous for every finite value of x and that

$$\lim_{|x|\to\infty} f(x) = 0, \quad \lim_{|x|\to\infty} \varphi(x) = 0. \tag{12}$$

If we assume that the transverse vibrations are very small compared to the length of the string, so that the change in the length of the string as it vibrates is negligible, it is easy to show (See, for example, Reddick and Miller, *Advanced Mathematics for Engineers*, p. 254) that $y(x, t)$ satisfies the equation

$$\frac{\partial^2 y}{\partial t^2} = V^2 \frac{\partial^2 y}{\partial x^2} \tag{13}$$

where V is the constant velocity of propagation. Moreover, for every t, we must have

$$(14_1) \qquad \lim_{|x| \to \infty} y(x, t) = 0, \qquad\qquad (14_2) \qquad \lim_{|x| \to \infty} \frac{\partial y(x, t)}{\partial x} = 0.$$

Assuming that a function $y(x, t)$ exists which satisfies (13), the initial conditions (10) and (11), and the conditions at infinity (14_1) and (14_2), multiplying (13) by e^{-ivx}, integrating with respect to x between $-\infty$ and $+\infty$, or, in other words, operating on (13) with the Fourier transform, we obtain formally

$$\int_{-\infty}^{\infty} \frac{\partial^2 y}{\partial t^2} e^{-ivx}\, dx = V^2 \int_{-\infty}^{\infty} \frac{\partial^2 y}{\partial x^2} e^{-ivx}\, dx,$$

and by (14_1), (14_2) and the theorem of Sec. 13.3

$$\frac{\partial^2}{\partial t^2} \int_{-\infty}^{\infty} e^{-ivx} y(x, t) dx = -V^2 v^2 \int_{-\infty}^{\infty} e^{-ivx} y(x, t) dt,$$

whence, writing

$$Y(v, t) = \frac{1}{\sqrt{2\pi}} \int_{-\infty}^{\infty} e^{-ivx} y(x, t) dx \tag{15}$$

we have

$$\int_{-\infty}^{\infty} e^{-ivx} \left[\frac{\partial^2 Y}{\partial t^2} + V^2 v^2 Y \right] dx = 0,$$

which is satisfied if $Y(v, t)$ is a solution of the ordinary differential equation of the second order:

$$\frac{\partial^2 Y}{\partial t^2} + V^2 v^2 Y = 0. \tag{16}$$

Denoting by $F(v)$ and $\Phi(v)$ the Fourier transforms of $f(x)$ and $\varphi(x)$, namely

$$F(v) = Y(v, 0) = \frac{1}{\sqrt{2\pi}}\int_{-\infty}^{\infty} e^{-ivx} f(x)dx \tag{17$_1$}$$

$$\Phi(v) = Y_t(v,0) = \frac{1}{\sqrt{2\pi}}\int_{-\infty}^{\infty} e^{-ivx}\varphi(x)dx, \tag{17$_2$}$$

the solution of (16) which satisfies the initial conditions (17_1) and (17_2) is given by

$$Y(v, t) = F(v) \cos (Vvt) + [\Phi(v)/Vv] [\sin (Vvt)],$$

or

$$Y(v, t) = \tfrac{1}{2}F(v) [e^{iVvt} + e^{-iVvt}] + [\Phi(v)/2iVv] [e^{iVvt} - e^{-iVvt}]. \tag{18}$$

By (18) and the Fourier formula of inversion, we derive from (15):

$$y(x, t) = \frac{1}{\sqrt{2\pi}}\int_{-\infty}^{\infty} e^{ivx}\, Y(v, t)dt$$

$$= \frac{1}{2}\left[\frac{1}{\sqrt{2\pi}}\int_{-\infty}^{\infty} F(v)\{e^{iv(x+Vt)} + e^{iv(x-Vt)}\}dv\right.$$

$$\left.+\frac{1}{V}\frac{1}{\sqrt{2\pi}}\int_{-\infty}^{\infty} \frac{\Phi(v)}{iv}\{e^{iv(x+Vt)} - e^{iv(x-Vt)}\}dv\right].$$

From (17_1) follows

$$f(x \pm Vt) = \frac{1}{\sqrt{2\pi}}\int_{-\infty}^{\infty} e^{iv(x\pm Vt)}F(v)dv,$$

and from (17_2)

$$\varphi(u) = \frac{1}{\sqrt{2\pi}}\int_{-\infty}^{\infty} e^{ivu}\, \Phi(v)\, dv,$$

and integrating on both sides of the last equation with respect to u between $x - Vt$ and $x + Vt$, we obtain

$$\int_{x-Vt}^{x+Vt} \varphi(u)du = \frac{1}{\sqrt{2\pi}}\int_{-\infty}^{\infty} \frac{\Phi(v)}{iv} [e^{iv(x+Vt)} - e^{iv(x-Vt)}]dv$$

whence, finally

$$(19)\quad y(x, t) = \tfrac{1}{2}[f(x + Vt) + f(x - Vt)] + \frac{1}{2V}\int_{x-Vt}^{x+Vt} \varphi(u)du.$$

It is easily verified that the function $Y(x, t)$ determined in this way satisfies the partial differential equation (13), the initial conditions (10) and (11) and the conditions (14_1) and (14_2) at infinity. (Cf. M. Picone, *Appunti di Analisi Superiore*, Napoli, 1940, p. 716, f.f.).

5. The so-called Fourier cosine and sine transforms can be defined analogously to the Fourier transform. If $f(x)$ is defined for $x > 0$, integrable in every finite interval interior to $(0, \infty)$ and defined in the interior of $(-\infty, 0)$ by the relation $f(-x) = f(x)$, $[f(-x) = -f(x)]$. Then if $f(x)$ is even

$$F(v) = \underset{(0,\,\infty)}{P.V.}\sqrt{\frac{2}{\pi}}\int_0^\infty f(\alpha)\cos v\alpha\, d\alpha$$

$$= \lim_{\substack{\sigma\to 0\\ n\to\infty}}\sqrt{\frac{2}{\pi}}\int_\sigma^n f(\alpha)\cos v\alpha\, d\alpha$$

and if $f(x)$ is odd

$$iF(v) = \underset{(0,\,\infty)}{P.V.}\sqrt{\frac{2}{\pi}}\int_0^\infty f(\alpha)\sin v\alpha\, d\alpha$$

$$= \lim_{\substack{\sigma\to 0\\ n\to\infty}}\sqrt{\frac{2}{\pi}}\int_\sigma^n f(\alpha)\sin\alpha\, d\alpha.$$

The two functions $F_c(v)$, $F_s(v)$ defined by the relations

$$(20_1)\qquad F_c(v) = \underset{(0,\,\infty)}{P.V.}\sqrt{\frac{2}{\pi}}\int_0^\infty f(\alpha)\cos v\alpha\, d\alpha;$$

$$(20_2)\qquad F_s(v) = \underset{(0,\,\infty)}{P.V.}\sqrt{\frac{2}{\pi}}\int_0^\infty f(\alpha)\sin v\alpha\, d\alpha$$

will be called respectively the Fourier cosine and sine transforms of $f(x)$.

In view of Sec. 13.1b, we have

THEOREM 61. If $h(x)$ and $l(x)$ are respectively the real and imaginary parts of $f(x) = h(x) + il(x)$, if $h(x)$ and $l(x)$ satisfy conditions (i), (ii), (iii), (iv) of Sec. 13.1b, if $\lim_{x\to+0} f(x) = f(0)$ in case of (20_1), or $\lim_{x\to+0} f(x) = f(0) = 0$ in case of (20_2), then (20_1) and (20_2) admit the following formulas of inversion (or reciprocity)

$$f(v) = \underset{(0,\,\infty)}{P.V.}\sqrt{\frac{2}{\pi}}\int_0^\infty F_c(\alpha)\cos v\alpha\, d\alpha; \tag{21$_1$}$$

$$f(v) = \underset{(0,\,\infty)}{P.V.}\sqrt{\frac{2}{\pi}}\int_0^\infty F(\alpha)\sin v\alpha\, d\alpha. \tag{21$_2$}$$

In Sec. 13.4 it was explained how certain problems may be solved by operating on the equation with the Fourier transform. In some cases it may be more convenient to operate with the cosine or sine transform.

6. Let $K(x, v)$ be a function of two variables x, v defined for $a \leqq x \leqq b$, $a \leqq v \leqq b$, $(a < b)$ where one or both of the numbers a and b may be infinite, and suppose that for the functions $f(x)$ of a certain class and for every value of v, the following integral exists

$$\int_a^b K(\alpha, v)\, f(\alpha) d\alpha.$$

Then the function $F(v)$ defined by

$$F(v) = \int_a^b K(\alpha, v) f(\alpha) d\alpha \tag{22}$$

will be called the transform of $f(x)$ with kernel $K(x, v)$. We shall call Fourier kernels those kernels $K(x, v)$ which have the property that there exists a corresponding kernel, $H(x, v)$, called the inverse kernel, such that (22) can be inverted by

$$f(x) = \int_a^b H(x, v) F(v) dv. \tag{23}$$

Under these conditions the transform (22) will be called a Fourier transform.

The results of Sec. 13.1 show that if $a = -\infty$, $b = \infty$, then $k(x, v) = e^{-ivx}/\sqrt{2\pi}$ is a Fourier kernel and the inverse kernel $H(x, v)$ is given by $H(x, v) = e^{ivx}/\sqrt{2\pi}$; or $H(x, v) = K(x, -v)$.

Similarly the results of Sec. 13.5 show that if $a = 0$, $b = \infty$,. $K(x, v) = \sqrt{2/\pi} \cos vx$ or $\sqrt{2/\pi} \sin vx$ is a Fourier kernel and $H(x, v) = K(x, v)$.

If $a = 0$, $b = \infty$ and $K(x, v) = e^{-xv}$, the transform (22) is the so-called Laplace transform (Cf., e.g., G. Sansone, *Equazioni Differenziali nel Campo Reale*, II, Bologna, 1948, pp. 227—249). If $a = 0$, $b = \infty$ and $K(x, v) = v^{x-1}$, the transform (22) is called the Mellin transform. For an extensive treatment of these and other transforms the reader is referred to treatises on Fourier Transforms (cf., e.g. Ian N. Sneddon, *Fourier Transforms*, New York, 1951, p.p. IX + 542.)

CHAPTER III

Expansions in Series of Legendre Polynomials and Spherical Harmonics

1. Legendre Polynomials

1. In the study of the attraction of spheroids and planetary motion, Legendre [68] was led to the consideration of the series expansion of the function $1/r$ defined by the relation

$$1/r = (1 - 2\varrho \cos \gamma + \varrho^2)^{-\frac{1}{2}}. \tag{1}$$

The remainder of this section will be devoted to deriving the expansion.

If we assume $|\varrho| < 1$, then by the binomial expansion we have

$$\begin{aligned} 1/r &= (1 - \varrho e^{i\gamma})^{-\frac{1}{2}} (1 - \varrho e^{-i\gamma})^{-\frac{1}{2}} \\ &= \sum_{n=0}^{\infty} (-1)^n \binom{-\frac{1}{2}}{n} \varrho^n e^{in\gamma} \sum_{n=0}^{\infty} (-1)^n \binom{-\frac{1}{2}}{n} \varrho^n e^{-in\gamma}, \end{aligned}$$

whence, multiplying the two series according to Cauchy's rule,

$$\begin{aligned} \frac{1}{r} = \sum_{n=0}^{\infty} \varrho^n \sum_{r=0}^{[n/2]} \binom{-\frac{1}{2}}{r} \binom{-\frac{1}{2}}{n-r} [(-1)^r (-1)^{n-r} e^{ir\gamma} e^{-i(n-r)\gamma} \\ + (-1)^r (-1)^{n-r} e^{-ir\gamma} e^{i(n-r)\gamma}] \end{aligned} \tag{1}$$

and consequently

$$\boxed{\frac{1}{r} = (1 - 2\varrho \cos \gamma + \varrho^2)^{-\frac{1}{2}} = \sum_{n=0}^{\infty} \varrho^n P_n (\cos \gamma)} \tag{2}$$

where

$$P_n(\cos \gamma) = (-1)^n \left[\binom{-\frac{1}{2}}{n} 2 \cos n\gamma \right. \tag{3}$$

$$\left. + \binom{-\frac{1}{2}}{n-1} \binom{-\frac{1}{2}}{1} 2 \cos (n-2)\gamma + \binom{-\frac{1}{2}}{n-2} \binom{-\frac{1}{2}}{2} 2 \cos (n-4)\gamma + \ldots \right]$$

or

$$(3') \quad \boxed{\begin{aligned} P_n(\cos\gamma) = &\frac{1\times3\times\ldots\times(2n-1)}{2\times4\times\ldots\times(2n)}\,2\cos n\gamma \\ &+\frac{1\times3\times\ldots\times(2n-3)}{2\times4\times\ldots\times(2n-2)}\,\frac{1}{2}\,2\cos(n-2)\gamma \\ &+\frac{1\times3\times\ldots\times(2n-5)}{2\times4\times\ldots\times(2n-4)}\,\frac{1\times3}{2\times4}\,2\cos(n-4)\gamma+\ldots \end{aligned}}$$

where, in the case n is even, the term which does not contain $\cos\gamma$ does not contain the factor 2.

The series (2), for fixed γ, converges for $|\varrho| < 1$, and since it is a power series it can be differentiated term by term with respect to ϱ.

In addition to the expansion (2), the expansion of $(1-\varrho^2)(1-2\varrho\cos\gamma+\varrho^2)^{-3/2}$ will also figure in the applications. We have

$$\frac{1-\varrho^2}{(1-2\varrho\cos\gamma+\varrho^2)^{3/2}} = \frac{1}{(1-2\varrho\cos\gamma+\varrho^2)^{1/2}}$$

$$+2\varrho\frac{d}{d\varrho}\frac{1}{(1-2\varrho\cos\gamma+\varrho^2)^{-1/2}} = \sum_{n=0}^{\infty}\varrho^n P_n(\cos\gamma)+2\varrho\sum_{n=0}^{\infty}n\varrho^{n-1}P(\cos\gamma)$$

and therefore

$$(4) \quad \boxed{\frac{1-\varrho^2}{(1-2\varrho\cos\gamma+\varrho^2)^{3/2}} = 1+\sum_{n=0}^{\infty}(2n+1)\varrho^n P_n(\cos\gamma)}$$

where, for fixed γ, the series on the right converges uniformly for ϱ varying in an interval interior to $(-1, 1)$.

2. If in (3′) we set $\cos\gamma = x$, and express the cosines of the multiples of γ in terms of powers of the $\cos\gamma$, we obtain $P_n(x)$ expressed as a polynomial in x. This expression can be derived easily as follows:

Let $2|x|+|\varrho| < 1$ (whence also $|\varrho(2x-\varrho)| < 1$]). Then

$$\begin{aligned}(1-2\varrho x+\varrho^2)^{-1/2} &= [1-\varrho(2x-\varrho)]^{-1/2}\\ &= \sum_{n=0}^{\infty}(-1)^n\binom{-\frac{1}{2}}{n}\varrho^n(2x-\varrho)^n\\ &= \sum_{n=0}^{\infty}\frac{1}{2^n}\frac{1\times3\times5\times\ldots\times(2n-1)}{n!}\varrho^n(2x-\varrho)^n\\ &= \sum_{n=0}^{\infty}\frac{1}{2^n}\frac{1\times3\times5\times\ldots\times(2n-1)}{n!}\varrho^n\sum_{r=0}^{\infty}(-1)^r\binom{n}{r}\varrho^r2^{n-r}x^{n-r}.\end{aligned}$$

The series of absolute values of this double series represents the series expansion of $[1-|\varrho|(2|x|+|\varrho|)]^{-1/2}$ and by our hypotheses the series of absolute values is convergent. Therefore it is permissible to collect the terms in ϱ^n to obtain

$$\begin{aligned}(1-2\varrho x+\varrho^2)^{-1/2} = 1+\sum_{n=0}^{\infty}\varrho^n\Bigg[&\frac{1\times3\times5\times\ldots\times(2n-1)}{n!}x^n\\ &-\frac{1}{2}\frac{1\times3\times5\times\ldots\times(2n-3)}{(n-1)!}\frac{n-1}{1}x^{n-2}\\ &+\frac{1}{2^2}\frac{1\times3\times5\times\ldots\times(2n-5)}{(n-2)!}\frac{(n-2)(n-3)}{1\times2}x^{n-4}\\ &-\frac{1}{2^3}\frac{1\times3\times5\times\ldots\times(2n-7)}{(n-3)!}\frac{(n-3)(n-4)(n-5)}{1\times2\times3}x^{n-6}+\ldots\Bigg]\end{aligned}$$

whence by (2)

$$(5_1)\qquad \boxed{\begin{aligned}P_n(x)=\frac{1\times3\times5\times\ldots\times(2n-1)}{n!}\Bigg[&x^n-\frac{n(n-1)}{2(2n-1)}x^{n-2}\\ &+\frac{n(n-1)(n-2)(n-3)}{2\times4\times(2n-1)(2n-3)}x^{n-4}-\ldots\Bigg]\end{aligned}}$$

$$(5_2)\qquad \boxed{P_0(x)=1}$$

The polynomials $P_n(x)$ defined by (5_1), and (5_2) are called Legendre polynomials. In particular

$$(5_3)\quad \begin{cases} P_0 = 1;\; P_1 = x;\; P_2 = \dfrac{3}{2}x^2 - \dfrac{1}{2};\; P_3 = \dfrac{5}{2}x^3 - \dfrac{3}{2}x; \\ P_4 = \dfrac{35}{8}x^4 - \dfrac{30}{8}x^2 + \dfrac{3}{8};\quad P_5 = \dfrac{63}{8}x^5 - \dfrac{70}{8}x^3 + \dfrac{15}{8}x; \\ P_6 = \dfrac{231}{16}x^6 - \dfrac{315}{16}x^4 + \dfrac{105}{16}x^2 - \dfrac{5}{16}; \\ P_7 = \dfrac{429}{16}x^7 - \dfrac{693}{16}x^5 + \dfrac{315}{16}x^3 - \dfrac{35}{16}x. \end{cases}$$

The graphs of these eight polynomials are given below.

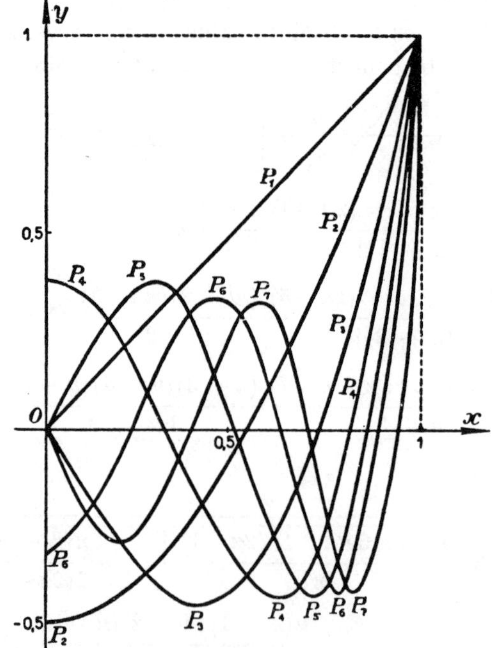

Fig. 12.

From (5_1) follows

$$(6_1)\quad \boxed{P_{2n}(0) = (-1)^n \frac{1\times 3\times 5\times \ldots \times (2n-1)}{2\times 4\times 6\times \ldots \times (2n)} = \binom{-\frac{1}{2}}{n}}$$

$$(n = 1, 2, \ldots);$$

$$(6_2) \qquad \boxed{P_{2n+1}(0) = 0,} \qquad (n = 0, 1, 2, \ldots);$$

$$(6_3) \qquad \boxed{P_{2n}(-x) = P_{2n}(x), \quad P_{2n+1}(-x) = -P_{2n+1}(x)}$$

$$(n = 0, 1, 2, \ldots).$$

The Legendre polynomials are a particular case of ultraspherical polynomials obtained from the series expansion of $(1 - 2\varrho \cos \gamma + {} + \varrho^2)^{-\lambda}$ (cf. e.g. G. Sansone: *Equazioni Differenziali nel Campo reale*, I, 2nd ed., Bologna, 1949, pp. 162—164, [98b]).

3. From the equalities

$$1\times 2\times 3\times 4\times \ldots \times (2n-1)2n = 1\times 3\times 5\times \ldots \times (2n-1)2\times 4\times 6\times \ldots \times 2n$$

$$n!(n+1)(n+2)\ldots(2n) = 1\times 3\times 5\times \ldots \times (2n-1)2^n n!$$

we have

$$1\times 3\times 5\times \ldots \times (2n-1) = \frac{2n(2n-1)\ldots(n+2)(n+1)}{2^n}$$

whence from (5_1)

$$2^n n!\, P_n(x) = 2n(2n-1)(2n-2)\ldots(n+1)\left[x^n - \frac{n(n-1)}{2(2n-1)}x^{n-2} + \frac{n(n-1)(n-2)(n-3)}{2\times 4\times (2n-1)(2n-3)}x^{n-4} - \ldots\right].$$

Consequently every term on the right side is of the form

$$\pm\frac{2n(2n-1)\ldots(n+1)n(n-1)\ldots(n-2m+1)}{2\times 4\times \ldots \times (2m)(2n-1)(2n-3)\ldots(2n-2m+1)}x^{n-2m}$$

$$= \pm\frac{2n(2n-1)(2n-2)\ldots(2n-2m+1)(2n-2m)(2n-2m-1)\ldots}{2\times 4\times 6\times \ldots \times (2m)(2n-1)(2n-3)\ldots}\,\frac{\ldots(n-2m+1)}{\ldots(2n-2m+1)}x^{n-2m}$$

$$= \pm\frac{n(n-1)\ldots(n-m+1)}{1\times 2\times \ldots \times m}(2n-2m)(2n-2m-1)\ldots \ldots(n-2m+1)x^{n-2m}$$

$$= \pm\binom{n}{m}\frac{d^n}{dx^n}x^{2(n-m)};$$

whence

$$2^n n!\,P_n(x) = \sum_{m=0}^{n} \binom{n}{m} (-1)^m \frac{d^n}{dx^n} x^{2(n-m)}$$

$$= \frac{d^n}{dx^n} \sum_{m=0}^{n} \binom{n}{m} (-1)^m x^{2(n-m)} = \frac{d^n}{dx^n} (x^2 - 1)^n,$$

and finally the familiar formula of Rodrigues [95]

$$(7) \qquad \boxed{P_n(x) = \frac{1}{2^n n!} \frac{d^n}{dx^n} (x^2 - 1)^n}.$$

4. We shall now derive a bound on the derivatives of the polynomials $P_n(x)$. Later in Sec. 17.6 this result will be improved. By the results of Sec. 1.3 we have

$$P_n(x) = \sum_{m=0}^{[n/2]} a_m x^{n-2m},$$

where

$$a_m = (-1)^m \frac{1}{2^n} \binom{n}{m} \binom{2n-2m}{n},$$

and by the identity

$$2^{n-1} = \binom{n}{1} + \binom{n}{3} + \ldots = 1 + \binom{n}{2} + \binom{n}{4} + \ldots$$

we obtain

$$|a_m| \leqq \frac{1}{2^n} 2^{n-1} 2^{2n-2m-1} \leqq 2^{2n-2}.$$

The number of terms of $P_n(x)$ for $n \geqq 1$ is $[n/2] + 1 \leqq n$. Also the kth derivative of every term of $P_n(x)$ will have a numerical factor not exceeding $n(n-1)\ldots(n-k+1)$. Therefore for $|x| \leqq 1$, $n \geqq 1$

$$(8_1) \qquad |P_n(x)| \leqq 2^{2n-2} n^2$$

$$(8_2) \qquad |P_n^{(k)}(x)| \leqq 2^{2n-2}(n-1)(n-2)\ldots(n-k+1), \quad (k = 2, 3, \ldots, n).$$

2. *Schläfli's Integral Formula*

We recall that if $f(z)$ is analytic in a domain D, if x is a point of D and C is a regular, simple, closed curve in D, surrounding x, then

$$f^{(n)}(x) = \frac{n!}{2\pi i}\oint_C \frac{f(t)}{(t-x)^{n+1}}\,dt.$$

Now let $f(z) = (z^2 - 1)^n$, $n = 0, 1, 2, \ldots$ Since $f(z)$ is analytic everywhere in the plane, we have by (7)

$$P_n(x) = \frac{1}{2\pi i}\int_C \frac{(t^2-1)^n}{2^n(t-x)^{n+1}}\,dt \tag{9}$$

where C is any regular, simple, closed curve surrounding x.

3. *Differential Equations of Legendre Polynomials*

We shall now show that the polynomial $P_n(x)$ is a solution of the differential equation

$$(1-x^2)\frac{d^2P_n}{dx^2} - 2x\frac{dP_n}{dx} + n(n+1)P_n = 0 \tag{10}$$

By virtue of Schläfli's formula (9), the left side of (10) can be written

$$(1-x^2)\frac{d^2P_n}{dx^2} - 2x\frac{dP_n}{dx} + n(n+1)P_n$$

$$= \frac{n+1}{2\pi i}\oint_C \frac{(t^2-1)^n dt}{2^n(t-x)^{n+3}}[n(t-x)^2 - 2(t-x)x + (n+2)(1-x^2)]$$

$$= \frac{n+1}{2\pi i}\oint_C \frac{(t^2-1)^n dt}{2^n(t-x)^{n+3}}[-(n+2)(t^2-1) + 2t(t-x)(n+1)]$$

$$= \frac{n+1}{2\pi i\cdot 2^n}\oint_C \frac{d}{dt}\left\{\frac{(t^2-1)^{n+1}}{(t-x)^{n+2}}\right\}dt.$$

Since n is an integer and the function $(t^2-1)^{n+1}(t-x)^{-n-2}$

returns to its original value as t describes C, the last integral clearly vanishes.

(10) can also be written in the form

$$\frac{d}{dx}\left[(1-x^2)\frac{dP_n}{dx}\right]+n(n+1)P_n=0. \tag{10'}$$

4. Recurrence Formulas for Legendre Polynomials

a. If C is a regular, simple, closed curve surrounding x, then

$$P_n(x)=\frac{1}{2^{n+1}\pi i}\oint_C\frac{(t^2-1)^n}{(t-x)^{n+1}}dt,\quad P_n'(x)=\frac{n+1}{2^{n+1}\pi i}\oint_C\frac{(t^2-1)^n}{(t-x)^{n+2}}dt.$$

Now

$$\frac{d}{dt}\frac{(t^2-1)^{n+1}}{(t-x)^{n+1}}=\frac{2(n+1)t(t^2-1)^n}{(t-x)^{n+1}}-\frac{(n+1)(t^2-1)^{n+1}}{(t-x)^{n+2}},$$

whence, integrating along C and dividing by $n+1$ we have

$$0=2\oint_C\frac{t(t^2-1)^n}{(t-x)^{n+1}}dt-\oint_C\frac{(t^2-1)^{n+1}}{(t-x)^{n+2}}dt;$$

which gives

$$\begin{aligned}\frac{1}{2^{n+2}\pi i}\oint_C\frac{(t^2-1)^{n+1}}{(t-x)^{n+2}}dt&=\frac{1}{2^{n+1}\pi i}\oint_C\frac{t(t^2-1)^n}{(t-x)^{n+1}}dt\\&=\frac{1}{2^{n+1}\pi i}\oint_C\frac{[(t-x)+x](t^2-1)^n}{(t-x)^{n+1}}dt\\&=\frac{1}{2^{n+1}\pi i}\oint_C\frac{(t^2-1)^n}{(t-x)^n}dt+\frac{x}{2^{n+1}\pi i}\oint_C\frac{(t^2-1)^n}{(t-x)^{n+1}}dt.\end{aligned}$$

Therefore

$$P_{n+1}(x)=\frac{1}{2^{n+1}\pi i}\oint_C\frac{(t^2-1)^n}{(t-x)^n}dt+xP_n(x), \tag{11}$$

whence, differentiating with respect to x,

$$P_{n+1}'(x)=\frac{n}{2^{n+1}\pi i}\oint_C\frac{(t^2-1)^n}{(t-x)^{n+1}}dt+P_n(x)+xP_n'(x)$$

and consequently

$$\boxed{P'_{n+1}(x) - xP'_n(x) = (n+1)P_n(x)}, \tag{12}$$

which is the first recurrence formula desired.

b. Performing the indicated differentiation under the integral sign in the identity

$$\oint_C \frac{d}{dt}\left[\frac{t(t^2-1)^n}{(t-x)^n}\right] dt = 0$$

we have

$$\oint_C \frac{(t^2-1)^n}{(t-x)^n}\,dt + 2n\oint_C \frac{t^2(t^2-1)^{n-1}}{(t-x)^n}\,dt - n\oint_C \frac{t(t^2-1)^n}{(t-x)^{n+1}}\,dt = 0.$$

But

$$\begin{aligned} 2n\oint_C \frac{t^2(t^2-1)^{n-1}}{(t-x)^n}\,dt &= 2n\oint_C \frac{(t^2-1)^n}{(t-x)^n}\,dt \\ + 2n\oint_C \frac{(t^2-1)^{n-1}}{(t-x)^n}\,dt &= 2n\oint_C \frac{(t^2-1)^n}{(t-x)^n}\,dt + 2^n\pi i\cdot 2nP_{n-1}(x), \\ -n\oint_C \frac{t(t^2-1)^n}{(t-x)^{n+1}}\,dt &= -n\oint_C \frac{(t^2-1)^n}{(t-x)^n}\,dt \\ -nx\oint_C \frac{(t^2-1)^n}{(t-x)^{n+1}}\,dt &= -n\oint_C \frac{(t^2-1)^n}{(t-x)^n}\,dt - 2^{n+1}\pi i\cdot nxP_n(x). \end{aligned}$$

Therefore

$$(n+1)\oint_C \frac{(t^2-1)^n}{(t-x)^n}\,dt + 2^n\pi i\cdot 2nP_{n-1}(x) - 2^{n+1}\pi i\cdot nxP_n(x) = 0,$$

whence by (11)

$$\begin{aligned} (n+1)2^{n+1}\pi i[P_{n+1}(x) - xP_n(x)] \\ + 2^n\pi i\cdot 2nP_{n-1}(x) - 2^{n+1}\pi i\cdot nxP_n(x) = 0, \\ (n+1)[P_{n+1}(x) - xP_n(x)] + nP_{n-1}(x) - nxP_n(x) = 0, \end{aligned}$$

and finally the recurrence formula between three consecutive

Legendre polynomials

$$(13)\quad \boxed{(n+1)P_{n+1}(x) - (2n+1)xP_n(x) + nP_{n-1}(x) = 0}\,.$$

From this formula follows readily

$$P_n(1) = 1, \qquad P_n(-1) = (-1)^n$$

(cf. Sec. 6.2 (20)).

c. Differentiating (13) we have

$$(n+1)P'_{n+1} - (2n+1)P_n - (2n+1)xP'_n + nP'_{n-1} = 0,$$
$$(n+1)[P'_{n+1} - xP'_n] - n[xP'_n - P'_{n-1}] - (2n+1)P_n = 0.$$

and by (12)

$$(n+1)^2 P_n - n[P'_{n+1} - (n+1)P_n - P'_{n-1}] - (2n+1)P_n = 0,$$
$$n(2n+1)P_n = n[P'_{n+1} - P'_{n-1}],$$

whence, finally

$$(14)\quad \boxed{P'_{n+1}(x) - P'_{n-1}(x) = (2n+1)P_n(x)}\,.$$

d. Substituting for n in (14) the values $n-1$, $n-2$, . . ., 2, 1, 0 and adding we obtain $[P'_1 = 1,\ P'_0 = 0,\ P_0 = 1]$

$$(15)\quad \boxed{\frac{dP_{n+1}(x)}{dx} + \frac{dP_n(x)}{dx} = P_0(x) + 3P_1(x) + 5P_2(x) + \cdots + (2n+1)P_n(x)}\,.$$

e. Subtracting (14) from (12) we have

$$(16)\quad \boxed{xP'_n(x) - P'_{n-1}(x) = nP_n(x)}\,.$$

f. Substituting $n-1$ for n in (12) gives

$$-P'_n + xP'_{n-1} = -nP_{n-1}.$$

Multiplying (16) through by x we have

$$x^2P'_n - xP'_{n-1} = nxP_n$$

whence adding

$$(17) \qquad \boxed{(x^2 - 1)P_n'(x) = nxP_n(x) - nP_{n-1}(x)}.$$

Formulas (12), (13), (14), (15), (16), (17) are called the recurrence formulas for Legendre polynomials.

5. *The Christoffel Formula of Summation*

We wish to prove the identity

$$(18) \quad \boxed{\sum_{r=0}^{n}(2r+1)P_r(x)P_r(y) = (n+1)\frac{P_n(x)P_{n+1}(y) - P_{n+1}(x)P_n(y)}{y - x}}.$$

From (13), substituting r for n and multiplying by $P_r(y)$

$$(r+1)P_{r+1}(x)P_r(y) - (2r+1)xP_r(x)P_r(y) + rP_{r-1}(x)P_r(y) = 0.$$

Interchanging x and y

$$(r+1)P_{r+1}(y)P_r(x) - (2r+1)yP_r(x)P_r(y) + rP_{r-1}(y)P_r(x) = 0$$

subtracting

$$\begin{aligned}(2r+1)(y-x)P_r(x)P_r(y) = (r+1)[P_{r+1}(y)P_r(x) - P_{r+1}(x)P_r(y)]\\ + r[P_{r-1}(y)P_r(x) - P_{r-1}(x)P_r(y)].\end{aligned}$$

Setting $r = 0, 1, 2, \ldots, n$ we obtain

$$\begin{aligned}(y - x)P_0(x)P_0(y) &= [P_1(y)P_0(x) - P_1(x)P_0(y)]\\ 3(y - x)P_1(x)P_1(y) &= 2[P_2(y)P_1(x) - P_2(x)P_1(y)]\\ &\quad + 1[P_0(y)P_1(x) - P_0(x)P_1(y)]\\ &\cdots\cdots\cdots\cdots\cdots\\ (2n+1)(y-x)P_n(x)P_n(y) &= (n+1)[P_{n+1}(y)P_n(x) - P_{n+1}(x)P_n(y)]\\ &\quad + n[P_{n-1}(y)P_n(x) - P_{n-1}(x)P_n(y)]\end{aligned}$$

whence (18) follows by addition.

Letting $y = 1$ in (18) and using (15), we obtain in particular

$$(18') \qquad \boxed{(n+1)\frac{P_n(x) - P_{n+1}(x)}{1 - x} = \frac{dP_{n+1}(x)}{dx} + \frac{dP_n(x)}{dx}}.$$

6. Laplace's Integral Formula for $P_n(x)$

1. Suppose $x \neq \pm 1$ and let C in Schläfli's formula (9) be a circle with center at x and radius $|x^2 - 1|^{1/2}$. It follows that

$$t = x + (x^2 - 1)^{1/2} e^{i\varphi}$$

where $-\pi \leqq \varphi \leqq \pi$ and $(x^2 - 1)^{1/2}$ denotes either branch of the function $\sqrt{x^2 - 1}$.

Then

$$(t - x)^{n+1} = (x^2 - 1)^{(n+1)/2} e^{i(n+1)\varphi},$$

$$\begin{aligned} t^2 - 1 &= (t+1)(t-1) \\ &= [(x+1) + (x^2-1)^{1/2} e^{i\varphi}][(x-1) + (x^2-1)^{1/2} e^{i\varphi}] \\ &= (x^2-1)^{1/2} e^{i\varphi}[2x + (x^2-1)^{1/2}\{e^{i\varphi} + e^{-i\varphi}\}] \\ &= 2(x^2-1)^{1/2} e^{i\varphi}[x + (x^2-1)^{1/2} \cos\varphi]; \end{aligned}$$

$$(t^2 - 1)^n = 2^n (x^2 - 1)^{n/2} e^{in\varphi}[x + (x^2 - 1)^{1/2} \cos\varphi]^n;$$

$$dt = (x^2 - 1)^{1/2} e^{i\varphi}\, i\, d\varphi,$$

whence

$$P_n(x) = \frac{1}{2\pi} \int_{-\pi}^{\pi} [x + (x^2 - 1)^{\frac{1}{2}} \cos\varphi]^n \, d\varphi,$$

and, finally, Laplace's formula

$$P_n(x) = \frac{1}{\pi} \int_0^{\pi} [x + (x^2 - 1)^{\frac{1}{2}} \cos\varphi]^n \, d\varphi \tag{19}$$

where the value of the integral is independent of the choice of the branch of $(x^2 - 1)^{1/2}$.

By continuity, (19) holds also for $x = \pm 1$.

2. From (19) for $x = \pm 1$, we obtain, as we have already observed in Sec. 4, b,

$$P_n(1) = 1, \quad P_n(-1) = (-1)^n. \tag{20}$$

For x real, $-1 < x < 1$, we have

$$| x + (x^2 - 1)^{1/2} \cos\varphi | = | x + i\sqrt{1-x^2} \cos\varphi |$$
$$= | x^2 + (1 - x^2) \cos^2\varphi |^{\frac{1}{2}} \leqq 1;$$

whence, except for $\varphi = 0$ or π, we have

$$| x + (x^2 - 1)^{1/2} \cos\varphi |^n < 1$$

and therefore by (19)

$$\boxed{| P_n(x) | < 1 \text{ for } -1 < x < 1}. \tag{21}$$

This bound on $P_n(x)$ can be improved as follows. From (19) we have, for $0 < \varepsilon < \pi$,

$$P_n(x) = \frac{1}{\pi}\int_0^{\varepsilon} [x + i(1 - x^2)^{1/2} \cos\varphi]^n \, d\varphi$$
$$+ \frac{1}{\pi}\int_{\varepsilon}^{\pi-\varepsilon} [x+i(1-x^2)^{1/2} \cos\varphi]^n \, d\varphi + \frac{1}{\pi}\int_{\pi-\varepsilon}^{\pi} [x+i(1-x^2)^{1/2} \cos\varphi]^n \, d\varphi.$$

Now

$$\left| \frac{1}{\pi}\int_0^{\varepsilon} [x + i(1 - x^2)^{1/2} \cos\varphi]^n \, d\varphi \right| < \frac{\varepsilon}{\pi},$$
$$\left| \frac{1}{\pi}\int_{\pi-\varepsilon}^{\pi} [x + i(1 - x^2)^{1/2} \cos\varphi]^n \, d\varphi \right| < \frac{\varepsilon}{\pi};$$

and for φ varying in $[\varepsilon, \pi - \varepsilon]$,

$$| x + i(1 - x^2)^{1/2} \cos\varphi | = | x^2 + (1 - x^2) \cos^2\varphi |^{1/2}$$
$$= | 1 - (1 - x^2) \sin^2\varphi |^{1/2} \leqq [1 - (1 - x^2) \sin^2\varepsilon]^{1/2},$$

whence

$$\left| \frac{1}{\pi}\int_{\varepsilon}^{\pi-\varepsilon} [x+i(1-x^2)^{1/2} \cos\varphi]^n \, d\varphi \right| \leqq \frac{1}{\pi}[1 - (1-x^2)\sin^2\varepsilon]^{n/2}(\pi - 2\varepsilon).$$

But for

$$-1 + \eta \leqq x \leqq 1 - \eta \qquad [1 > \eta > 0]$$

we have

$$1 - x^2 \geqq 1 - (1 - \eta)^2 = \eta(2 - \eta) > \eta,$$

whence

$$| P_n(x) | < \frac{2\varepsilon}{\pi} + \frac{\pi - 2\varepsilon}{\pi} [1 - \eta \sin^2 \varepsilon]^{n/2}.$$

From this it follows that for a fixed $\sigma > 0$, there exists an integer $n_0 > 0$ such that for $n > n_0$, and for all x in $(-1 + \eta,\ 1 - \eta)$, we have

$$| P_n(x) | < \sigma. \tag{22}$$

In Sec. 10 this property will figure more prominently; for the present it will suffice to observe that at every point x interior to $[-1, 1]$

$$\lim_{n\to\infty} P_n(x) = 0, \qquad (-1 < x < 1)$$

and that $P_n(x)$ tends uniformly to zero in every interval $[-1 + \eta, 1 - \eta]$, interior to $[-1, 1]$.

7. *Mehler's Formulas*

LEMMA. If a and b are real or complex constants, $a \neq 0$, $b \neq 0$, and if $| b/a | < 1$, when b/a is real, then

$$\int_0^\pi \frac{d\varphi}{a + b \cos \varphi} = \frac{\pi}{\sqrt{a^2 - b^2}}, \tag{23}$$

where the sign of the radical is chosen so that

$$| (-a + \sqrt{a^2 - b^2})/b | < 1.$$

Proof. $bz^2 + 2az + b = b(z - \alpha)(z - \beta)$ where $\alpha = (-a - \sqrt{a^2 - b^2})/b$, $\beta = (-a + \sqrt{a^2 - b^2})/b$, $\alpha\beta = 1$, and the sign of the radical is chosen so that $| \alpha | > 1$, $| \beta | < 1$. This is always possible since otherwise we would have $\alpha = e^{-i\vartheta}$, $\beta = e^{+i\vartheta}$, whence $-b/a = 2/(e^{-i\vartheta} + e^{i\vartheta}) = 1/\cos \vartheta$, and then $| b/a |$ would be real and greater than or equal to 1, contrary to hypothesis.

Thus

$$\frac{1}{a+b\cos\varphi}=\frac{2}{2a+b(e^{i\varphi}+e^{-i\varphi})}=\frac{2e^{i\varphi}}{be^{2i\varphi}+2ae^{i\varphi}+b}$$

$$=-\frac{2e^{i\varphi}}{b(\alpha-e^{i\varphi})(e^{i\varphi}-\beta)}=\frac{2}{b(\beta-\alpha)}\left(\frac{\alpha}{\alpha-e^{i\varphi}}+\frac{\beta}{e^{i\varphi}-\beta}\right)$$

$$=\frac{1}{\sqrt{a^2-b^2}}\left[\frac{1}{1-\beta e^{i\varphi}}+\frac{\beta e^{-i\varphi}}{1-\beta e^{-i\varphi}}\right]$$

$$=\frac{1}{\sqrt{a^2-b^2}}\left[\beta e^{-i\varphi}\sum_{n=0}^{\infty}\beta^n e^{-in\varphi}+\sum_{n=0}^{\infty}\beta^n e^{in\varphi}\right],$$

$$\frac{1}{a+b\cos\varphi}=\frac{1}{\sqrt{a^2-b^2}}\left[1+2\sum_{n=1}^{\infty}\beta^n\cos n\varphi\right],$$

where the series in parentheses converges uniformly for φ in $[0, \pi]$. This justifies integrating term by term on the right side whence, integrating both sides between 0 and π, and observing that

$$\int_0^\pi \cos n\varphi\, d\varphi = 0 \quad \text{for} \quad n = 1, 2, \ldots$$

we obtain (23).

Writing

$$a = 2[1-\varrho(\mu+\mu')+\varrho^2],\ b = 2\varrho(\mu-\mu')$$

and determining ξ so that

$$2\xi = \mu+\mu'-(\mu-\mu')\cos\varphi \tag{24}$$

where

$$1 \geqq \mu' > \mu \geqq -1, \quad 0 \leqq \varrho < 1;$$

we have

$$a+b\cos\varphi = 2[1-2\varrho\xi+\varrho^2],$$

$$a+b = 2[1-2\varrho\mu'+\varrho^2], \qquad a-b = 2(1-2\varrho\mu+\varrho^2)$$

whence by (23)

$$\int_0^\pi \frac{d\varphi}{1-2\varrho\xi+\varrho^2} = \frac{\pi}{\sqrt{(1-2\varrho\mu+\varrho^2)(1-2\varrho\mu'+\varrho^2)}}. \tag{25}$$

Clearly

$$2(\xi - \mu) = (\mu' - \mu)(1 + \cos\varphi),$$
$$2(\mu' - \xi) = (\mu' - \mu)(1 - \cos\varphi),$$
$$4(\xi - \mu)(\mu' - \xi) = (\mu' - \mu)^2 \sin^2\varphi,$$
$$(\mu' - \mu)\sin\varphi = 2\sqrt{(\mu' - \xi)(\xi - \mu)}.$$

By (24) ξ decreases from μ' to μ as φ varies from 0 to π. Therefore, if we assume ξ to be the independent variable on the left side of (25), and observe that

$$(\mu' - \mu)\cos\varphi\, d\varphi = \frac{\mu' + \mu - 2\xi}{\sqrt{(\mu'-\xi)(\xi-\mu)}}\, d\xi = \frac{(\mu - \mu')\cos\varphi}{\sqrt{(\mu' - \xi)(\xi - \mu)}}\, d\xi$$

we obtain from (25)

$$(26)\quad \int_{\mu}^{\mu'} \frac{d\xi}{\sqrt{(\mu' - \xi)(\xi - \mu)(1 - 2\varrho\xi + \varrho^2)}} = \frac{\pi}{\sqrt{(1 - 2\varrho\mu + \varrho^2)(1 - 2\varrho\mu' + \varrho^2)}}.$$

If we let $\mu' = 1$, we get

$$\int_{\mu}^{1} \frac{(1 - \varrho)d\xi}{\sqrt{(1 - \xi)(\xi - \mu)(1 - 2\varrho\xi + \varrho^2)}} = \frac{\pi}{\sqrt{1 - 2\varrho\mu + \varrho^2}}.$$

If we also let $\mu = \cos\gamma$, $0 < \gamma \leqq \pi$, and $\xi = \cos\varphi$ we have

$$\int_0^{\gamma} \frac{(1-\varrho)\sin\varphi d\varphi}{(1-2\varrho\cos\varphi+\varrho^2)\sqrt{(\cos\varphi-\cos\gamma)(1-\cos\varphi)}} = \frac{\pi}{\sqrt{1-2\varrho\cos\gamma+\varrho^2}},$$

whence, by (2) of Sec. 1.1, follows

$$(27)\quad \frac{2}{\pi}\int_0^{\gamma} \frac{(1 - \varrho)\cos\frac{\varphi}{2}\, d\varphi}{(1 - 2\varrho\cos\varphi + \varrho^2)\sqrt{2(\cos\varphi - \cos\gamma)}} = \frac{1}{\sqrt{1 - 2\varrho\cos\gamma + \varrho^2}} = \sum_{n=0}^{\infty} \varrho^n P_n(\cos\gamma).$$

Clearly

$$\frac{1}{1-2\varrho\cos\varphi+\varrho^2}=\frac{1}{(1-\varrho e^{i\varphi})(1-\varrho e^{-i\varphi})}=\sum_{n=0}^{\infty}\varrho^n e^{in\varphi}\sum_{n=0}^{\infty}\varrho^n e^{-in\varphi}$$

$$=1+\sum_{n=1}^{\infty}\varrho^n[e^{in\varphi}+e^{i(n-2)\varphi}+\ldots+e^{-i(n-2)\varphi}+e^{-in\varphi}]$$

$$=1+\sum_{n=1}^{\infty}\varrho^n e^{-in\varphi}\frac{e^{2i(n+1)\varphi}-1}{e^{2i\varphi}-1}=1+\sum_{n=1}^{\infty}\varrho^n\frac{e^{i(n+1)\varphi}-e^{-i(n+1)\varphi}}{e^{i\varphi}-e^{-i\varphi}}$$

$$=\sum_{n=0}^{\infty}\varrho^n\frac{\sin(n+1)\varphi}{\sin\varphi},$$

whence

$$\frac{(1-\varrho)\cos\frac{\varphi}{2}}{1-2\varrho\cos\varphi+\varrho^2}=\frac{1}{2\sin\frac{\varphi}{2}}\sum_{n=0}^{\infty}\varrho^n[\sin(n+1)\varphi-\sin n\varphi],$$

and consequently

$$\frac{(1-\varrho)\cos\frac{\varphi}{2}}{1-2\varrho\cos\varphi+\varrho^2}=\sum_{n=0}^{\infty}\varrho^n\cos(n+\tfrac{1}{2})\varphi,\qquad(0<\varrho<1),\tag{28}$$

where the series on the right side is uniformly convergent with respect to φ and therefore the partial sums are uniformly bounded as φ varies in an arbitrary interval.

If the two sides of (28) are multiplied by $d\varphi/\sqrt{2(\cos\varphi-\cos\gamma)}$, we may integrate term by term, since the partial sums of the new series are in absolute value less than a constant multiplied by the integrable function $1/\sqrt{2(\cos\varphi-\cos\gamma)}$ (cf. Appendix, Ths. 8a and 8), to obtain

$$\frac{2}{\pi}\int_0^{\gamma}\frac{(1-\varrho)\cos\frac{\varphi}{2}\,d\varphi}{(1-2\varrho\cos\varphi+\varrho^2)\sqrt{2(\cos\varphi-\cos\gamma)}}$$

$$=\sum_{n=0}^{\infty}\varrho^n\frac{2}{\pi}\int_0^{\gamma}\frac{\cos(n+\frac{1}{2})\varphi}{\sqrt{2(\cos\varphi-\cos\gamma)}}\,d\varphi.$$

Comparing this result with (27) we obtain Mehler's formula

$$P_n(\cos\gamma) = \frac{2}{\pi}\int_0^\gamma \frac{\cos(n+\frac{1}{2})\varphi}{\sqrt{2(\cos\varphi - \cos\gamma)}}\,d\varphi, \tag{29}$$

$$(n=0, 1, 2, \ldots; 0<\gamma\leqq\pi).$$

If we substitute $\pi - \gamma$ for γ and $\pi - \varphi$ for φ in (29) and take account of (6_3) we obtain

$$(-1)^n P_n(\cos\gamma) = \frac{2}{\pi}(-1)^n \int_\gamma^\pi \frac{\sin(n+\frac{1}{2})\varphi}{\sqrt{2(\cos\gamma - \cos\varphi)}}\,d\varphi,$$

which gives Mehler's second formula

$$P_n(\cos\gamma) = \frac{2}{\pi}\int_\gamma^\pi \frac{\sin(n+\frac{1}{2})\varphi}{\sqrt{2(\cos\gamma - \cos\varphi)}}\,d\varphi, \tag{30}$$

$$(n=0, 1, 2, \ldots; 0\leqq\gamma<\pi).$$

(cf. F. G. Mehler [75]). The proof in the text is due to Hermite.

8. Zeros of the Legendre Polynomials: Bruns' Inequalities

1a. In Sec. 8.3, use will be made of the following theorem due to Sturm as formulated by G. Szego [107e].

Given the two differential equations

$$(31_1) \quad \frac{d^2y}{dx^2} - Q(x)y(x) = 0, \qquad (31_2) \quad \frac{d^2z}{dx^2} - Q_1(x)z(x) = 0,$$

and suppose $Q(x)$, $Q_1(x)$ are continuous and $Q(x) \geqq Q_1(x)$ for $a < x \leqq b$, and that $Q(x)$ and $Q_1(x)$ are not identically equal. Suppose further that (31_1) has a solution which satisfies the conditions

$$y(x) > 0 \quad \text{for} \quad a < x < b, \qquad y(b) = 0. \tag{32}$$

Then, if $z(x)$ is a nonvanishing solution of (31_2) for which

$$\lim_{x \to a+0} (y'z - yz') = 0,$$

$z(x)$ has at least one zero in the open interval (a, b).

Proof. The proof will proceed by reductio ad absurdum. Suppose $z(x)$ never vanishes in the open interval (a, b) and assume $z(x) > 0$ for $a < x < b$.

Multiplying (31_1) by z and (31_2) by y and subtracting we obtain

$$\frac{d}{dx}[zy' - yz'] - (Q - Q_1)yz = 0,$$

whence, by integration between $a + \varepsilon$ and b, $\varepsilon > 0$, we have

$$(33) \quad z(b)y'(b) - [z(a + \varepsilon)y'(a + \varepsilon) - y(a + \varepsilon)z'(a + \varepsilon)] = \int_{a+\varepsilon}^{b} (Q - Q_1)yz\,dx.$$

Now $z(b) \geqq 0$. On the other hand, $y'(b) \neq 0$, for if $y'(b) = 0$, then by the uniqueness of the solution of equation (31_1) satisfying the boundary condition $y(b) = 0$ it would follow that $y(x) \equiv 0$. Since $y(b) = 0$ and $y(x) > 0$ for $a < x < b$, $y'(b) < 0$. Therefore as $\varepsilon \to +0$, the limit of the left side of (33) is negative or zero while the limit on the right side is positive. We are thus led to a contradiction and, consequently $z(x)$ vanishes in the open interval (a, b).

1b. The same conclusions follow if we suppose $Q(x)$, $Q_1(x)$ continuous in the closed interval $[a, b]$ and

$$y(a) = y(b) = 0, \quad y(x) > 0 \quad \text{for} \quad a < x < b.$$

In fact, the preceding argument shows that if $z(x) > 0$ for $a < x < b$, then

$$z(b)y'(b) - z(a)y'(a) > 0$$

while

$$y'(a) > 0,\ y'(b) < 0;\ z(a) \geqq 0,\ z(b) \geqq 0.$$

2. We proceed to show that the zeros of polynomials $P_n(x)$ are real, simple, and in the open interval $(-1, 1)$. (This theorem will

reappear as a particular case of a more general theorem which will be proved in Ch. IV, Sec. 8.2.)

First we observe that $(x^2 - 1)^n$ has two zeros of order n at $x = -1$ and $x = 1$. Consequently its derivative, which is of degree $2n - 1$, has two zeros of order $n - 1$ at $x = -1$ and $x = 1$ and hence by Rolle's theorem exactly one zero in the open interval $(-1, 1)$. The second derivative, which is of degree $2n - 2$, has two zeros of order $n - 2$ at $x = -1$ and $x = 1$ and consequently exactly two distinct zeros in the open interval $(-1, 1)$. Continuing in this way, we see that the derivative of order n of $(x^2 - 1)^n$, and consequently $P_n(x)$ (cf. (7)) has n real, simple zeros in the open interval $(-1, 1)$.

3. The distribution of the n zeros of $P_n(x)$ according to Bruns' inequalities will now be established. (Bruns [13])

If in equation (10) of 3, we set

$$x = \cos \varphi, \qquad z(\varphi) = (\sin \varphi)^{1/2} P_n(\cos \varphi)$$

we obtain for z the equation

$$z'' + \left[(n + \tfrac{1}{2})^2 + \frac{1}{4 \sin^2 \varphi} \right] z = 0. \tag{34}$$

If we compare (34) with the equation

$$y'' + (n + \tfrac{1}{2})^2 y = 0,$$

which has the solution $y = \sin (n + \frac{1}{2})(\varphi - \varphi_0)$, $[\varphi_0 = \text{constant}]$, it follows by Sturm's theorem Sec. 8.1b) that $P_n(\cos \varphi)$ has a zero in every interval of length $\pi/(n + \frac{1}{2})$, and, therefore, there exists at least one zero of z in each of the open intervals

$$((\nu - 1)\pi/(n + \tfrac{1}{2}), \quad \nu\pi/(n + \tfrac{1}{2})), \quad (\nu = 2, 3, \ldots, n).$$

If we let $y = \sin (n + \frac{1}{2})\varphi$, then $\lim_{\varphi \to 0} \{y'z - yz'\} = 0$, whence, by the result of Sec. 8.1a), $P_n(\cos \varphi)$ has a zero in the open interval $(0, \pi/(n + \frac{1}{2}))$, and thus we have shown that $P_n(\cos \varphi)$ has n distinct zeros $\varphi_1, \varphi_2, \ldots, \varphi_n$ belonging to the interval $(0, \pi)$ and satisfying the inequalities

$$\frac{\nu-1}{n+\frac{1}{2}}\pi<\varphi_\nu<\frac{\nu}{n+\frac{1}{2}}\pi, \qquad (\nu=1,2,\ldots,n).$$

Now $P_n(-x)=(-1)^nP_n(x)$, [Sec. 1, (6_3)] whence

$$\varphi_\nu=\pi-\varphi_{n+1-\nu}>\pi-\frac{(n+1)-\nu}{n+\frac{1}{2}}\pi=\frac{\nu-\frac{1}{2}}{n+\frac{1}{2}}\pi,$$

and, finally, Bruns' inequalities

$$\boxed{\frac{\nu-\frac{1}{2}}{n+\frac{1}{2}}\pi<\varphi_\nu<\frac{\nu}{n+\frac{1}{2}}\pi}.$$

9. *The Complete Orthonormal System* $\{[\frac{1}{2}(2n+1)]^{\frac{1}{2}}P_n(x)\}$

1. From the two equations [cf. Sec. 3, (10')],

$$\frac{d}{dx}\left[(1-x^2)\frac{dP_n}{dx}\right]=-n(n+1)P_n, \frac{d}{dx}\left[(1-x^2)\frac{dP_m}{dx}\right]=-m(m+1)P_m$$

multiplying the first by P_m, the second by P_n and subtracting we have

$$\frac{d}{dx}\left[(1-x^2)\left\{P_m\frac{dP_n}{dx}-P_n\frac{dP_m}{dx}\right\}\right]=\{m(m+1)-n(n+1)\}P_nP_m,$$

whence integrating between -1 and 1,

$$[m(m+1)-n(n+1)]\int_{-1}^{1}P_n(x)P_m(x)dx=0,$$

and therefore for $n\neq m$

$$(35_1) \qquad \boxed{\int_{-1}^{1}P_n(x)P_m(x)\,dx=0}.$$

The integral $\int_{-1}^{1}P_n^2(x)dx$ can be easily evaluated as follows. By (13) of Sec. 4, we have

$$nP_n(x)=(2n-1)xP_{n-1}(x)-(n-1)P_{n-2}(x).$$

If now we multiply by $P_n(x)dx$, integrate between -1 and 1, and take into account (35_1), we obtain

$$(36_1) \quad n\int_{-1}^{1}[P_n(x)]^2dx = (2n-1)\int_{-1}^{1} xP_n(x)P_{n-1}(x)dx.$$

If, on the other hand, we multiply (13) by $P_{n-1}(x)\,dx$ and integrate between -1 and 1, we have

$$(36_2) \quad (2n+1)\int_{-1}^{1} xP_n(x)P_{n-1}(x)dx = n\int_{-1}^{1}[P_{n-1}(x)]^2\,dx.$$

Now from (36_1) and (36_2) follows

$$\int_{-1}^{1}[P_n(x)]^2\,dx = \frac{2n-1}{2n+1}\int_{-1}^{1}[P_{n-1}(x)]^2\,dx.$$

If we substitute for n in this formula the values $n-1$, $n-2$, . . ., 1, we obtain by multiplication

$$(35_2) \quad \boxed{\int_{-1}^{1}[P_n(x)]^2\,dx = \frac{2}{2n+1}}.$$

2a. From (35_1), (35_2) follows that the system $\left\{\sqrt{\frac{2n+1}{2}}\,P_n(x)\right\}$ is orthonormal. We now prove that it is complete with respect to square integrable functions.

Vitali's condition for completeness for the interval $[-1, x]$ (Ch. 1, Sec. 6.6), becomes here

$$1 + x = \sum_{r=0}^{\infty}\frac{2r+1}{2}\left[\int_{-1}^{x}P_r(x)\,dx\right]^2$$

or, since $P_0 = 1$,

$$(37) \quad 1-x^2 = \sum_{r=1}^{\infty}(2r+1)\left[\int_{-1}^{x}P_r(x)\,dx\right]^2$$

Consequently, by (14) of Sec. 4,

$$(2r+1)P_r(x) = P'_{r+1}(x) - P'_{r-1}(x),$$

$$(2r+1)\int_{-1}^{x} P_r(x)dx = [P_{r+1}(x) - P_{r-1}(x)]_{-1}^{x} = P_{r+1}(x) - P_{r-1}(x);$$

$$\begin{aligned}(2r+1)\left[\int_{-1}^{x} P_r(x)\,dx\right]^2 &= \frac{1}{2r+1}[P_{r+1}(x) - P_{r-1}(x)]^2 \\ &= \frac{1}{2r+1}\int_{-1}^{x}\frac{d}{dx}[P_{r+1}(x) - P_{r-1}(x)]^2\,dx \\ &= \frac{2}{2r+1}\int_{-1}^{x}[P_{r+1}(x) - P_{r-1}(x)][P'_{r+1}(x) - P'_{r-1}(x)]\,dx \\ &= 2\int_{-1}^{x}[P_{r+1}(x) - P_{r-1}(x)]P_r(x)\,dx,\end{aligned}$$

and therefore we obtain for the sum $S_n(x)$ of the first n terms of the series of the right side of (37)

$$\begin{aligned}(38)\quad S_n(x) &= 2\int_{-1}^{x}\sum_{r=1}^{n}[P_{r+1}(x) - P_{r-1}(x)]P_r(x)dx \\ &= 2\int_{-1}^{x}[P_{n+1}(x)P_n(x) - x]\,dx.\end{aligned}$$

The terms of the sequence $\{P_{n+1}(x)P_n(x) - x\}$ are less than or equal to 2 in absolute value in $[-1, 1]$. Moreover, we have in $(-1, 1)$

$$\lim_{n\to\infty}[P_{n+1}(x)P_n(x) - x] = -x.$$

Therefore, we may pass to the limit under the integral sign in (38) to obtain

$$\lim_{n\to\infty} S_n(x) = -2\int_{-1}^{x} x\,dx = 1 - x^2,$$

which is precisely (37).

2b. By the argument of Ch. II, Sec. 2.3, the system $\left\{\sqrt{\frac{2n+1}{2}}P_n(x)\right\}$ can be proved complete with respect to functions integrable in $[-1, 1]$.

Indeed, let $\theta(x)$ be a function integrable in $[-1, 1]$ such that

$$\int_{-1}^{1}\theta(x)dx = 0; \qquad \int_{-1}^{1}\theta(x)P_n(x)dx = 0, \qquad n = 1, 2, \ldots.$$

If we also let

$$\omega(x) = c + \int_0^x \theta(x)dx, \quad c = -\tfrac{1}{2}\int_{-1}^{1} dx \int_0^x \theta(x)dx,$$

then

$$\int_{-1}^{1} \omega(x)dx = 0,$$

and

$$\int_{-1}^{1} \omega(x)P_n(x)dx = \left[\omega(x)\int_{-1}^{x} P_n(x)dx\right]_{-1}^{1}$$

$$- \frac{1}{2n+1}\int_{-1}^{1} \theta(x)[P_{n+1}(x) - P_{n-1}(x)]dx = 0,$$

whence $\omega(x)$ vanishes identically, and thus $\theta(x)$ vanishes almost everywhere.

3. Let $f(x)$ be integrable in $[-1, 1]$ and consider its Fourier coefficients a_n' with respect to the system $\left\{\sqrt{\frac{2n+1}{2}}\, P_n(x)\right\}$, namely

$$a_n' = \sqrt{\frac{2n+1}{2}}\int_{-1}^{1} f(x)P_n(x)\, dx.$$

The Fourier series of $f(x)$ with respect to this system is therefore

$$\sum_{n=0}^{\infty} \sqrt{\frac{2n+1}{2}}\, a_n' P_n(x)$$

or, if we let

$$\boxed{a_n = \frac{2n+1}{2}\int_{-1}^{1} f(x)P_n(x)\, dx}$$

we have

$$f(x) \sim \sum_{n=0}^{\infty} a_n P_n(x). \tag{39}$$

The series $\sum_{n=0}^{\infty} a_n P_n(x)$ will be called *the series of Legendre polynomials* or simply *the Legendre series of* $f(x)$, and by the results of Ch. I, Sec. 6.5, we can assert that if $f(x)$ is square integrable in $[-1, 1]$, its Legendre series converges in the mean in $[-1, 1]$ to $f(x)$.

Evidently if $f(x)$ is even, the series (27) will contain only the terms $P_n(x)$ of even index; if $f(x)$ is odd, the terms of odd index only.

We can also affirm that if $f(x)$ is continuous in $[-1, 1]$ and its Legendre series is uniformly convergent there, then

$$f(x) = \sum_{n=0}^{\infty} a_n P_n(x).$$

In Sec. 14 we shall consider pointwise convergence of series (39) in general. In the next section we consider specifically two expansions in Legendre series which can be obtained by elementary considerations.

4a. Consider the Legendre series of x^m, where m is a positive integer. Then $x^m = \sum a_n P_n(x)$, and since x^m is a linear combination of $P_m(x)$, x^{m-2}, x^{m-4}, . . .; x^{m-2} is a linear combination of P_{m-2}, x^{m-4}, . . . ; it follows that x^m can be expressed in the form

$$x^m = a_m P_m + a_{m-2} P_{m-2} + \cdots$$

and

$$\text{(40)} \quad a_s = \frac{2s+1}{2} \int_{-1}^{1} x^m P_s(x)dx \text{ for } m-s \geqq 0,\ m-s \text{ even,}$$

while

$$\int_{-1}^{1} x^m P_s(x)dx = 0 \quad \text{for} \quad m-s \text{ odd.}$$

To evaluate the integral (40) we proceed as follows. If we let

$$P_s(x) = \alpha x^s + \beta x^{s-2} + \gamma x^{s-4} + \ldots, \ [\alpha+\beta+\gamma+\ldots = P_s(1) = 1],$$

then

$$\frac{1}{2}\int_{-1}^{1} x^m\, P_s(x)dx = \frac{\alpha}{m+s+1} + \frac{\beta}{m+s-1} + \frac{\gamma}{m+s-3} + \cdots$$

$$= \frac{f(m)}{(m+s+1)(m+s-1)(m+s-3)\ldots}$$

where $f(m)$ is a polynomial in m of degree $s/2$ or $(s-1)/2$ according as s is even or odd. By virtue of the orthogonality of P_n, $f(m)$ vanishes for $m = s-2, s-4, s-6, \ldots$ and since the coefficient of the highest power of $f(m)$ is $\alpha + \beta + \gamma + \ldots = 1$ it follows that

$$(41_1) \quad \tfrac{1}{2}\int_{-1}^{1} x^m P_s(x)dx = \frac{m(m-2)\ldots(m-s+2)}{(m+s+1)(m+s-1)\ldots(m+1)}$$

for m and s even, and

$$(41_2) \quad \tfrac{1}{2}\int_{-1}^{1} x^m P_s(x)dx = \frac{(m-1)(m-3)\ldots(m-s+2)}{(m+s+1)(m+s-1)\ldots(m+2)}$$

for m and s odd. Multiplying the numerator and denominator of the fraction in (41_1) by $1\times3\times5\times\ldots\times(m-1)$ and $2\times4\times6\times\ldots\times(m-s)$, and in (41_2) by $1\times3\times5\times\ldots\times m$ and $2\times4\times6\times\ldots\times(m-s)$ we obtain

$$(42) \qquad \tfrac{1}{2}\int_{-1}^{1} x^m P_s(x)dx = \frac{m!}{(m-s)!!\,(m+s+1)!!}$$

if $m \geqq s$, $m-s=0$ (mod. 2), where $(m-s)!! = (m-s)(m-s-2)\times\ldots\times6\times4\times2$ and $(m+s+1)!! = (m+s+1)\times(m+s-1)\times\ldots\times5\times3\times1$. Consequently, for m a positive integer, we have the identity

$$(43) \quad \boxed{\begin{aligned} x^m = \frac{m!}{(2m+1)!!}\Big[&(2m+1)P_m(x) + (2m-3)\frac{2m+1}{2}P_{m-2}(x) \\ &+ (2m-7)\frac{(2m+1)(2m-1)}{2\cdot4}P_{m-4}(x) \\ &+ (2m-11)\frac{(2m+1)(2m-1)(2m-3)}{2\cdot4\cdot6}P_{m-6}(x) + \ldots\Big] \end{aligned}}$$

4b. The Legendre series expansion for $P'_n(x)$ can be derived as follows.

Arguing as in 4a we see that $P'_n(x)$ can be expressed as a linear combination with constant coefficients of P_{n-1}, P_{n-3}, . . ., namely

$$P'_n = a_{n-1} P_{n-1} + a_{n-3} P_{n-3} + a_{n-5} P_{n-5} + \ldots$$

where

$$a_s = \frac{2s+1}{2} \int_{-1}^{1} P'_n P_s \, dx, \quad s < n, \; n - s \text{ odd}.$$

Now

$$\int_{-1}^{1} P'_n P_s \, dx = [P_n P_s]_{-1}^{1} - \int_{-1}^{1} P_n P'_s \, dx = 2 - \int_{-1}^{1} P_n P'_s \, dx;$$

but by the orthogonality of P_n and P'_s we have

$$\int_{-1}^{1} P_n P'_s \, dx = 0,$$

whence $a_s = 2s + 1$ and, finally,

$$(44) \qquad \boxed{P'_n = (2n-1)P_{n-1} + (2n-5)P_{n-3} + (2n-9)P_{n-5} + \cdots}$$

This formula could also be deduced from (15) of Sec. 4 by equating the even and odd terms on both sides of the equation.

10. Stieltjes' Bounds for Legendre Polynomials

The proofs in this section are substantially those given by L. Fejér in his memoir [33e].

1. We recall first the Brunacci-Abel transformation from which we derive an immediate consequence.

Let $z_1, z_2, \ldots, z_n$ be n constants, real or complex, and define s_k

$$(1) \qquad z_1 + z_2 + \ldots + z_k = s_k, \qquad (k = 1, 2, \ldots, n).$$

Let

$$(2) \qquad |s_k| \leqq M \quad \text{for} \quad k = 1, 2, \ldots, n.$$

Finally, let $\varepsilon_1, \varepsilon_2, \ldots, \varepsilon_n$ be a monotonic sequence. We wish to show that if

$$(3) \qquad \varepsilon_1 \geqq \varepsilon_2 \geqq \ldots \geqq \varepsilon_n \geqq 0$$

then

$$(4) \qquad |\,\varepsilon_1 z_1 + \varepsilon_2 z_2 + \ldots + \varepsilon_n z_n\,| \leqq \varepsilon_1 M;$$

and if, on the other hand,

$$(3') \qquad 0 \leqq \varepsilon_1 \leqq \varepsilon_2 \leqq \ldots \leqq \varepsilon_n$$

then

$$(4') \qquad |\,\varepsilon_1 z_1 + \varepsilon_2 z_2 + \ldots + \varepsilon_n z_n\,| \leqq 2\varepsilon_n M.$$

Proof. Clearly

$$\sum_{k=1}^{n} \varepsilon_k z_k = \varepsilon_1 s_1 + \varepsilon_2(s_2 - s_1) + \ldots + \varepsilon_n(s_n - s_{n-1}) = s_1(\varepsilon_1 - \varepsilon_2) + s_2(\varepsilon_2 - \varepsilon_3) + \ldots + s_{n-1}(\varepsilon_{n-1} - \varepsilon_n) + \varepsilon_n s_n$$

whence

$$\left|\sum_{k=1}^{n} \varepsilon_k z_k\right| \leqq M[\,|\,\varepsilon_1 - \varepsilon_2\,| + |\,\varepsilon_2 - \varepsilon_3\,| + \ldots + |\,\varepsilon_{n-1} - \varepsilon_n\,| + |\,\varepsilon_n\,|\,]$$

and depending on whether (3) or (3′) holds, (4) or (4′) follows.

2. Let

$$(5) \qquad \alpha_k = (-1)^k \binom{-\frac{1}{2}}{k} = \frac{(2k-1)!!}{(2k)!!}, \quad k = 1, 2, \ldots,; \ \alpha_0 = 1,$$

where $(2k)!! = 2k\,(2k-2)\,(2k-4)\ldots 4\cdot 2$ and $(2k-1)!! = (2k-1)(2k-3)\ldots 3\cdot 1$. The following inequality will be established

$$(6) \qquad |\,s_n(z)\,| = |\,\alpha_0 + \alpha_1 z + \alpha_2 z^2 + \ldots + \alpha_n z^n\,| \leqq \frac{\sqrt{2}}{\sqrt{|1-z|}};$$

$$n = 0, 1, 2, \ldots,; \quad |\,z\,| \leqq 1; \quad z \neq 1.$$

Indeed

$$[s_n(z)]^2 = \sum_{r=0}^{2n} q_r z^r$$

where q_r is given by the following expression for $0 < r \leqq n$

$$(-1)^r q_r = \binom{-\frac{1}{2}}{0}\binom{-\frac{1}{2}}{r} + \binom{-\frac{1}{2}}{1}\binom{-\frac{1}{2}}{r-1} + \dots + \binom{-\frac{1}{2}}{r-1}\binom{-\frac{1}{2}}{1}$$
$$+ \binom{-\frac{1}{2}}{r}\binom{-\frac{1}{2}}{0} = \binom{-1}{r} = (-1)^r$$

and therefore

$$q_r = 1, \qquad (r = 0, 1, 2, \dots, n).$$

Consequently, if $r = n + k$ and $k \geqq 0$

$$q_{n+k} = \sum_{\nu=0}^{n-k} \alpha_{k+\nu}\,\alpha_{n-\nu}$$

whence

$$q_{n+k} - q_{n+k+1} = \sum_{\nu=0}^{(n-k-1)} (\alpha_{k+\nu} - \alpha_{k+1+\nu})\alpha_{n-\nu} + \alpha_n\alpha_k.$$

But for $k = 0, 1, 2, \dots, \alpha_k > 0$ and

$$\alpha_k/\alpha_{k+1} = (2k + 2)/(2k + 1) > 1, \tag{7}$$

whence $q_{n+k} > q_{n+k+1}$. This implies

$$[s_n(z)]^2 = 1 + z + z^2 + \dots + z^n + q_{n+1}z^{n+1} + \dots + q_{2n}z^{2n}$$

where

$$1 = q_n > q_{n+1} > \dots > q_{2n}$$

and since clearly

$$|\,1 + z + z^2 + \dots + z^\nu\,| = |\,(1 - z^{\nu+1})/(1 - z)\,| \leqq 2/|\,1 - z)\,|,$$
$$|\,z\,| \leqq 1, \quad z \neq 1$$

from (4) follows precisely (6).

3. FIRST THEOREM OF STIELTJES [104b]. The following bound holds for $0 < \gamma < \pi$, and $n = 1, 2, 3, \dots$

$$\boxed{|\,P_n(\cos\gamma)\,| \leqq \sqrt{2}\,\frac{4}{\sqrt{\pi}}\,\frac{1}{\sqrt{n}\,\sqrt{\sin\gamma}}}, \tag{8}$$

$$n = 1, 2, \dots, \quad 0 < \gamma < \pi.$$

Proof. According to Sec. 1.1, (3) we have

$$P_n(\cos\gamma) = \sum_{k=0}^{n} (-1)^n \binom{-\frac{1}{2}}{k}\binom{-\frac{1}{2}}{n-k} e^{ik\gamma}\, e^{-i(n-k)\gamma}$$

or

$$P_n(\cos\gamma) = \sum_{k=0}^{n} \alpha_k \alpha_{n-k}\, e^{ik\gamma}\, e^{-i(n-k)\gamma} = e^{-in\gamma} \sum_{k=0}^{n} \alpha_k \alpha_{n-k}\, e^{2ik\gamma}. \tag{9}$$

If we write

$$e^{2i\gamma} = z, \qquad e^{-2i\gamma} = z_0$$

then for $0 < h < n$ we have

$$|P_n(\cos\gamma)| \leqq \Big|\sum_{k=0}^{h} \alpha_k \alpha_{n-k} z^k\Big| + \Big|\sum_{k=h+1}^{n} \alpha_k \alpha_{n-k} z^k\Big|.$$

But

$$\Big|\sum_{k=h+1}^{n} \alpha_k \alpha_{n-k} z^k\Big| = \Big|z_0^{-n} \sum_{k=h+1}^{n} \alpha_k \alpha_{n-k} z_0^{n-k}\Big| = \Big|\sum_{l=0}^{n-h-1} \alpha_l \alpha_{n-l} z_0^l\Big|$$

whence

$$|P_n(\cos\gamma)| \leqq \Big|\sum_{k=0}^{h} \alpha_k \alpha_{n-k} z^k\Big| + \Big|\sum_{l=0}^{n-h-1} \alpha_l \alpha_{n-l} z_0^l\Big|. \tag{10}$$

However,

$$|\alpha_0|,\ |\alpha_0 + \alpha_1 z|,\ |\alpha_0 + \alpha_1 z + \alpha_2 z^2|, \ldots$$
$$\leqq \frac{\sqrt{2}}{\sqrt{|1-z|}}, \quad |z| \leqq 1, \quad z \neq 1,$$

and since by (7) the sequence

$$\alpha_n,\ \alpha_{n-1},\ \alpha_{n-2}, \ldots,$$

is increasing, it follows from (4′) that

$$\begin{aligned} \Big|\sum_{k=0}^{h} \alpha_k \alpha_{n-k} z^k\Big| &\leqq \frac{2\sqrt{2}}{\sqrt{|1-z|}}\, \alpha_{n-h}, \\ \Big|\sum_{l=0}^{n-h-1} \alpha_l \alpha_{n-l} z_0^l\Big| &\leqq \frac{2\sqrt{2}}{\sqrt{|1-z_0|}}\, \alpha_{h+1}. \end{aligned} \tag{11}$$

But

$$|1 - z| = |1 - z_0| = |1 - \cos 2\gamma - i \sin 2\gamma|$$
$$= \sqrt{2(1 - \cos 2\gamma)} = 2 \sin \gamma$$

and therefore by (10) we have

$$|P_n(\cos \gamma)| \leqq \frac{2}{\sqrt{\sin \gamma}} (\alpha_{n-h} + \alpha_{h+1}), \quad 0 < h < n.$$

On the other hand, Wallis' formula gives

$$\alpha_k = \frac{(2k-1)!!}{(2k)!!} = \frac{1}{\sqrt{\pi k}} \sqrt{\frac{2k}{2k+\theta}} < \frac{1}{\sqrt{\pi k}}, \quad 0 < \theta < 1, \tag{12}$$

and therefore

$$|P_n(\cos \gamma)| < \frac{2}{\sqrt{\pi}\sqrt{\sin \gamma}} \left(\frac{1}{\sqrt{n-h}} + \frac{1}{\sqrt{h+1}} \right). \tag{13}$$

For n even, we write $h = n/2$ to obtain

$$1/\sqrt{n-h} + 1/\sqrt{h+1} = \sqrt{2/n} + \sqrt{2/(n+2)} < 2\sqrt{2/n},$$

for n odd, we write $h = (n-1)/2$ to obtain

$$1/\sqrt{n-h} + 1/\sqrt{h+1} = \sqrt{2}[1/\sqrt{2n-2h} + 1/\sqrt{2h+2}]$$
$$= 2\sqrt{2}/\sqrt{n+1} < 2\sqrt{2/n}.$$

Consequently in either case (8) follows from (13).

If we put $\cos \gamma = x$ in (8), we obtain the formula

$$\boxed{\left|\sqrt{n}\sqrt[4]{1-x^2}P_n(x)\right| \leqq 4\sqrt{\frac{2}{\pi}}}, \quad (-1 \leqq x \leqq 1,\ n = 0, 1, 2, \ldots). \tag{14}$$

4. SECOND THEOREM OF STIELTJES. For $-1 \leqq x \leqq 1$,

$$\boxed{|P_n(x) - P_{n+2}(x)| < \frac{4}{\sqrt{\pi}\sqrt{n+2}}} \tag{15}$$

Proof. From (3) of Sec. 1.1 we have

$$P_{2\nu-1}(\cos\gamma) = 2\sum_{k=0}^{\nu-1} \alpha_k \alpha_{n-k} \cos(n-2k)\gamma, \qquad (n = 2\nu - 1)$$

$$P_{2\nu}(\cos\gamma) = 2\sum_{k=0}^{\nu-1} \alpha_k \alpha_{n-k} \cos(n-2k)\gamma + \alpha_\nu^2, \quad (n = 2\nu),$$

whence

$$(16_1) \quad P_{2\nu-1}(\cos\gamma) - P_{2\nu+1}(\cos\gamma) = -2\alpha_{n+2}\cos(n+2)\gamma + 2\sum_{k=0}^{\nu-1}(\alpha_k\alpha_{n-k} - \alpha_{k+1}\alpha_{n-k+1})\cos(n-2k)\gamma, \quad (n = 2\nu - 1),$$

$$(16_2) \quad P_{2\nu}(\cos\gamma) - P_{2\nu+2}(\cos\gamma) = \alpha_\nu^2 - \alpha_{\nu+1}^2 - 2\alpha_{n+2}\cos(n+2)\gamma + 2\sum_{k=0}^{\nu-1}(\alpha_k\alpha_{n-k} - \alpha_{k+1}\alpha_{n-k+1})\cos(n-2k)\gamma, \quad (n = 2\nu).$$

If we let $\gamma = 0$ [cf. Sec. 6.2, (20)], we obtain

$$2\sum_{k=0}^{\nu-1}(\alpha_k\alpha_{n-k} - \alpha_{k+1}\alpha_{n-k+1}) = \begin{cases} 2\alpha_{n+2} \text{ for } n = 2\nu - 1 \\ 2\alpha_{n+2} - (\alpha_\nu^2 - \alpha_{\nu+1}^2) \text{ for } n = 2\nu. \end{cases}$$

But

$$\alpha_k > \alpha_{k+1}, \quad \alpha_{n-k} > \alpha_{n-k+1}$$

whence

$$\alpha_k\alpha_{n-k} - \alpha_{k+1}\alpha_{n-k+1} > 0, \quad \alpha_\nu^2 - \alpha_{\nu+1}^2 > 0,$$

and consequently from (16_1), (16_2) and (12), follows

$$|P_n(\cos\gamma) - P_{n+2}(\cos\gamma)| < 4\alpha_{n+2} < \frac{4}{\sqrt{\pi}\sqrt{n+2}};$$

which is precisely (15).

5. Integrating between -1 and x in (14) of Sec. 4, we have

$$(2n+1)\int_{-1}^{x} P_n(\xi)\,d\xi = P_{n+1}(x) - P_{n-1}(x),$$

whence by (15)

$$(17) \qquad \left|\int_{-1}^{x} P_n(\xi)d\xi\right| < \frac{4}{\sqrt{\pi}\sqrt{n+1}\,(2n+1)} < \frac{2}{\sqrt{\pi}\, n^{3/2}}$$

$$\text{for } |x| \leqq 1, \ n = 1, 2, \ldots .$$

Integrating between -1 and x in (10′) of Sec. 3, we have

$$(1-x^2)\frac{dP_n(x)}{dx} = -n(n+1)\int_{-1}^{x} P_n(\xi)d\xi,$$

whence by (17) we obtain for $|x|<1$

$$\left|\frac{dP_n(x)}{dx}\right| < \frac{4n(n+1)}{\sqrt{\pi}\sqrt{n+1}(2n+1)}\,\frac{1}{1-x^2}.$$

However, since $2n(n+1)/\sqrt{n+1}(2n+1) < \sqrt{n}$, we have

$$\text{(18)}\qquad \boxed{\left|\frac{dP_n(x)}{dx}\right| < \frac{2}{\sqrt{\pi}}\,\frac{\sqrt{n}}{1-x^2}} \quad \text{for } |x|<1,\ n=1, 2, \ldots.$$

Integrating between -1 and x in (12) of Sec. 4, we have

$$P_{n+1}(x) + (-1)^n - \int_{-1}^{x} \xi P_n'(\xi)\,d\xi = (n+1)\int_{-1}^{x} P_n(\xi)\,d\xi,$$

$$P_{n+1}(x) - xP_n(x) = n\int_{-1}^{x} P_n(\xi)\,d\xi,$$

$$P_{n+1}(x) + P_n(x) = (1+x)\,P_n(x) + n\int_{-1}^{x} P_n(\xi)\,d\xi,$$

whence by (14) and (17)

$$|\,P_{n+1}(x) + P_n(x)\,| < \frac{4\sqrt{2}}{\sqrt{n}\sqrt{\pi}}\,\frac{1+x}{\sqrt{1-x^2}} + \frac{2}{\sqrt{\pi}\sqrt{n}}$$

$$= \frac{2}{\sqrt{n\pi}}\,\frac{\sqrt{1-x}+2\sqrt{2}\sqrt{1+x}}{\sqrt{1-x}}$$

and since $\sqrt{1-x}+2\sqrt{2}\sqrt{1+x} \leqq 3\sqrt{2}$ for $|x| \leqq 1$, we have finally

$$\text{(19)}\qquad \boxed{|\,P_{n+1}(x) + P_n(x)\,| < \frac{6\sqrt{2}}{\sqrt{\pi}}\,\frac{1}{\sqrt{n}}\,\frac{1}{\sqrt{1-x}}}, \quad \text{for } |x|<1.$$

11. Series of Legendre Polynomials for Functions of Bounded Variation: Picone's and Jackson's Theorems

1. THEOREM. Let $F(x)$ be of bounded variation in $[-1, 1]$ and consider its Legendre series

$$F(x) \sim \sum_{n=0}^{\infty} a_n P_n(x)$$

where

$$a_n = \frac{2n+1}{2} \int_{-1}^{1} F(x) P_n(x)\, dx.$$

If $| F(x) | \leqq M$ and if V denotes the total variation of $F(x)$ for $| x | \leqq 1$, then, for $n \geqq 1$, we have

$$(1_1) \qquad | a_n P_n(x) | < \frac{16\sqrt{2}}{\pi} \frac{M+V}{\sqrt[4]{1-\delta^2}} \frac{1}{n} \text{ for } | x | < \delta < 1$$

$$(1_1) \qquad | a_n P_n(x) | < \frac{4}{\sqrt{\pi}} (M+V) \frac{1}{\sqrt{n}} \text{ for } | x | \leqq 1$$

(See Jackson [51a], p. 73, Picone [88b], p. 259).

Proof. Let $P(x)$ and $N(x)$ denote the positive and negative variation of $F(x)$ in $[-1, x]$. If $F(-1) \geqq 0$, we set

$$F(x) = F_1(x) - F_2(x)$$

where

$$F_1(x) = F(-1) + P(x), \quad F_2(x) = N(x).$$

Then

$$0 \leqq F(-1) + P(x) \leqq M + P(x) \leqq M + V$$

and

$$0 \leqq F_2(x) = N(x) \leqq V.$$

Similarly if $F(-1) < 0$, we set

$$F(x) = P(x) - [N(x) - F(-1)].$$

In either case we can suppose

$$F(x) = F_1(x) - F_2(x)$$

where $F_1(x)$ and $F_2(x)$ are nonnegative and nondecreasing and

$$0 \leqq F_1(x) \leqq M + V, \; 0 \leqq F_2(x) \leqq M + V.$$

Now, by the second theorem of the mean

$$\int_{-1}^{1} F_1(x)P_n(x)\,dx = (M + V)\int_{\xi_1}^{1} P_n(x)\,dx,$$

$$\int_{-1}^{1} F_2(x)P_n(x)\,dx = (M + V)\int_{\xi_2}^{1} P_n(x)\,dx,$$

where $-1 \leqq \xi_1 \leqq 1$, $-1 \leqq \xi_2 \leqq 1$. Subtracting gives

$$\int_{-1}^{1} F(x)P_n(x)\,dx = (M + V)\int_{\xi_1}^{\xi_2} P_n(x)\,dx$$

whence by (17) of Sec. 10.5, we have

$$\left|\int_{-1}^{1} F(x)P_n(x)\,dx\right| < \frac{8(M + V)}{\sqrt{\pi}\sqrt{n+1}(2n + 1)}.$$

Therefore

$$(2_1) \qquad |a_n| < \frac{4(M + V)}{\sqrt{\pi}\sqrt{n+1}}$$

$$(2_2) \qquad |a_nP_n(x)| < \frac{4(M + V)}{\sqrt{\pi}\sqrt{n+1}}|P_n(x)|.$$

Moreover, by (14) of Sec. 10.3 we have

$$(3_1) \qquad |P_n(x)| \leqq \frac{4\sqrt{2}}{\sqrt{\pi}}\frac{1}{\sqrt{n}\sqrt[4]{1 - x^2}},$$

also

$$(3_2) \qquad |P_n(x)| \leqq 1,$$

and clearly, (1_1) and (1_2) follow from (2_2), (3_1), (3_2).

2. PICONE'S THEOREM [[88b], p. 260]. Let $f(x)$ be integrable in $[-1, 1]$, let $(1 - x^2)\,f(x)$ be of bounded variation in $[-1, 1]$, let M' denote the least upper bound of $|f(x)(1 - x^2)|$ in $[-1, 1]$ and, finally, let V' denote the total variation of $f(x)\ (1 - x^2)$ in $[-1, 1]$.

Given the function

$$(4) \qquad F(x) = F(-1) + \int_{1}^{x} f(x)dx$$

then the terms of its Legendre series

$$F(x) \sim \sum_{n=0}^{\infty} a_n P_n(x), \quad a_n = \frac{2n + 1}{2}\int_{-1}^{1} F(x)\,P_n(x)\,dx$$

satisfy the following inequalities for $[n \geqq 1]$

$$(5_1) \qquad | a_n P_n(x) | < \frac{8\sqrt{2}}{\sqrt{\pi}} \frac{M' + V'}{\sqrt[4]{1 - \delta^2}} \frac{1}{n^{3/2}} \quad \text{for } |x| \leqq \delta < 1,$$

$$(5_2) \qquad | a_n P_n(x) | < 2(M' + V') \frac{1}{n} \text{ for } | x | \leqq 1.$$

Moreover, the Legendre series of $F(x)$ is uniformly and absolutely convergent in every interval interior to $[-1, 1]$ and converges to $F(x)$ in each such interval.

The remainder of the series beginning with the $(n + 1)$st term satisfies the inequality

$$(6) \qquad | R_{n+1}(x) | < \frac{16\sqrt{2}}{\sqrt{\pi}} \frac{M' + V'}{\sqrt[4]{(1-\delta^2)}} \frac{1}{\sqrt{n}} \quad \text{for } | x | \leqq \delta < 1.$$

Proof. By (10′) of Sec. 3 we have

$$\int_{-1}^{1} F(x)P_n(x)dx = -\frac{1}{n(n+1)}\int_{-1}^{1} F(x)\frac{d}{dx}\left[(1-x^2)\frac{dP_n}{dx}\right]dx$$

$$= \frac{1}{n(n+1)}\int_{-1}^{1} f(x)(1-x^2)\frac{dP_n}{dx}\,dx,$$

and reasoning as in Sec. 11.1

$$\int_{-1}^{1} F(x)P_n(x)dx = \frac{M' + V'}{n(n+1)}\int_{\xi_1}^{\xi_2} \frac{dP_n}{dx}\,dx.$$

Consequently

$$|a_n| = \frac{2n+1}{2}\left|\int_{-1}^{1} F(x)P_n(x)dx\right| \leqq \frac{2n+1}{n(n+1)}(M'+V') < 2\frac{M'+V'}{n},$$

whence (5_1) and (5_2) follow with the help of (3_1) and (3_2).

By (5_1) we have for $| x | \leqq \delta < 1$

$$\left|\sum_{n=1}^{\infty} a_n P_n(x)\right| \leqq \sum_{n=1}^{\infty}\left| a_n P_n(x)\right| < \frac{8\sqrt{2}}{\sqrt{\pi}} \frac{M' + V'}{\sqrt[4]{1 - \delta^2}} \sum_{n=1}^{\infty} \frac{1}{n^{3/2}},$$

and from the convergence of the series $\sum_{n=1}^{\infty} n^{-3/2}$ follows the uniform and absolute convergence of $\sum_{n=1}^{\infty} a_n P_n(x)$ in the interval $I = (-\delta, +\delta)$. But this series converges in the mean in I to the continuous function $F(x)$. Therefore in I,

$$F(x) = \sum_{n=1}^{\infty} a_n P_n(x).$$

We shall see in Sec. 14.7, in view of the bounded variation of $F(x)$, that the series $\sum_{n=1}^{\infty} a_n P_n(x)$ is convergent also at $x = -1$ and $x = 1$ to the values $F(-1)$ and $F(+1)$.

Finally, we have

$$| R_{n+1}(x) | = | \sum_{k=n+1}^{\infty} a_k P_k(x) | < \frac{8\sqrt{2}}{\sqrt{\pi}} \frac{M' + V'}{\sqrt[4]{1 - \delta^2}} \sum_{k=n+1}^{\infty} k^{-3/2},$$

and since for $\alpha > 0$

$$\frac{1}{n}\left[\frac{n^{1+\alpha}}{(n+1)^{1+\alpha}} + \frac{n^{1+\alpha}}{(n+2)^{1+\alpha}} + \dots\right] = \frac{1}{n}\sum_{k=1}^{\infty}\left(1 + \frac{k}{n}\right)^{-(1+\alpha)}$$

$$< \int_0^{+\infty} (1 + x)^{-(\alpha+1)}\, dx = \frac{1}{\alpha},$$

whence for $\alpha > 0$

$$\frac{1}{(n+1)^{1+\alpha}} + \frac{1}{(n+2)^{1+\alpha}} + \dots < \frac{1}{\alpha n^{\alpha}}.$$

Therefore

$$| R_{n+1}(x) | < \frac{16\sqrt{2}}{\sqrt{\pi}} \frac{M' + V'}{\sqrt[4]{1 - \delta^2}} \frac{1}{\sqrt{n}}, \qquad \text{Q.E.D.}$$

3. JACKSON'S THEOREM [[51a] p. 76]. Let $f(x)$ be of bounded variation in $[-1, 1]$ and let M' and V' denote respectively the least upper bound of $| f(x) |$ and the total variation of $f(x)$ in $[-1, 1]$.

Given the function

$$F(x) = F(-1) + \int_{-1}^{x} f(x)dx,$$

then the coefficients a_n

$$a_n = \frac{2n+1}{2}\int_{-1}^{1} F(x)\, P_n(x)\, dx$$

of its Legendre series satisfy the inequalities

$$(7_1) \qquad | a_n | < \frac{6}{\sqrt{\pi}} (M' + V') \frac{1}{n^{3/2}} \text{ for } n \geqq 1;$$

$$(7_2) \qquad | a_n | < \frac{4}{\sqrt{\pi}} (M' + V') \frac{1}{n^{3/2}} \text{ for } n \geqq 2.$$

Moreover, the Legendre series of $F(x)$ converges uniformly and absolutely to $F(x)$ in $[-1, 1]$.

The remainder of the series beginning with the $(n + 1)$st term satisfies the inequalities

$$(8_1) \quad | R_{n+1}(x) | < \frac{8}{\sqrt{\pi}} (M' + V') \frac{1}{\sqrt{n}} \qquad \text{for } | x | \leqq 1, \ n \geqq 1,$$

$$(8_2) \quad | R_{n+1}(x) | < \frac{16\sqrt{2}}{\pi} \frac{M' + V'}{\sqrt[4]{1 - \delta^2}} \frac{1}{n} \qquad \text{for } | x | \leqq \delta < 1, \ n \geqq 1.$$

Proof. Proceeding as in Sec. 10.5, from, Sec. 4 (14), if we integrate between -1 and x we have

$$\omega_n(x) = (2n + 1) \int_{-1}^{x} P_n(\xi) d\xi = P_{n+1}(x) - P_{n-1}(x)$$

whence

$$a_n = \frac{2n + 1}{2} \int_{-1}^{1} F(x) P_n(x) dx = \tfrac{1}{2} \int_{-1}^{1} F(x) \omega_n'(x) dx$$

$$= \tfrac{1}{2} [F(x)\omega_n(x)]_{-1}^{1} - \tfrac{1}{2} \int_{-1}^{1} \omega_n(x) f(x) dx.$$

But $\omega_n(-1) = 0$, and for $n \geqq 1$, $\omega_n(1) = 0$, whence, reasoning as in Sec. 11.1, we find

$$a_n = - \tfrac{1}{2} \int_{-1}^{1} \omega_n(x) f(x) dx = - \tfrac{1}{2} (M' + V') \int_{\xi_1}^{\xi_2} \omega_n(x) dx$$

$$= - \tfrac{1}{2} (M' + V') \int_{\xi_1}^{\xi_2} [P_{n+1}(x) - P_{n-1}(x)] dx,$$

and by (17) of Sec. 10.5

$$| a_n | < \tfrac{1}{2} (M' + V') \frac{8}{\sqrt{\pi}} \left[\frac{1}{\sqrt{n + 2}(2n + 3)} + \frac{1}{\sqrt{n}(2n - 1)} \right]$$

$$(9) \quad |a_n| < \frac{4}{\sqrt{\pi}}(M' + V')\frac{1}{n^{3/2}}\left[\sqrt{\frac{n}{n+2}}\frac{n}{2n+3} + \frac{n}{2n-1}\right].$$

The factor in square brackets is less than 3/2 for $n \geqq 1$, and less than 1 for $n \geqq 2$. Therefore (7_1) and (7_2) follow from (9).

In $[-1, 1]$ we have $|P_n(x)| \leqq 1$, whence

$$\left|\sum_{n=1}^{\infty} a_n P_n(x)\right| \leqq \sum_{n=1}^{\infty} |a_n P_n(x)| < \frac{6}{\sqrt{\pi}}(M' + V') \sum_{n=1}^{\infty} n^{-3/2}$$

and, therefore, the Legendre series of $F(x)$ converges uniformly and absolutely in $[-1, 1]$ and consequently converges there to $F(x)$.

Now for $n \geqq 1$

$$\left|F(x) - \sum_{k=0}^{n} a_k P_k(x)\right| = \left|\sum_{k=n+1}^{\infty} a_k P_k(x)\right| < \frac{4}{\sqrt{\pi}}(M' + V') \sum_{k=n+1}^{\infty} n^{-3/2}.$$

Since, however, $\sum_{k=n+1}^{\infty} n^{-3/2} < 2n^{-1/2}$, we have for $n \geqq 1$

$$|R_{n+1}(x)| = \left|F(x) - \sum_{k=0}^{n} a_k P_k(x)\right| < \frac{8}{\sqrt{\pi}}(M' + V')\frac{1}{\sqrt{n}};$$

$$|x| \leqq 1, \quad n \geqq 1.$$

If, on the other hand, we restrict x by the relation $|x| \leqq \delta < 1$ then by (7_2) and (3_1) we have

$$\left|F(x) - \sum_{k=0}^{n} a_k P_k(x)\right| < \frac{16\sqrt{2}}{\pi}\frac{M' + V'}{\sqrt[4]{1-\delta^2}} \sum_{k=n+1}^{\infty} n^{-2}$$

and, consequently,

$$|R_{n+1}(x)| < \frac{16\sqrt{2}}{\pi}\frac{M' + V'}{\sqrt[4]{1-\delta^2}}\frac{1}{n} \quad \text{for } |x| \leqq \delta < 1, n \geqq 1.$$

Thus the theorem is proved.

In Sec. 14 we shall derive criteria for pointwise convergence of series of Legendre polynomials. To this end we shall establish Stieltjes' formula for the asymptotic approximation of Legendre polynomials [Sec. 12], and shall prove two theorems on limits of certain integrals [Sec. 13].

12. Formulas and Series for Asymptotic Approximation of Legendre Polynomials

1. We list below the important approximation formulas for Legendre polynomials:

a) Laplace's formula [65e]:

$$(1)\qquad P_n(\cos\gamma) = \sqrt{\frac{2}{\pi n \sin\gamma}}\cos\left[(n+\tfrac{1}{2})\gamma - \frac{\pi}{4}\right] + O(n^{-3/2}),$$

$$\varepsilon \leqq \gamma \leqq \pi - \varepsilon, \quad 0 < \varepsilon < \pi/2.$$

b) Bonnet-Heine's formula [O. Bonnet [11]]

$$(2)\qquad P_n(\cos\gamma) = \sqrt{\frac{2}{\pi n \sin\gamma}}\left\{\left(1 - \frac{1}{4n}\right)\cos\left[(n+\tfrac{1}{2})\gamma - \frac{\pi}{4}\right]\right.$$

$$\left. + \frac{1}{8n}\cot\gamma\sin\left[(n+\tfrac{1}{2})\gamma - \frac{\pi}{4}\right]\right\} + O(n^{-5/2}),$$

$$\varepsilon \leqq \gamma \leqq \pi - \varepsilon, \quad 0 < \varepsilon < \pi/2.$$

This result was first stated incorrectly by Bonnet with $(-1)^n/4n$ in place of $-1/4n$. The correction is due to Heine [[45]; pp. 171—187].

c) Darboux' formula and series [26]:

$$(3)\qquad P_n(\cos\gamma) = 2\alpha_n\left\{\frac{\cos\left[(n+\frac{1}{2})\gamma - \frac{\pi}{4}\right]}{(2\sin\gamma)^{1/2}}\right.$$

$$- \alpha_1\frac{1}{2n-1}\,\frac{\cos[(n-\frac{1}{2})\gamma + \frac{1}{4}\pi]}{(2\sin\gamma)^{3/2}}$$

$$\left. + \alpha_2\frac{1\times 3}{(2n-1)(2n-3)}\,\frac{\cos[(n-\frac{3}{2})\gamma + \frac{3}{4}\pi]}{(2\sin\gamma)^{5/2}} - \cdots\right\}$$

where

$$(3') \qquad \alpha_n = \frac{(2n-1)!!}{(2n)!!}, \qquad \varepsilon \leqq \gamma \leqq \pi - \varepsilon,\ 0 < \varepsilon < \pi/2.$$

We recall that

$$(2n-1)!! = 1\times 3\times 5\times \ldots \times (2n-1);\ (2n)!! = 2\times 4\times 6\times \ldots \times (2n).$$

The series is convergent for $\pi/6 < \gamma < 5\pi/6$. If the sum of the first r terms is taken as an approximation for $P_n(\cos\gamma)$ for $0 < \gamma < \pi$, the error is of the order $O(n^{-r-\frac{1}{2}})$.

d) Stieltjes' formula [104c]:

$$\begin{aligned} P_n(\cos\gamma) = \frac{4}{\pi}\frac{1}{(2n+1)\alpha_n}\Bigg[&\frac{\cos\left[(n+\frac{1}{2})\gamma - \frac{\pi}{4}\right]}{(2\sin\gamma)^{\frac{1}{2}}} \\ (4) \qquad &+ \alpha_1 \frac{1}{2n+3}\frac{\cos\left[(n+\frac{3}{2})\gamma - \frac{3}{4}\pi\right]}{(2\sin\gamma)^{3/2}} \\ &+ \alpha_2 \frac{1\times 3}{(2n+3)(2n+5)}\frac{\cos\left[(n+\frac{5}{2})\gamma - \frac{5}{4}\pi\right]}{(2\sin\gamma)^{5/2}} + \ldots\Bigg]. \end{aligned}$$

The series is convergent for $\pi/6 < \gamma < 5\pi/6$. If the sum of the first r terms is taken as an approximation of $P_n(\cos\gamma)$ for $0 < \gamma < \pi$, the error $p_{n,r}(\gamma)$ satisfies the inequality

$$(5) \qquad |\, p_{n,r}(\gamma)\,| < \frac{4}{\pi}\frac{1}{(2n+1)\alpha_n} \cdot \alpha_r \frac{(2r-1)!!}{(2n+3)(2n+5)\ldots(2n+2r+1)}\frac{2}{(2\sin\gamma)^{r+\frac{1}{2}}}.$$

In other words, the error is less than twice the absolute value of the first term omitted if the cosine term is replaced by 1.

e) Additional approximation formulas are due to Watson and Szegö [107c].

2. In this section we shall give essentially Stieltjes' proof of formula (4).

The point of departure is the Laplace integral formula [Sec. 6.1, (19)]

$$P_n(x) = \frac{1}{\pi}\int_0^\pi [x + i\sqrt{1-x^2}\cos\varphi]^n d\varphi, \qquad -1 < x < 1,$$

where the radical $\sqrt{1-x^2}$ is to be interpreted to mean positive square root.

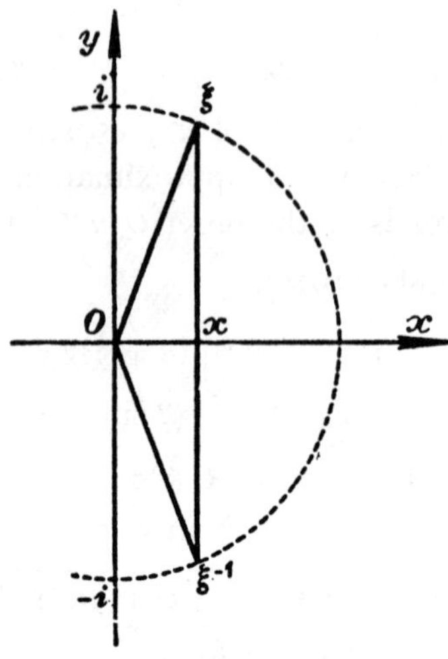

Fig. 13.

Let

$$\text{(6)} \qquad \begin{aligned} \xi &= x + i\sqrt{1-x^2}, \\ \xi^{-1} &= x - i\sqrt{1-x^2}, \end{aligned}$$

and make the following change in variables

$$x + i\sqrt{1-x^2}\cos\varphi = z.$$

Then

$$\begin{aligned} \sin\varphi &= \sqrt{1-2xz+z^2}/\sqrt{1-x^2}, \\ d\varphi &= -\,i^{-1}dz/\sqrt{1-2xz+z^2}, \end{aligned}$$

where $\sqrt{1-2xz+z^2}$ denotes the branch of the function which assumes the value 1 for $z = 0$ for values of z such that $|z| \leqq 1$. It follows that

$$P_n(x) = \frac{1}{\pi i}\int_{\xi^{-1}}^{\xi} \frac{z^n}{\sqrt{1-2xz+z^2}}\,dz.$$

Since the integral $\int \frac{z^n}{\sqrt{1-2xz+z^2}}\,dz$ vanishes on the contour consisting of the triangle whose vertices are the points 0, ξ^{-1}, ξ, we have

$$P_n(x) = \frac{1}{\pi i}\left[\int_0^{\xi} \frac{z^n}{\sqrt{1-2xz+z^2}}\,dz - \int_0^{\xi^{-1}} \frac{z^n}{\sqrt{1-2xz+z^2}}\,dz\right].$$

If, now, we write

$$(7_1) \qquad A = \frac{1}{\pi i}\int_0^{\xi} \frac{z^n}{\sqrt{1-2xz+z^2}}\,dz,$$

$$(7_2) \qquad B = \frac{1}{\pi}\int_0^{\xi^{-1}} \frac{z^n}{\sqrt{1-2xz+z^2}}\,dz,$$

then

$$(8) \qquad P_n(x) = A - B.$$

If in A we let $z = \xi(1-u)$, $0 \leqq u \leqq 1$, and $k = \xi^2/(\xi^2-1)$, then

$$z^2 - 2xz + 1 = (1-\xi z)(1-\xi^{-1}z) = u(1-\xi z) = \frac{1}{1-k}u(1-ku).$$

Finally, if we let $\sqrt{u}$ denote the positive square root and let $\sqrt{1-k}$ denote the value of $\sqrt{1-ku}$ for $u = 1$, we have

$$A = \frac{1}{\pi i}\int_1^0 \xi^n(1-u)^n\sqrt{1-k}\;\frac{-\xi\,du}{\sqrt{u(1-ku)}}$$
$$= \frac{1}{\pi i}\,\xi^{n+1}\sqrt{1-k}\int_0^1 \frac{(1-u)^n du}{\sqrt{u(1-ku)}}.$$

Similarly for B, we have

$$(9) \qquad P_n(x) = \frac{1}{\pi i}\left[\xi^{n+1}\sqrt{1-k}\int_0^1 \frac{(1-u)^n du}{\sqrt{u(1-ku)}} - \xi^{-(n+1)}\sqrt{1-k_1}\int_0^1 \frac{(1-u)^n du}{\sqrt{u(1-k_1u)}}\right]$$

where

(10) $$k = \frac{\xi^2}{\xi^2 - 1}, \qquad k_1 = \frac{\xi^{-2}}{\xi^{-2} - 1}.$$

If we write $x = \cos \gamma$ where $0 < \gamma < \pi$, then

$$\xi = e^{i\gamma}, \quad \xi^{-1} = e^{-i\gamma}$$

$$k = \frac{\xi}{\xi - \xi^{-1}} = \frac{\cos \gamma + i \sin \gamma}{2i \sin \gamma} = \frac{e^{i\left(\gamma - \frac{\pi}{2}\right)}}{2 \sin \gamma}$$

or

(11) $$k = \frac{e^{i\alpha}}{2 \sin \gamma}, \qquad \alpha = \gamma - \frac{\pi}{2}.$$

Now

$$1 - ku = 1 - u/2 + iu/2 + iu/2 \cot \gamma,$$

whence the real part of $\sqrt{1 - ku}$ can never vanish as u varies from 0 to 1, and, consequently, if we agree that $\sqrt{1 - ku} = 1$ for $u = 0$, then we must choose the branch of $\sqrt{1 - k}$ with positive real part.

Moreover

$$Re\, \sqrt{1 - k} = Re \sqrt{-\frac{k}{\xi^2}} = Re\, \frac{i\sqrt{k}}{\xi} = Re\, \frac{ie^{\frac{1}{2}\alpha i}}{\xi\sqrt{2 \sin \gamma}}$$

$$= Re\, \frac{ie^{(\frac{1}{2}\alpha - \gamma)i}}{\sqrt{2 \sin \gamma}} = -\frac{\sin\left(\frac{1}{2}\alpha - \gamma\right)}{\sqrt{2 \sin \gamma}} = \frac{\sin\left(\frac{\gamma}{2} + \frac{\pi}{4}\right)}{\sqrt{2 \sin \gamma}}.$$

It follows from our conventions that $\sqrt{2 \sin \gamma}$ denotes the positive square root.

Now

$$\xi^{n+1}\sqrt{1 - k} = \frac{ie^{(n\gamma + \frac{1}{2}\alpha)i}}{\sqrt{2 \sin \gamma}}, \qquad \xi^{-(n+1)}\sqrt{1 - k_1} = \frac{-ie^{-(n\gamma + \frac{1}{2}\alpha)i}}{\sqrt{2 \sin \gamma}}$$

and consequently (9) can be written

$$(12)\quad P_n(\cos\gamma)=\frac{e^{(n\gamma+\frac{1}{2}\alpha)i}}{\pi\sqrt{2\sin\gamma}}\int_0^1\frac{(1-u)^n\,du}{\sqrt{u(1-ku)}}+\frac{e^{-(n\gamma+\frac{1}{2}\alpha)i}}{\pi\sqrt{2\sin\gamma}}\int_0^1\frac{(1-u)^n\,du}{\sqrt{u(1-k_1u)}}$$

whence

$$(12')\qquad P_n(\cos\gamma)=\frac{2}{\pi\sqrt{2\sin\gamma}}\,Re\;e^{(n\gamma+\frac{1}{2}\alpha)i}\int_0^1\frac{(1-u)^n\,du}{\sqrt{u(1-ku)}}.$$

Since $|k|=|e^{i\alpha}/(2\sin\gamma)|=1/(2\sin\gamma)$, it follows that for $\sin\gamma>\frac{1}{2}$, namely for $\pi/6<\gamma<5\pi/6$, the following series expansion is uniformly convergent to the function in the left member

$$\frac{1}{\sqrt{1-ku}}=1+\sum_{r=1}^{\infty}\frac{(2r-1)!!}{(2r)!!}k^r u^r=1+\sum_{r=1}^{\infty}\frac{(2r-1)!!}{(2r)!!}\frac{e^{ir\alpha}}{2^r\sin^r\gamma}u^r,$$

whence

$$\frac{e^{(n\gamma+\frac{1}{2}\alpha)i}(1-u)^n}{\sqrt{u(1-ku)}}=e^{(n\gamma+\frac{1}{2}\alpha)i}u^{-\frac{1}{2}}(1-u)^n+\sum_{r=1}^{\infty}\frac{(2r-1)!!}{(2r)!!}\frac{e^{(n\gamma+\frac{1}{2}\alpha+r\alpha)i}}{2^r\sin^r\gamma}u^{r-\frac{1}{2}}(1-u)^n,$$

and

$$(13)\quad Re\,\frac{e^{(n\gamma+\frac{1}{2}\alpha)i}(1-u)^n}{\sqrt{u(1-ku)}}=\cos\left(n\gamma+\tfrac{1}{2}\alpha\right)u^{-\frac{1}{2}}(1-u)^n+\sum_{r=1}^{\infty}\frac{(2r-1)!!}{(2r)!!}\frac{\cos\left(n\gamma+\frac{1}{2}\alpha+r\alpha\right)}{2^r\sin^r\gamma}(1-u)^n u^{r-\frac{1}{2}}.$$

But by the well-known properties of the B function, we have

$$\frac{(2r-1)!!}{(2r)!!}\int_0^1(1-u)^n u^{r-\frac{1}{2}}\,du=\frac{(2r-1)!!}{(2r)!!}B(r+\tfrac{1}{2};n+1)$$

$$=\frac{(2r-1)!!}{(2r)!!}\frac{n!}{(r+\frac{1}{2})(r+\frac{3}{2})\ldots(r+\frac{1}{2}+n)}=\frac{[(2r-1)!!]^2(2n)!!2}{(2r)!!(2n+2r+1)!!}$$

$$=\frac{2}{\alpha_n(2n+1)}\alpha_r\frac{(2r-1)!!}{(2n+3)(2n+5)\ldots(2n+2r+1)},$$

and, therefore, term by term integration of (13) between 0 and 1 gives Stieltjes' series (4) which converges for $\pi/6 < \gamma < 5\pi/6$ as previously stated.

3a. Following the arguments used by Stieltjes we shall show that whether or not γ satisfies the inequality $\pi/6 < \gamma < 5\pi/6$ if the sum of the first r terms of (4) is taken as an approximation for $P_n(\cos\gamma)$, the error $p_{n,r}(\gamma)$ is less than (5).

First we observe that

$$\frac{1}{\pi}\int_0^\pi \frac{dv}{1-ku\sin^2 v} = \frac{2}{\pi}\int_0^{\pi/2}\frac{dv}{\cos^2 v + (1-ku)\sin^2 v}$$

$$= \frac{2}{\pi}\int_0^{\pi/2}\frac{d\tan v}{1+(1-ku)\tan^2 v} = \frac{2}{\pi}\int_0^{+\infty}\frac{dt}{1+(1-ku)t^2} = \frac{1}{\sqrt{1-ku}},$$

where $\sqrt{1-ku}$ denotes the branch whose real part is positive and therefore (12′) can be written

$$P_n(\cos\gamma) = \frac{2}{\pi^2\sqrt{2\sin\gamma}}\, Re\, e^{(n\gamma+\frac{1}{2}\alpha)i}\int_0^1 u^{-\frac{1}{2}}(1-u)^n du \int_0^\pi \frac{dv}{1-ku\sin^2 v}. \tag{14}$$

However,

$$\frac{1}{1-ku\sin^2 v} = 1 + ku\sin^2 v + \ldots + (ku\sin^2 v)^{r-1} + \frac{(ku\sin^2 v)^r}{1-ku\sin^2 v},$$

and since

$$\int_0^\pi \sin^{2s} v\, dv = \pi\frac{(2s-1)!!}{(2s)!!},$$

and $k^r = e^{ir\alpha}/(2\sin\gamma)^r$ it follows that if we take the sum of the first r terms as an approximation for $P_n(\cos\gamma)$ the error $p_{n,r}(\gamma)$ is

$$\begin{aligned} p_{n,r}(\gamma) &= \frac{2}{\pi^2\sqrt{(2\sin\gamma)^{2r+1}}}\, Re\, e^{\left[(n+\gamma+\frac{1}{2})\gamma-(r+\frac{1}{2})\frac{\pi}{2}\right]i} \\ &\quad\times\int_0^1 u^{r-\frac{1}{2}}(1-u)^n du\int_0^\pi \frac{\sin^{2r} v}{1-ku\sin^2 v}dv, \end{aligned} \tag{15}$$

and therefore

$$|p_{n,r}(\gamma)| \leq \frac{2}{\pi\sqrt{(2\sin\gamma)^{2r+1}}}\left|\frac{1}{\pi}\int_0^1 u^{r-\frac{1}{2}}(1-u)^n du\int_0^\pi\frac{\sin^{2r} v}{1-ku\sin^2 v}dv\right|.$$

Also

$$|1 - ku \sin^2 v|^{-1} = \left|1 - \tfrac{1}{2}(1 - i \cot \gamma)u \sin^2 v\right|^{-1}$$
$$= 2 \sin \gamma / \sqrt{4 \sin^2 \gamma \cos^2 \gamma + (2 \sin^2 \gamma - u \sin^2 v)^2}$$

whence, for $\sin^2 \gamma \leqq \frac{1}{2}$, we have

$$|1 - ku \sin^2 v|^{-1} \leqq 2 \sin \gamma / \sqrt{4 \sin^2 \gamma \cos^2 \gamma} = 1/|\cos \gamma| < 2.$$

While for $\sin^2 \gamma > \frac{1}{2}$, we have

$$(2 \sin^2 \gamma - u \sin^2 v)^2 \geqq (2 \sin^2 \gamma - 1)^2 = \cos^2 2\gamma,$$
$$4 \sin^2 \gamma \cos^2 \gamma + (2 \sin^2 \gamma - u \sin^2 v)^2 \geqq \sin^2 2\gamma + \cos^2 2\gamma = 1.$$

Consequently

$$|1 - ku \sin^2 v|^{-1} < 2 \sin \gamma$$

and, therefore, in every case

$$|1 - ku \sin^2 v|^{-1} < 2.$$

Finally, from (16) we infer

$$|p_{n,r}(\gamma)| < \frac{4}{\pi\sqrt{(2 \sin \gamma)^{2r+1}}} \int_0^1 (1-u)^n u^{r-\frac{1}{2}} du \frac{1}{\pi} \int_0^\pi \sin^{2r} v \, dv$$

or

$$|p_{n,r}(\gamma)| < \frac{4}{\pi} \frac{1}{(2n+1)\alpha_n} \alpha_r \frac{(2r-1)!!}{(2n+3)(2n+5)..(2n+2r+1)} \frac{2}{(2 \sin \gamma)^{r+\frac{1}{2}}},$$

which is precisely (5).

For r fixed, since $1/\alpha_n = (\pi n)^{1/2}(1 + \theta/2n)^{-\frac{1}{2}}$ [Sec. 10. 3, (12)] it follows that $1/\alpha_n = O(n^{1/2})$ and therefore

$$\text{(17)} \qquad \boxed{|p_{n,r}(\gamma)| < \frac{A}{(n \sin \gamma)^{r+\frac{1}{2}}}}, \quad 0 < \gamma < \pi, \quad n \geqq 1,$$

where A is a constant independent of n and γ.

3b. We now derive an inequality for $dp_{n,r}(\gamma)/d\gamma$ which will be needed in Sec. 14.

If we take the derivative of both sides of (15) with respect to γ, observe that $n+r+\frac{1}{2}$ comes out as a factor in $\frac{d}{d\gamma}e^{\left[(n+r+\frac{1}{2})\gamma-(r+\frac{1}{2})\frac{\pi}{2}\right]i}$ and that

$$\left|\frac{d}{d\gamma}\frac{1}{1-ku\sin^2 v}\right| = \frac{|u\sin^2 v|}{|1-ku\sin^2 v|^2}\left|\frac{d}{d\gamma}\left(\tfrac{1}{2}-\frac{i}{2}\cot\gamma\right)\right| < \frac{2}{\sin^2\gamma},$$

it follows that for r fixed and $\varepsilon \leqq \gamma \leqq \pi-\varepsilon$, $0<\varepsilon<\pi/2$

$$\boxed{\frac{dp_{n,r}(\gamma)}{d\gamma} = O(n^{-r+\frac{1}{2}})}, \qquad \varepsilon \leqq \gamma \leqq \pi-\varepsilon. \tag{18}$$

13. *Limits of Integrals: Singular Integrals*

The following theorems concerning the expansion in series of integrals of a given function will be needed in the discussion in Sec. 14 of the expansion of a function in series of Legendre polynomials.

THEOREM 1. Let $\Phi(x', x, \lambda_n)$ be defined for $x' \in [a, b]$ where $[a, b]$ is a finite interval, $x \in g$ where g is any subset of the x-axis, and $\lambda_n \in \{\lambda_n\}$ where $\{\lambda_n\}$ is any sequence of numbers, real or complex. Now suppose the following conditions are satisfied:

a) For fixed $x \in g$ and $\lambda_n \in \{\lambda_n\}$ then for all $x' \in [a, b]$, except at most a set of measure zero,

$$|\Phi(x', x, \lambda_n)| < k \tag{1}$$

where k is a constant independent of x and λ_n.

b) For fixed $x \in g$ and $\lambda_n \in \{\lambda_n\}$, let the function $\Phi(x', x, \lambda_n)$ be integrable in every subinterval $[\alpha, \beta]$ of $[a, b]$, then

$$\lim_{n\to\infty}\int_\alpha^\beta \Phi(x', x, \lambda_n)dx' = 0 \tag{2}$$

uniformly for all $x \in g$.

Under these hypotheses, we shall show that, if $f(x')$ is integrable in $[a, b]$ then

$$\lim_{n\to\infty}\int_a^b f(x')\Phi(x', x, \lambda_n)dx' = 0, \tag{3}$$

uniformly for all $x \in g$.

Proof. For $\varepsilon > 0$, there exists a continuous function $\varphi(x')$ such that

$$\int_a^b | f(x') - \varphi(x') | \, dx' < \frac{\varepsilon}{k} \tag{4}$$

(Ch. II, Sec. 2.5). Now subdivide $[a, b]$ into subintervals $[a_{s-1}, a_s]$, $s = 1, 2, \ldots, r$, where $a = a_0 < a_1 < \ldots < a_{r-1} < a_r = b$ such that the oscillation of $\varphi(x')$ in every subinterval is less than $\varepsilon/[k(b-a)]$. Denote by $\psi(x')$ the function defined as follows:

$$\psi(a_s) = 0, \qquad (s = 0, 1, 2, \ldots, r)$$

$$\psi(x') = \varphi\left(\frac{a_{s-1} + a_s}{2}\right) = c_s \text{ for } a_{s-1} < x' < a_s, \quad (s = 1, 2, \ldots, r).$$

$\psi(x')$ assumes the values $0, c_1, c_2, \ldots c_r$, in $[a, b]$, and since $| \varphi(x') - \psi(x') | < \varepsilon/[k(b-a)]$, except possibly at the points $x' = a_s$, it follows that

$$\int_a^b | \varphi(x') - \psi(x') | \, dx' < \frac{\varepsilon}{k} \tag{5}$$

whence by (4) and (5)

$$\int_a^b | f(x') - \psi(x') | \, dx' < \frac{2\varepsilon}{k}. \tag{6}$$

By hypothesis $f(x') \Phi(x', x, \lambda_n)$ is integrable with respect to x' in $[a, b]$, and we have, therefore,

$$\left| \int_a^b f(x') \Phi(x', x, \lambda_n) dx' \right| \leqq \left| \int_a^b \{f(x') - \psi(x')\} \Phi(x', x, \lambda_n) dx' \right|$$
$$+ \sum_{s=1}^r |c_s| \left| \int_{a_{s-1}}^{a_s} \Phi(x', x, \lambda_n) dx' \right| \leqq 2\varepsilon + \sum_{s=1}^r |c_s| \left| \int_{a_{s-1}}^{a_s} \Phi(x', x, \lambda_n) dx' \right|,$$

and since, in accordance with the hypothesis, there exists a positive integer n_ε, such that for all $n > n_\varepsilon$ we have

$$\left| \int_{a_{s-1}}^{a_s} \Phi(x', x, \lambda_n) dx' \right| < \varepsilon / \sum_{s=1}^r | c_s |; \quad (s = 1, 2, \ldots, r; n > n_\varepsilon),$$

then for all $n > n_\varepsilon$, and for every x in g we have

$$\left| \int_a^b f(x')\Phi(x', x, \lambda_n)dx' \right| < 3\varepsilon, \qquad \text{Q.E.D.}$$

THEOREM 2. Let $F(x', x, \lambda_n)$ be defined for $x' \in [a, b]$, where $[a, b]$ is a finite interval, $x \in g$ where g is contained in $[a + \mu, b - \mu]$, $0 < \mu < (b - a)/2$, and for $\lambda_n \in \{\lambda_n\}$.

Suppose that for the function $F(x', x, \lambda_n)$ the following hypotheses are satisfied:

(a) There exists a positive number d, $\mu \geqq d > 0$, such that for a given x in g and a given λ_n of $\{\lambda_n\}$ and for all the values x' of $[a, b]$ where $|x' - x| \geqq d$, except possibly for a set of measure zero, we have

$$|F(x', x, \lambda_n)| < k \tag{7}$$

where k is a constant independent of x and λ_n.

(b) For every subinterval $[\alpha, \beta]$ of $[a, b]$, $a \leqq \alpha < \beta \leqq b$, and for every x in g not in the open interval $I' = (\alpha - d, \beta + d)$, let $F(x', x, \lambda_n)$ be integrable with respect to x' and let

$$\lim_{n\to\infty} \int_\alpha^\beta F(x', x, \lambda_n)\, dx' = 0 \tag{8}$$

uniformly for all x in g not in I'.

Under the preceding hypotheses, we proceed to prove that if $f(x')$ is integrable in $[a, b]$ and μ_x and μ_x' are two functions of x such that $d \leqq \mu_x \leqq \mu$, $d \leqq \mu_x' \leqq \mu$, and otherwise arbitrary, we have

$$\lim_{n\to\infty} \int_a^{x-\mu_x} f(x')F(x', x, \lambda_n)dx' = 0, \tag{9$_1$}$$

$$\lim_{n\to\infty} \int_{x+\mu_x'}^b f(x')F(x', x, \lambda_n)dx' = 0, \tag{9$_2$}$$

uniformly with respect to all x in g.

Proof. For a given x in g and a given λ_n of $\{\lambda_n\}$ we define $\Phi(x', x, \lambda_n)$ as follows:

$$\Phi(x', x, \lambda_n) = F(x', x, \lambda_n) \text{ if } a \leqq x' \leqq x - \mu_x, \tag{10$_1$}$$

$$\Phi(x', x, \lambda_n) = 0 \qquad \text{if } x - \mu_x < x' \leqq b. \tag{10$_2$}$$

Then for a given pair of values of x and λ_n and for all values of x', except possibly a set of measure zero, we have by (7)

$$|\,\Phi(x', x, \lambda_n)\,| < k.$$

Now let $[\alpha, \beta]$ be a subinterval of $[a, b]$ and let us consider the integral

$$J = \int_\alpha^\beta \Phi(x', x, \lambda_n)dx'.$$

If $x < \alpha + \mu_x$ $[x - \mu_x < \alpha \leqq x']$ we have

$$J = 0;$$

if $x > \beta + \mu_x$ $[x' \leqq \beta < x - \mu_x]$ we have

$$J = \int_\alpha^\beta F(x', x, \lambda_n)dx'$$

which, by hypothesis, tends uniformly to zero as $n \to \infty$ for all x in g such that $x > \beta + d$ and therefore also for $x > \beta + \mu_x$.

When $\alpha + \mu_x \leqq x \leqq \beta + \mu_x$ we have

$$J = \int_\alpha^{x-\mu_x} F(x', x, \lambda_n)dx'$$

and therefore, again, provided that $x - (x - \mu_x) = \mu_x \geqq d$ we have $J \to 0$ as $n \to \infty$.

We now prove that even when x satisfies the inequality

$$\alpha + \mu_x \leqq x \leqq \beta + \mu_x$$

the convergence of J to zero is uniform with respect to x.

In fact, given an $\varepsilon > 0$, determine an integer $m > 0$ such that $(\beta - \alpha)/m < \varepsilon/2k$ and consider the points

$$\alpha_r = \alpha + r(\beta - \alpha)/m, \quad (r = 0, 1, 2, \ldots, m).$$

Then by the hypotheses (a) and (b) there exists an integer n_ε such that for all $n > n_\varepsilon$, we have

$$\left|\int_\alpha^{\alpha_r} F(x', x, \lambda_n)dx'\right| < \varepsilon/2 \text{ for } r=0, 1, 2, \ldots, m;\ x \geqq \alpha_r + d;\ n > n_\varepsilon;$$

and, consequently, for $\alpha_r \leqq x - \mu_x < \alpha_{r+1}$ $[x \geqq \alpha_r + \mu_x \geqq \alpha_r + d]$ we have

$$|\,J\,| \leqq \left|\int_\alpha^{\alpha_r} F(x', x, \lambda_n)dx'\right| + \int_{\alpha_r}^{x-\mu_x} |\,F(x', x, \lambda_n)\,|\,dx' < \varepsilon \text{ for } n > n_\varepsilon.$$

We conclude that the function $\Phi(x', x, \lambda_n)$ satisfies the conditions of Theorem 1, and therefore

$$\lim_{n\to\infty} \int_a^b f(x')\Phi(x', x, \lambda_n)dx' = \lim_{n\to\infty} \int_\alpha^{x-\mu_x} f(x')F(x', x, \lambda_n)dx' = 0$$

uniformly with respect to x in g.

Similar reasoning holds for the integral (9_2).

DEFINITION. An integral of the type

$$\int_a^b f(x')F(x', x, \lambda_n)dx'$$

where $f(x')$ is integrable in $[a, b]$ and $F(x', x, \lambda)$ satisfies the conditions of Theorem 2 for a number $d > 0$ and which does not satisfy the conditions for $d = 0$, is called a *singular integral*. For example, the integral

$$\int_{-\pi}^{\pi} f(x') \frac{\sin n(x' - x)}{x' - x} dx', -\pi < x < \pi,$$

is a singular integral.

As we shall see in Sec. 14, and as we have already seen for Fourier expansions, the convergence of the series are intimately connected with the singular integrals.

14. Convergence of Series of Legendre Polynomials: Hobson's Theorem

1. Let $f(x)$ be integrable in $[-1, 1]$ and let

$$(1) \quad f(x) \sim \sum_{n=0}^{\infty} a_n P_n(x), \qquad a_n = \frac{2n+1}{2}\int_{-1}^{1} f(x')P_n(x')dx'.$$

Then the sum $S_n(x)$ of the first $n + 1$ terms of the series is given by the expression

$$S_n(x) = \int_{-1}^{1} \tfrac{1}{2} \sum_{k=0}^{n} \{(2k + 1)P_k(x)P_k(x')\} f(x')\, dx'$$

and by Christoffel's summation formula [Sec. 5, (18)]

$$(2) \quad \boxed{S_n(x) = \tfrac{1}{2}(n+1)\int_{-1}^{1} \frac{P_n(x)P_{n+1}(x') - P_{n+1}(x)P_n(x')}{x' - x} f(x')dx'}\,.$$

We suppose here that $f(x)(1-x^2)^{-\frac{1}{4}}$ is integrable in $[-1, 1]$ which is equivalent to assuming that $f(x)(1-x)^{-\frac{1}{4}}$ and $f(x)(1+x)^{-\frac{1}{4}}$ are integrable in intervals to the left of $+1$ and to the right of -1 respectively.

Letting $x = \cos\gamma$, $0 \leqq \gamma \leqq \pi$, our hypothesis is equivalent to supposing $f(\cos\gamma)\sin^{1/2}\gamma$ is integrable in $[0, \pi]$.

2. Let s be a positive number < 1, and suppose s is expressed as the sum of two positive numbers μ and ε, $\mu + \varepsilon = s$; also let d be a positive number, $0 < d \leqq \mu$, and let μ_x and μ'_x be two functions of x defined in $[-1, 1]$ that satisfy the inequalities

$$d \leqq \mu_x \leqq \mu; \quad d \leqq \mu'_x \leqq \mu.$$

Now consider the sum of the two integrals

$$I_n(x) = \tfrac{1}{2}(n+1)\int_{-1}^{x-\mu_x} \frac{P_n(x)P_{n+1}(x') - P_{n+1}(x)P_n(x')}{x'-x} f(x')dx'$$

$$+ \tfrac{1}{2}(n+1)\int_{x+\mu'_x}^{1} \frac{P_n(x)P_{n+1}(x') - P_{n+1}(x)P_n(x')}{x'-x} f(x')dx'$$

for all values of x in the interval $[-1+s, 1-s]$. Then we wish to prove that for a given positive number σ, there exists a positive integer n_σ, such that for $n > n_\sigma$ we have

$$|I_n(x)| < \sigma \text{ for } n > n_\sigma, \text{ and } -1+s \leqq x \leqq 1-s.$$

2a. We begin by proving that there exists a positive number ε_1 less than ε, such that

$$(3_1) \quad \left| \tfrac{1}{2}(n+1)\int_{-1}^{-1+\varepsilon_1} \frac{P_n(x)P_{n+1}(x') - P_{n+1}(x)P_n(x')}{x'-x} f(x')dx' \right| < \frac{\sigma}{4},$$

$$(3_2) \quad \left| \tfrac{1}{2}(n+1)\int_{1-\varepsilon_1}^{1} \frac{P_n(x)P_{n+1}(x') - P_{n+1}(x)P_n(x')}{x'-x} f(x')dx' \right| < \frac{\sigma}{4},$$

for

$$n = 0, 1, 2, \ldots; \quad -1+\varepsilon+\mu \leqq x \leqq 1-\varepsilon-\mu; \quad \varepsilon_1 < \varepsilon.$$

We give the proof of (3_1). Similar reasoning will establish (3_2).

By (14) of Sec. 10.3, we have

(a) $$| \sqrt{n+1}\, \sqrt[4]{1-x^2}\, P_n(x) | = | \sqrt{n}\, \sqrt[4]{1-x^2}\, P_n(x) \sqrt{1+1/n} \,| < 8/\sqrt{\pi},$$
$$-1 \leqq x \leqq 1, \quad n = 0, 1, 2, \ldots;$$

(b) $$| \sqrt{n+1}\, \sqrt[4]{1-x'^2}\, P_{n+1}(x') | < 4\sqrt{2}/\sqrt{\pi},$$
$$-1 \leqq x' \leqq 1, \quad n = 0, 1, 2, \ldots.$$

(c) For

$$0 < \varepsilon_1 < \varepsilon, -1 \leqq x' \leqq -1 + \varepsilon_1 < -1 + \varepsilon, -1 + \varepsilon + \mu \leqq x$$

we have

$$1/|x' - x| < 1/\mu.$$

(d) For

$$-1 + \varepsilon + \mu \leqq x \leqq 1 - \varepsilon - \mu$$

we have

$$1/\sqrt[4]{1-x^2} \leqq 1/\sqrt[4]{1-(1-\varepsilon-\mu)^2} < 1/\sqrt[4]{1-(1-\mu)^2};$$

therefore

$$\left| \tfrac{1}{2}(n+1) \int_{-1}^{-1+\varepsilon_1} \frac{P_n(x)P_{n+1}(x')}{x'-x} f(x')dx' \right| < \frac{16\sqrt{2}}{\pi\mu \sqrt[4]{1-(1-\mu)^2}} \int_{-1}^{-1+\varepsilon_1} \frac{|f(x')|}{\sqrt[4]{1-x'^2}} dx',$$

and by the integrability of $f(x')/\sqrt[4]{1-x'^2}$ in $[-1, 1]$ there exists a positive $\varepsilon_1 < \varepsilon$ such that

$$\left| \tfrac{1}{2}(n+1) \int_{-1}^{-1+\varepsilon_1} \frac{P_n(x)P_{n+1}(x')}{x'-x} f(x')dx' \right| < \frac{\sigma}{8}.$$

We can deduce similarly, by taking ε_1, smaller if necessary, that

$$\left| \tfrac{1}{2}(n+1) \int_{-1}^{-1+\varepsilon_1} \frac{P_{n+1}(x)P_n(x')}{x'-x} f(x')dx' \right| < \frac{\sigma}{8},$$

whence (3_1) follows.

2b. Let us now prove that there exists a positive integer n such that for $n > n_\sigma$, where

$$-1+\varepsilon+\mu \leqq x \leqq 1-\varepsilon-\mu,\ 0 < d \leqq \mu_x \leqq \mu,\ 0 < d \leqq \mu'_x \leqq \mu$$

we have

$$(4_1)\quad \left| \tfrac{1}{2}(n+1)\int_{-1+\varepsilon_1}^{x-\mu_x} \frac{P_n(x)P_{n+1}(x') - P_n(x')P_{n+1}(x)}{x'-x} f(x')dx' \right| < \frac{\sigma}{4},$$

$$(4_2)\quad \left| \tfrac{1}{2}(n+1)\int_{x+\mu'_x}^{1-\varepsilon_1} \frac{P_n(x)P_{n+1}(x') - P_{n+1}(x)P_n(x')}{x'-x} f(x')dx' \right| < \frac{\sigma}{4}.$$

To this end we define a function $F(x', x, n)$ by the following: for x in $[-1+\varepsilon_1+\mu, 1-\varepsilon_1-\mu]$ and n a positive integer, let

$$F(x', x, n) = \tfrac{1}{2}(n+1)\,\frac{P_n(x)P_{n+1}(x') - P_{n+1}(x)P_n(x')}{x'-x}$$

for

$$-1+\varepsilon_1 \leqq x' \leqq x-\mu_x;$$

let $F(x', x, n) = 0$ for $-1 \leqq x' < -1+\varepsilon_1$, or

$$x-\mu_x < x' \leqq 1.$$

We now prove that, as x' varies in $[-1, 1]$, and x in $[-1+\varepsilon_1+\mu, 1-\varepsilon_1-\mu]$, the function $F(x', x, n)$ has the properties of the $F(x', x, n)$ of Theorem 2 of Sec. 13.

In fact there exists a constant λ such that for $-1+\varepsilon_1 \leqq x' \leqq 1-\varepsilon_1$ we have $|P_n(x')| < \lambda n^{-\frac{1}{2}}$ [Sec. 10, (8)], and since $-1+\varepsilon_1+\mu \leqq x \leqq 1-\varepsilon_1-\mu$ we also have $|P_n(x)| < \lambda n^{-\frac{1}{2}}$, and when $F(x', x, n)$ is not zero we have $|x'-x| \geqq d$. Therefore, in all cases we have

$$|F(x', x, n)| < \tfrac{1}{2}(n+1)\,\frac{2\lambda^2}{dn^{1/2}(n+1)^{1/2}} \leqq \frac{\sqrt{2}\lambda^2}{d}$$

that is, for given ε_1 and μ, $|F(x', x, n)|$ is less than a constant ndependent of x', x and n.

If (α_1, β_1) is an interval of $[-1+\varepsilon_1, 1-\varepsilon_1]$ we next prove that

$$\lim_{n\to\infty} \int_{\alpha_1}^{\beta_1} F(x', x, n)dx' = 0 \tag{5}$$

uniformly for all values of x in $g = [-1+\varepsilon_1+\mu, 1-\varepsilon_1-\mu]$ not interior to $[\alpha_1 - d, \beta_1 + d]$.

Evidently we may suppose that $\beta_1 \leqq x - \mu_x$, for the upper limit of the integral

$$\int_{\alpha_1}^{\beta_1} F(x', x, n)\, dx',$$

and then since $1/(x'-x)$ is monotonically decreasing for x' in $[-1+\varepsilon_1, x-\mu_x]$ we get by the second theorem of the mean,

$$\begin{aligned}
&\int_{\alpha_1}^{\beta_1} F(x', x, n)dx' \\
&\quad = \frac{n+1}{2}\left\{P_n(x)\int_{\alpha_1}^{\beta_1} \frac{P_{n+1}(x')}{x'-x}\,dx' - P_{n+1}(x)\int_{\alpha_1}^{\beta_1} \frac{P_n(x')}{x'-x}\,dx'\right\} \\
&\quad = \frac{n+1}{2}\left[\frac{P_n(x)}{\alpha_1 - x}\int_{\alpha_1}^{\xi_1} P_{n+1}(x')dx' + \frac{P_n(x)}{\beta_1 - x}\int_{\xi_1}^{\beta_1} P_{n+1}(x')dx'\right] \\
&\quad\quad - \frac{n+1}{2}\left[\frac{P_{n+1}(x)}{\alpha_1 - x}\int_{\alpha_1}^{\xi_2} P_n(x')dx' + \frac{P_{n+1}(x)}{\beta_1 - x}\int_{\xi_2}^{\beta_1} P_n(x')dx'\right],
\end{aligned} \tag{6}$$

where $\alpha_1 < \xi_1,\ \xi_2 < \beta_1$.

But if $[\gamma_1, \gamma_2]$ is an interval of $[-1, 1]$ we have [Sec. 10, (17)]

$$\left|\int_{\gamma_1}^{\gamma_2} P_n(x')dx'\right| < \frac{4}{\sqrt{\pi n}\sqrt{n}}, \quad \left|\int_{\gamma_1}^{\gamma_2} P_{n+1}(x')dx'\right| < \frac{4}{\sqrt{\pi n}\sqrt{n}}.$$

We also have for all points x not interior to $[\alpha_1 - d, \beta_1 + d]$ simultaneously $|\alpha_1 - x| \geqq d$, $|\beta_1 - x| \geqq d$, and since for x in $[-1+\varepsilon_1, 1-\varepsilon_1]$

$$|P_n(x)| < \lambda n^{-\frac{1}{2}}, \qquad |P_{n+1}(x)| < \lambda n^{-\frac{1}{2}}$$

from (6) we get

$$\left|\int_{\alpha_1}^{\beta_1} F(x', x, n)dx'\right| < \frac{16\lambda}{\sqrt{\pi d}}\,\frac{1}{n}.$$

Consequently (5) holds uniformly for all values of x in g not interior to $[\alpha_1 - d, \beta_1 + d]$.

Applying Theorem 2 of the preceding section we get (4_1) [and in the same way (4_2)] for all values of x of the interval $[-1 + \varepsilon_1 + \mu, 1 - \varepsilon_1 - \mu]$ which has in its interior the interval $(-1 + \varepsilon + \mu, 1 - \varepsilon - \mu) = (-1 + s, 1 - s)$.

Adding (3_1), (3_2), (4_1), (4_2) we get the theorem which was to be proved.

3. If $T_n(x)$ is defined by

$$(7) \quad T_n(x) = \tfrac{1}{2}(n+1) \int_{x-\mu_x}^{x+\mu'_x} \frac{P_n(x)P_{n+1}(x') - P_{n+1}(x)P_n(x')}{x' - x} f(x')dx'$$

where

$$0 < \mu_x \leqq \mu, \quad 0 \leqq \mu'_x \leqq \mu, \quad -1 < x - \mu < x + \mu < 1,$$

then, for a fixed point x interior to $[-1, 1]$, we have, by Sec. 14.2,

$$S_n(x) = I_n(x) + T_n(x),$$

and for $\sigma > 0$ there exists an n_σ such that for $n > n_\sigma$ we have

$$| S_n(x) - T_n(x) | < \sigma \text{ for } n > n_\sigma.$$

Therefore, if one of the two limits $\lim_{n\to\infty} S_n(x)$ and $\lim_{n\to\infty} T_n(x)$ exists, the other one also exists and they have the same value; moreover, if the convergence to the limit of $T_n(\cos\gamma)$ when $n\to\infty$ is uniform for all γ in an interval $[\varrho+\tau, \pi-\varrho-\tau]$, for $\varrho > 0$, $\tau > 0$, $\varrho + \tau < \pi$, the convergence of $S_n(\cos\gamma)$, when $n \to \infty$, to the same limit, is uniform in the same interval and conversely.

4. In order to determine $\lim_{n\to\infty} T_n(x)$, we let $x = \cos\gamma$ and substitute for P_n and P'_n their asymptotic expressions.

Consider formula (7) for $T_n(x)$ for x varying in the interval $[-1 + s, 1 - s]$, and where

$$s = \mu + \varepsilon, \quad 0 < d \leqq \mu_x \leqq \mu, \quad 0 < d \leqq \mu'_x < \mu$$

for any positive $\mu < s$ and any $\varepsilon > 0$.

Let

$$\gamma = \text{arc cos } x,$$

$$\eta'_x = \text{arc cos } x - \text{arc cos } (x + \mu),\ \eta_x = \text{arc cos}(x - \mu) - \text{arc cos } x$$

$$x = \cos \gamma, \quad x + \mu = \cos (\gamma - \eta'_x), \quad x - \mu = \cos (\gamma + \eta_x)$$

where γ, η'_x, η_x are positive angles less than π.

The functions η'_x and η_x are continuous and positive for x in $[-1 + s,\ 1 - s]$. Therefore if $\eta(\eta > 0)$ denotes the minimum of arc cos u — arc cos $(u + \mu)$ for u in $[-1,\ 1 - \mu]$, we have

$$\gamma > \gamma - \eta \geqq \gamma - \eta'_x, \quad \gamma < \gamma + \eta \leqq \gamma + \eta_x$$

$$x < \cos (\gamma - \eta) \leqq x + \mu, \quad x > \cos (\gamma + \eta) \geqq x - \mu.$$

Now let

$$(8_1) \quad \begin{cases} \mu'_x = \cos (\gamma - \eta) - x = \cos [\text{arc cos } x - \eta] - x, \\ \mu_x = x - \cos (\gamma + \eta) = x - \cos [\text{arc cos } x + \eta] \end{cases}$$

$$(8_2) \quad x + \mu'_x = \cos (\gamma - \eta), \quad x - \mu_x = \cos (\gamma + \eta).$$

Then μ'_x and μ_x are continuous and positive for x in $[-1 + s, 1-s]$ and, consequently, have a positive minimum $d > 0$, so that

$$0 < d \leqq \mu'_x \leqq \mu, \quad 0 < d \leqq \mu_x \leqq \mu.$$

The μ_x and μ'_x given by (8_1) satisfy the conditions imposed on the μ_x and μ'_x in (7). Then, if we let

$$x = \cos \gamma, \quad x' = \cos \gamma', \quad 0 < \gamma < \pi, \quad 0 < \gamma' < \pi$$

$$1 - \varepsilon = \cos \varrho, \quad 1 - (\varepsilon + \mu) = 1 - s = \cos (\varrho + \tau),$$

$$0 < \varrho < \pi/2, \quad 0 < \tau, \quad \varrho + \tau < \pi/2$$

the expression (7) for $T_n(\cos \gamma)$ becomes

$$(9) \quad T_n(\cos \gamma)$$

$$= \tfrac{1}{2}(n+1) \int_{\gamma-\eta}^{\gamma+\eta} \frac{P_n(\cos \gamma) P_{n+1}(\cos \gamma') - P_{n+1}(\cos \gamma)\, P_n(\cos \gamma')}{\cos \gamma' - \cos \gamma} \sin \gamma' f(\cos \gamma') d\gamma'$$

$$\varrho > 0, \quad \tau > 0, \quad \varrho + \tau \leqq \gamma \leqq \pi - (\varrho + \tau), \quad [0 < \eta \leqq \tau],$$

and, in view of the results of Sec. 14.3, it follows that $S_n(\cos \gamma)$ and $T_n(\cos \gamma)$ either both converge to the same limit, or neither converges, for a given value of γ in $(0, \pi)$. Moreover, if $T_n(\cos \gamma)$

converges to a limiting value uniformly as $n \to \infty$ for all γ in an interval $[\varrho + \tau, \pi - \varrho - \tau]$, then $S_n(\cos \gamma)$ converges uniformly to the same limiting value as $n \to \infty$ in the same interval, and conversely.

5. We shall now obtain an alternative expression for $T_n(\cos \gamma)$ by introducing in (9) the asymptotic expressions for P_n and P_{n+1}.

By (4) and (5) of Sec. 12 we have the following formula for $P_n(\cos \gamma)$ valid for γ in $[\varrho, \pi - \varrho]$, and, therefore, also in $[\varrho + \tau, \pi - \varrho - \tau]$

$$(10) \quad P_n(\cos \gamma) = \frac{4}{\pi} \frac{1}{(2n+1)\alpha_n} \frac{\cos\left[(n+\frac{1}{2})\gamma - \frac{\pi}{4}\right]}{(2 \sin \gamma)^{1/2}} + p_{n,1}(\gamma),$$

where

$$(11_1) \quad p_{n,1}(\gamma) = \frac{2}{\pi} \frac{1}{(2n+1)(2n+3)\alpha_n} \frac{\cos\left[(n+\frac{3}{2})\gamma - \frac{3\pi}{4}\right]}{(2 \sin \gamma)^{3/2}} + p_{n,2}(\gamma)$$

$$(11_2) \quad p_{n,2}(\gamma) = \frac{\alpha(n, \gamma)}{n^{5/2}}, \quad \frac{dp_{n,2}}{d\gamma} = \frac{\beta(n, \gamma)}{n^{3/2}}$$

and where $\alpha(n, \gamma)$, $\beta(n, \gamma)$ are less in absolute value than a constant independent of n and γ. Replacing P_n and P_{n+1} in (9) in accordance with (10) and making use of the identity

$$\cos\left[(n+\tfrac{1}{2})\gamma - \frac{\pi}{4}\right] \cos\left[(n+\tfrac{3}{2})\gamma' - \frac{\pi}{4}\right]$$
$$- \cos\left[(n+\tfrac{3}{2})\gamma - \frac{\pi}{4}\right] \cos\left[(n+\tfrac{1}{2})\gamma' - \frac{\pi}{4}\right]$$
$$= \sin[(n+1)(\gamma - \gamma')] \sin\frac{\gamma+\gamma'}{2} - \cos[(n+1)(\gamma+\gamma')] \sin\frac{\gamma - \gamma'}{2}$$

we have

$$(12) \quad T_n(\cos \gamma) = U_n(\gamma) - A_n(\gamma) + B_n(\gamma) + C_n(\gamma) + D_n(\gamma)$$

where

$$(13)\quad U_n(\gamma) = \frac{2}{\pi^2}\frac{n+1}{(2n+1)(2n+3)\alpha_n\alpha_{n+1}}\frac{1}{(\sin\gamma)^{1/2}}$$
$$\times\int_{\gamma-\eta}^{\gamma+\eta}\frac{\sin[(n+1)(\gamma-\gamma')]}{\sin\frac{1}{2}(\gamma-\gamma')}\sin^{1/2}\gamma' f(\cos\gamma')\,d\gamma',$$

$$(13_1)\quad A_n(\gamma) = \frac{2}{\pi^2}\frac{n+1}{(2n+1)(2n+3)\alpha_n\alpha_{n+1}}\frac{1}{(\sin\gamma)^{1/2}}$$
$$\times\int_{\gamma-\eta}^{\gamma+\eta}\frac{\cos[(n+1)(\gamma+\gamma')]}{\sin\frac{1}{2}(\gamma+\gamma')}\sin^{1/2}\gamma' f(\cos\gamma')\,d\gamma'$$

$$(13_2)\quad B_n(\gamma) = \frac{2}{\pi}\frac{n+1}{(2n+1)\alpha_n}\int_{\gamma-\eta}^{\gamma+\eta}\frac{1}{\cos\gamma'-\cos\gamma}\left[\frac{\cos\left[(n+\frac{1}{2})\gamma-\dfrac{\pi}{4}\right]}{(2\sin\gamma)^{1/2}}p_{n+1,1}(\gamma')\right.$$
$$\left.-\frac{\cos\left[(n+\frac{1}{2})\gamma'-\dfrac{\pi}{4}\right]}{(2\sin\gamma')^{1/2}}p_{n+1,1}(\gamma)\right]\sin\gamma' f(\cos\gamma')\,d\gamma',$$

$$(13_3)\quad C_n(\gamma) = \frac{2}{\pi}\frac{n+1}{(2n+3)\alpha_{n+1}}\int_{\gamma-\eta}^{\gamma+\eta}\frac{1}{\cos\gamma'-\cos\gamma}\left[\frac{\cos\left[(n+\frac{3}{2})\gamma'-\dfrac{\pi}{4}\right]}{(2\sin\gamma')^{1/2}}p_{n,1}(\gamma)\right.$$
$$\left.-\frac{\cos\left[(n+\frac{3}{2})\gamma-\dfrac{\pi}{4}\right]}{(2\sin\gamma)^{1/2}}p_{n,1}(\gamma')\right]\sin\gamma' f(\cos\gamma')\,d\gamma',$$

$$(13_4)\quad D_n(\gamma) = \tfrac{1}{2}(n+1)\int_{\gamma-\eta}^{\gamma+\eta}\frac{1}{\cos\gamma'-\cos\gamma}[p_{n,1}(\gamma)p_{n+1,1}(\gamma')$$
$$-p_{n+1,1}(\gamma)p_{n,1}(\gamma')]\sin\gamma' f(\cos\gamma')\,d\gamma'.$$

We now prove that

$$(14)\quad \lim_{n\to\infty}A_n(\gamma) = \lim_{n\to\infty}B_n(\gamma) = \lim_{n\to\infty}C_n(\gamma) = \lim_{n\to\infty}D_n(\gamma) = 0$$

and that the convergence to zero is uniform for γ in $[\varrho+\tau, \pi-\varrho-\tau]$.

5a. *Uniform convergence to zero of* $A_n(\gamma)$ in $[\varrho + \tau, \pi - \varrho - \tau]$.

Making use of (12) of Sec. 10, namely,

$$\alpha_k = \sqrt{2}/\sqrt{\pi(2k+\theta)}, \quad 0 < \theta < 1,$$

we have the following expression for the numerical coefficient on the right side of (13_1):

$$\frac{2}{\pi^2}\frac{n+1}{(2n+1)(2n+3)\alpha_n\alpha_{n+1}} = \frac{2}{\pi^2}\frac{n+1}{(2n+1)(2n+3)}\frac{\pi}{2}\sqrt{(2n+\theta)(2n+2+\theta_1)},$$

and consequently

$$\lim_{n\to\infty}\frac{2}{\pi^2}\frac{n+1}{(2n+1)(2n+3)\alpha_n\alpha_{n+1}} = \frac{1}{2\pi}. \tag{15}$$

We shall now prove that as $n \to \infty$ the following integral

$$\int_{\gamma+\eta}^{\gamma+\eta}\frac{\cos[(n+1)(\gamma+\gamma')]}{\sin\frac{1}{2}(\gamma+\gamma')}\sin^{1/2}\gamma' f(\cos\gamma')d\gamma'$$

converges uniformly to zero for γ in $[\varrho + \tau, \pi - \varrho - \tau]$.

For a fixed γ in $[\varrho + \tau, \pi - \varrho - \tau]$, let

$$(16_1)\ F(\gamma',\gamma,n) = \frac{\cos[(n+1)(\gamma+\gamma')]}{\sin\frac{1}{2}(\gamma+\gamma')} \text{ for } \gamma-\eta \leqq \gamma' \leqq \gamma+\eta, [0<\eta\leqq\tau]$$

$$(16_2)\ F(\gamma',\gamma,n) = 0, \text{ for } \varrho \leqq \gamma' < \gamma-\eta, \text{ or } \gamma+\eta < \gamma' \leqq \pi-\varrho.$$

Then

$$\varrho + \frac{\tau}{2} = \tfrac{1}{2}(\varrho+\tau+\varrho) \leqq \tfrac{1}{2}(\gamma+\gamma') \leqq \tfrac{1}{2}(\pi-\varrho-\tau+\pi-\varrho) = (\pi-\varrho)-\frac{\tau}{2}.$$

Consequently, there exists a constant $\delta > 0$ such that $|\sin\frac{1}{2}(\gamma+\gamma')| \geqq \delta$, and, therefore,

$$|F(\gamma',\gamma,n)| < 1/\delta$$

where δ is independent of γ', γ, n.

For γ in $[\varrho + \tau, \pi - \varrho - \tau]$, consider the integral

$$J_n(\gamma) = \int_\alpha^\beta F(\gamma',\gamma,n)\,d\gamma'$$

taken over any subinterval $[\alpha, \beta]$ of $[\varrho, \pi - \varrho]$. In view of (16_1) and (16_2), this integral is equal to the following integral taken over a subinterval $[\alpha_1, \beta_1]$ of $[\gamma - \eta, \gamma + \eta]$,

$$\int_{\alpha_1}^{\beta_1} F(\gamma', \gamma, n) d\gamma' = \int_{\alpha_1}^{\beta_1} \frac{\cos[(n+1)(\gamma+\gamma')]}{\sin\frac{1}{2}(\gamma+\gamma')} d\gamma'.$$

If as γ' varies in $[\alpha_1, \beta_1]$, $\sin(\gamma+\gamma')/2$ is not monotone, we can break up the integral into the sum of two integrals for which $\sin(\gamma+\gamma')/2$ is monotone. Therefore, we may suppose for simplicity that $\sin(\gamma+\gamma')/2$ is monotone in $[\alpha_1, \beta_1]$ and consequently, by the second theorem of the mean, we have $[\alpha_1 < \xi < \alpha_2]$

$$\left|\int_{\alpha_1}^{\beta_1} \frac{\cos[(n+1)(\gamma+\gamma')]}{\sin\frac{1}{2}(\gamma+\gamma')} d\gamma'\right| = \left|\frac{1}{\sin\frac{1}{2}(\gamma+\alpha_1)} \int_{\alpha_1}^{\xi} \cos[(n+1)(\gamma+\gamma')] d\gamma'\right.$$

$$\left. + \frac{1}{\sin\frac{1}{2}(\gamma+\beta_1)} \int_{\xi}^{\beta_1} \cos[(n+1)(\gamma+\gamma')] d\gamma'\right| < 4/(n+1)\delta.$$

Therefore, $\lim_{n\to\infty} J_n(\gamma) = 0$ uniformly for γ in $[\varrho + \tau, \pi - \varrho - \tau]$.

Then by Theorem 1 of Sec. 13 we have

$$\lim_{n\to\infty} \int_{\rho}^{\pi-\rho} F(\gamma', \gamma, n) \sin^{1/2}\gamma' f(\cos\gamma') d\gamma'$$

$$= \lim_{n\to\infty} \int_{\gamma-\eta}^{\gamma+\eta} \frac{\cos[(n+1)(\gamma+\gamma')]}{\sin\frac{1}{2}(\gamma+\gamma')} \sin^{1/2}\gamma' f(\cos\gamma') d\gamma' = 0$$

uniformly for all γ in $[\varrho + \tau, \pi - \varrho - \tau]$.

5b. *Uniform convergence to zero of* $B_n(\gamma)$, $C_n(\gamma)$ *in* $[\varrho + \tau, \pi - \varrho - \tau]$.

If in the expression (13_2) of $B_n(\gamma)$ we substitute for $p_{n+1,1}(\gamma)$, $p_{n+1,1}(\gamma')$ the expressions given by (11_1) we get

$$B_n(\gamma) = B_n'(\gamma) + B_n''(\gamma), \tag{17}$$

where

$$(18_1)\quad B_n'(\gamma) = \frac{4}{\pi^2}\frac{n+1}{(2n+1)(2n+3)(2n+5)\alpha_n\alpha_{n+1}}$$

$$\times \int_{\gamma-\eta}^{\gamma+\eta}\left[\frac{\cos\left[(n+\frac{1}{2})\gamma-\frac{\pi}{4}\right]\cos\left[(n+\frac{5}{2})\gamma'-\frac{3\pi}{4}\right]}{(2\sin\gamma)^{1/2}(2\sin\gamma')^{3/2}}\right.$$

$$\left.-\frac{\cos\left[(n+\frac{1}{2})\gamma'-\frac{\pi}{4}\right]\cos\left[(n+\frac{5}{2})\gamma-\frac{3\pi}{4}\right]}{(2\sin\gamma')^{1/2}(2\sin\gamma)^{3/2}}\right]\frac{\sin\gamma' f(\cos\gamma')}{\cos\gamma'-\cos\gamma}d\gamma',$$

$$(18_2)\quad B_n''(\gamma) = \frac{2}{\pi}\frac{n+1}{(2n+1)\alpha_n}\int_{\gamma-\eta}^{\gamma+\eta}\left[\frac{\cos\left[(n+\frac{1}{2})\gamma-\frac{\pi}{4}\right]}{(2\sin\gamma)^{1/2}}p_{n+1,2}(\gamma')\right.$$

$$\left.-\frac{\cos\left[(n+\frac{1}{2})\gamma'-\frac{\pi}{4}\right]}{(2\sin\gamma')^{1/2}}p_{n+1,2}(\gamma)\right]\frac{\sin\gamma' f(\cos\gamma')}{\cos\gamma'-\cos\gamma}d\gamma'.$$

For the numerical factor which figures in $B_n'(\gamma)$ we have

$$\frac{4}{\pi^2}\frac{n+1}{(2n+1)(2n+3)(2n+5)\alpha_n\alpha_{n+1}}$$

$$=\frac{4}{\pi^2}\frac{n+1}{(2n+1)(2n+3)(2n+5)}O(\sqrt{n})O(\sqrt{n}) = O\left(\frac{1}{n+1}\right).$$

Moreover,

$$\frac{1}{n+1}\frac{1}{\cos\gamma'-\cos\gamma}\frac{1}{(2\sin\gamma)^{3/2}(2\sin\gamma')^{3/2}}$$

$$\left\{\cos\left[(n+\tfrac{1}{2})\gamma-\frac{\pi}{4}\right]\cos\left[(n+\tfrac{5}{2})\gamma'-\frac{3\pi}{4}\right]2\sin\gamma\right.$$

$$\left.-\cos\left[(n+\tfrac{1}{2})\gamma'-\frac{\pi}{4}\right]\cos\left[(n+\tfrac{5}{2})\gamma-\frac{3\pi}{4}\right]2\sin\gamma'\right\}$$

$$=\frac{1}{n+1}\frac{(2\sin\gamma)^{-3/2}(2\sin\gamma')^{-3/2}}{2\sin\frac{1}{2}(\gamma-\gamma')\sin\frac{1}{2}(\gamma+\gamma')}$$

$$\{\sin\tfrac{1}{2}(\gamma-\gamma')\cos[(n+2)(\gamma+\gamma')]$$

$$-\sin\tfrac{3}{2}(\gamma-\gamma')\cos[(n+1)(\gamma+\gamma')]$$

$$+\sin(\gamma-\gamma')\sin\tfrac{1}{2}(\gamma+\gamma')\cos[(n+1)(\gamma-\gamma')]$$

$$+\sin\tfrac{1}{2}(\gamma+\gamma')[3\sin\gamma\sin\gamma'-\cos\gamma\cos\gamma'-1]\sin[(n+1)(\gamma-\gamma')]\}.$$

When we divide the first three terms in the braces by $\sin 2^{-1}(\gamma - \gamma')$ we see at once that their contribution to B'_n as $n \to \infty$ is zero uniformly for all γ in $[\varrho + \tau, \pi - \varrho - \tau]$; therefore, it remains to consider the limit as $n \to \infty$ of:

$$\bar{B}'_n(\gamma) = \frac{1}{2^4 \sin^{3/2}\gamma} \frac{1}{n+1}$$
$$\times \int_{\gamma-\eta}^{\gamma+\eta} \frac{\sin[(n+1)(\gamma-\gamma')]}{\sin \frac{1}{2}(\gamma-\gamma')} \frac{3\sin\gamma\sin\gamma' - \cos\gamma\cos\gamma' - 1}{\sin^{1/2}\gamma'} f(\cos\gamma')d\gamma'.$$

Since

$$\varrho \leqq \varrho + \tau - \eta \leqq \gamma - \eta \leqq \gamma' \leqq \gamma + \eta \leqq \pi - \varrho - \tau + \eta \leqq \pi - \varrho$$

we have $|\sin\gamma'| \geqq \sin\varrho$. Moreover,

$$|3\sin\gamma\sin\gamma' - \cos\gamma\cos\gamma' - 1|$$
$$= |\cos(\gamma - \gamma') - 2\cos(\gamma + \gamma') - 1| \leqq 4,$$

whence

$$\left| \frac{3\sin\gamma\sin\gamma' - \cos\gamma\cos\gamma' - 1}{\sin^{1/2}\gamma'} f(\cos\gamma') \right| \leqq \frac{4}{\sin^{1/2}\varrho} |f(\cos\gamma')|.$$

On the other hand, since $|(\gamma - \gamma')/2| \leqq \eta/2 < \pi/2$, we have

$$\tfrac{1}{2}\left| \frac{\sin[(n+1)(\gamma-\gamma')]}{(n+1)\sin\frac{1}{2}(\gamma-\gamma')} \right| = \left| \frac{\sin(n+1)(\gamma-\gamma')}{(n+1)(\gamma-\gamma')} \frac{(\gamma-\gamma')/2}{\sin\frac{1}{2}(\gamma-\gamma')} \right| < \frac{\pi}{2}$$

and, therefore, for any positive number $\omega < \eta$ we have

$$|\bar{B}'_n(\gamma)| \leqq \frac{\pi}{2^2\sin^2\varrho} \int_{\gamma-\omega}^{\gamma+\omega} |f(\cos\gamma')|\,d\gamma'$$
$$+ \frac{1}{n+1} \frac{1}{2^2\sin^2\varrho\sin\frac{\omega}{2}} \left[\int_{\gamma-\eta}^{\gamma-\omega} |f(\cos\gamma')|\,d\gamma' + \int_{\gamma+\omega}^{\gamma+\eta} |f(\cos\gamma')|\,d\gamma' \right],$$

and from the integrability of $f(\cos\gamma')$ in $[\varrho, \pi - \varrho]$ it follows that $\lim_{n\to\infty} B'_n(\cos\gamma) = 0$, uniformly for all γ in $[\varrho + \tau, \pi - \varrho - \tau]$.

We shall now consider the $\lim_{n\to\infty} B''_n(\gamma)$. For its numerical coefficient we have

$$\frac{2}{\pi}\frac{n+1}{(2n+1)\alpha_n} = \frac{2}{\pi}\frac{n+1}{2n+1}O(n^{1/2}) = O(n^{1/2}).$$

The integrand has as a factor

$$\frac{1}{\cos\gamma' - \cos\gamma}\left[\frac{\cos\left[(n+\frac{1}{2})\gamma - \frac{\pi}{4}\right]p_{n+1,2}(\gamma')}{(2\sin\gamma)^{1/2}} - \frac{\cos\left[(n+\frac{1}{2})\gamma' - \frac{\pi}{4}\right]p_{n+1,2}(\gamma)}{(2\sin\gamma')^{1/2}}\right]$$

By applying Cauchy's theorem of the mean, this expression becomes

$$-\frac{1}{\sin\xi}\left[\frac{\cos\left[(n+\frac{1}{2})\gamma - \frac{\pi}{4}\right]}{(2\sin\gamma)^{1/2}}p'_{n+1,2}(\xi) - \frac{d}{d\xi}\frac{\cos\left[(n+\frac{1}{2})\xi - \frac{\pi}{4}\right]}{(2\sin\xi)^{1/2}}p_{n+1,2}(\gamma)\right]$$

where $\gamma - \eta < \xi < \gamma + \eta$.

But we have $p'_{n+1,2}(\xi) = O(n^{-3/2})$, $p_{n+1,2}(\gamma) = O(n^{-5/2})$, and, therefore

$$\frac{2}{\pi}\frac{n+1}{(2n+1)\alpha_n}\frac{1}{\cos\gamma' - \cos\gamma}\left[\frac{\cos\left[(n+\frac{1}{2})\gamma - \frac{\pi}{4}\right]}{(2\sin\gamma)^{1/2}}p_{n+1,2}(\gamma') - \frac{\cos\left[(n+\frac{1}{2})\gamma' - \frac{\pi}{4}\right]}{(2\sin\gamma')^{1/2}}p_{n+1,2}(\gamma)\right] = O\left(\frac{1}{n}\right)$$

uniformly for all γ' in $[\varrho, \pi - \varrho]$ and for all γ in $[\varrho + \tau, \pi - \varrho - \tau]$. Finally, from the integrability of $f(\cos\gamma')\sin\gamma'$ in $[\varrho, \pi - \varrho]$ it follows that $\lim_{n\to\infty} B''_n(\gamma) = 0$ uniformly for ali γ in $[\varrho + \tau, \pi - \varrho - \tau]$.

Analogous conclusions hold for $C_n(\gamma)$.

5c. *Uniform convergence to zero of* $D_n(\gamma)$ in $[\varrho + \tau, \pi - \varrho - \tau]$. Again, by Cauchy's theorem of the mean we have

$$\frac{1}{2}\frac{n+1}{\cos\gamma'-\cos\gamma}[p_{n,1}(\gamma)p_{n+1,1}(\gamma')-p_{n+1,1}(\gamma)p_{n,1}(\gamma')]$$
$$=-\frac{1}{2}\frac{n+1}{\sin\xi}[p_{n,1}(\gamma)p'_{n+1,1}(\xi)-p_{n+1,1}(\gamma)p'_{n,1}(\xi)],$$

where $\gamma-\eta<\xi<\gamma+\eta$. However, $p_{n,1}(\gamma)=O(n^{-3/2})$, $p'_{n,1}(\gamma)=O(n^{-1/2})$, whence

$$\frac{1}{2}\frac{n+1}{\cos\gamma'-\cos\gamma}[p_{n,1}(\gamma)p_{n+1,1}(\gamma')-p_{n+1,1}(\gamma)p_{n,1}(\gamma')]=O\left(\frac{1}{n}\right)$$

uniformly for all γ' in $[\varrho,\pi-\varrho]$ and for all γ in $[\varrho+\tau,\pi-\varrho-\tau]$, and therefore, $\lim_{n\to\infty} D_n(\gamma)=0$ uniformly for all γ in $[\varrho+\tau,\pi-\varrho-\tau]$.

6. HOBSON'S THEOREM

6a. By Secs. 14.1 to 14.5, it follows that for any positive number σ it is possible to determine a positive integer n_σ such that for all $n>n_\sigma$

$$(19)\quad |S_n(\cos\gamma)-U_n(\cos\gamma)|=\Bigg|S_n(\cos\gamma)-\frac{2}{\pi^2}\frac{n+1}{(2n+1)(2n+3)\alpha_n\alpha_{n+1}}$$
$$\times\frac{1}{(\sin\gamma)^{1/2}}\int_{\gamma-\eta}^{\gamma+\eta}\frac{\sin[(n+1)(\gamma-\gamma')]}{\sin\frac{1}{2}(\gamma-\gamma')}\sin^{1/2}\gamma' f(\cos\gamma')d\gamma'\Bigg|<\sigma$$

uniformly for all γ in $[\varrho+\pi,\pi-\varrho-\tau]$, where n is a positive constant not greater than τ and independent of σ.

Now, in view of (15) and the fact that the existence of

$$\lim_{n\to\infty}\frac{1}{2\pi}\int_{\gamma-\eta}^{\gamma+\eta}\frac{\sin[(n+1)(\gamma-\gamma')]}{\sin\frac{1}{2}(\gamma-\gamma')}\sin^{1/2}\gamma' f(\cos\gamma')d\gamma'$$

expresses the necessary and sufficient condition for the convergence of the Fourier series of $f(\cos\gamma)\sin^{1/2}\gamma$ at point γ (Cf. Ch. II, Sec. 4.4), we have that if $f(x)(1-x^2)^{-\frac{1}{4}}$ is integrable in $[-1,1]$, then a necessary and sufficient condition for the convergence of the series of Legendre polynomials of $f(x)$ at a point x interior to the interval $[-1,1]$ is that the Fourier series of $f(\cos\gamma)\sin^{1/2}\gamma$ is convergent at the point $\gamma=\arccos x$, $[0<\gamma<\pi]$.

6b. If we let $\cos\gamma_1 = x_1$, $\cos\gamma_2 = x_2$ then

$$\begin{aligned} f(\cos\gamma_1)\sin^{1/2}\gamma_1 - f(\cos\gamma_2)\sin^{1/2}\gamma_2 \\ &= \sin^{1/2}\gamma_1[f(\cos\gamma_1) - f(\cos\gamma_2)] + f(\cos\gamma_2)[\sin^{1/2}\gamma_1 - \sin^{1/2}\gamma_2] \\ &= (1-x_1^2)^{1/4}[f(x_1) - f(x_2)] + f(x_2)[(1-x_1^2)^{1/4} - (1-x_2^2)^{1/4}] \end{aligned}$$

and, therefore, if $f(x)$ is of bounded variation in an interval of $[-1, 1]$, then $f(\cos\gamma)\sin^{1/2}\gamma$ is of bounded variation in the corresponding interval of $[0, \pi]$.

Clearly, then, if at a point x_1 interior to $[-1, 1]$, $f(x)$ satisfies a Lipschitz condition of order α,

$$0 < \alpha \leqq 1, \quad | f(x_1) - f(x_2) | < L | x_1 - x_2 |^{\alpha},$$

in view of the fact that the function $(1 - x^2)^{1/4}$ has a finite derivative in an interval interior to $[-1, 1]$, and therefore satisfies a Lipschitz condition of the first order there, it follows that $f(\cos\gamma)\sin^{1/2}\gamma$ also satisfies a Lipschitz condition of order $\alpha > 0$ at the point γ_1 corresponding to x_1.

By the results of Ch. II we have thus established Hobson's theorem [cf. U. Dini [28c, e, f]; E. W. Hobson, [48b c, d,]]:

If $f(x')/(1 - x^2)^{1/4}$ is integrable in the interval $[-1, 1]$, then a necessary and sufficient condition for the convergence of its series of Legendre polynomials

$$\sum_{n=0}^{\infty} (n + \tfrac{1}{2}) P_n(x) \int_{-1}^{1} f(x') P_n(x') dx' \tag{20}$$

at a point x interior to $[-1, 1]$ is that the trigonometric Fourier series of $f(\cos\gamma)\sin^{1/2}\gamma$ be convergent at the point $\gamma = \text{arc} \cos x$, $0 < \gamma < \pi$, and under this condition if the sum of the first series is S, then the sum of the second series is $S \sin^{1/2}\gamma$.

In particular:

1) Series (20) converges to $[f(x+) + f(x-)]/2$ if $f(x)$ is of bounded variation in a neighborhood of x.

2) Series (20) converges to $f(x)$, if $f(x)$ satisfies a Lipschitz condition of order $\alpha > 0$ at x, in other words if we have $| f(x \pm h) - f(x) | < Lh^{\alpha}$, for $0 < h < \varepsilon$.

Moreover, under the hypothesis that $f(x)/(1-x^2)^{1/4}$ is integrable in $[-1, 1]$, we have:

If series (20) is uniformly convergent in an interval I interior to $[-1, 1]$, then the Fourier series of $f(\cos\gamma)\sin^{1/2}\gamma$ is uniformly convergent in the interval I' of $[0, \pi]$ corresponding to I by the transformation $x = \cos\gamma$, and conversely.

In particular, if $f(x)/(1-x^2)^{1/4}$ is integrable in $[-1, 1]$ and if $f(x)$ is continuous and of bounded variation in an interval I_1 of $[-1, 1]$, then series (20) is uniformly convergent to $f(x)$ in any interval interior to I_1.

7. Finally, we consider the convergence of the series

$$(1) \qquad \begin{cases} \displaystyle\sum_{n=0}^{\infty} a_n P_n(x)\,, \\ \displaystyle a_n = \frac{2n+1}{2}\int_{-1}^{1} f(x')P_n(x')dx' \end{cases}$$

at the points -1 and $+1$. We shall prove that if $f(x)$ is of bounded variation in the interval $[-1, 1]$, then its series of Legendre polynomials converges at points -1 and $+1$ to $f(-1+), f(1-)$ respectively.

By formula (2) for the sum $S_n(1)$ of the first n terms of (1) we have

$$(21) \quad S_n(1) = \tfrac{1}{2}(n+1)\int_{-1}^{1}\frac{P_{n+1}(x') - P_n(x')}{x'-1}f(x')dx' = I_{1,n} + I_{2,n} + I_{3,n}$$

where

$$I_{1,n} = \tfrac{1}{2}(n+1)\int_{-1}^{-1+\epsilon}\frac{P_n(x') - P_{n+1}(x')}{1-x'}f(x')dx',$$

$$I_{2,n} = \tfrac{1}{2}(n+1)\int_{1-\epsilon}^{1}\frac{P_n(x') - P_{n+1}(x')}{1-x'}f(x')dx',$$

$$I_{3,n} = \tfrac{1}{2}(n+1)\int_{-1+\epsilon}^{1-\epsilon}\frac{P_n(x') - P_{n+1}(x')}{1-x'}f(x')dx'.$$

If we substitute $x' = \cos \gamma'$, $0 \leqq \gamma' \leqq \pi$ in $I_{3,n}$ and let $1 - \varepsilon = \cos \varrho$, then in view of (10) and the fact that

$$\cos\left[(n + \tfrac{1}{2})\gamma' - \frac{\pi}{4}\right] - \cos\left[(n + \tfrac{3}{2})\gamma' - \frac{\pi}{4}\right] = 2 \sin\left[(n + 1)\gamma' - \frac{\pi}{4}\right] \sin \tfrac{1}{2}\gamma'$$

we get

$$I_{3,n} = \frac{2}{\pi} \frac{n+1}{2n+1} \frac{1}{\alpha_n} \int_\rho^{\pi-\rho} \sin\left[(n+1)\gamma' - \frac{\pi}{4}\right] \times \cot^{1/2}(\tfrac{1}{2}\gamma') f(\cos \gamma')\, d\gamma' + \frac{n+1}{\pi}\left[\frac{1}{(2n+1)\alpha_n} - \frac{1}{(2n+3)\alpha_{n+1}}\right] \int_\rho^{\pi-\rho} \cos\left[(n + \tfrac{3}{2})\gamma' - \frac{\pi}{4}\right] \frac{\cos \frac{1}{2}\gamma'}{(\sin \frac{1}{2}\gamma')^{3/2}} f(\cos \gamma')\, d\gamma'$$
$$+ \tfrac{1}{2}(n+1) \int_\rho^{\pi-\rho} [p_{n,1}(\gamma') - p_{n+1,1}(\gamma')] \operatorname{ctg}(\tfrac{1}{2}\gamma') f(\cos \gamma')\, d\gamma'.$$

Since

$$\frac{n+1}{\pi}\left[\frac{1}{(2n+1)\alpha_n} - \frac{1}{(2n+3)\alpha_{n+1}}\right] = O(n^{-1/2}),$$
$$\tfrac{1}{2}(n+1)[p_{n,1}(\gamma') - p_{n+1,1}(\gamma')] = O(n^{-1/2}),$$

for γ' in $[\varrho, \pi - \varrho]$, the second and third terms of $I_{3,n}$ approach zero as $n \to \infty$. Moreover $2(n+1)/[\pi(2n+1)\alpha_n] = O(\sqrt{n})$. Consequently, if we let $F(\gamma') = \cot^{1/2}(\frac{1}{2}\gamma') f(\cos \gamma')$ it is sufficient to consider the limit as $n \to \infty$ of

$$\sqrt{n} \int_\rho^{\pi-\rho} \sin\left[(n+1)\gamma' - \frac{\pi}{4}\right] F(\gamma')\, d\gamma'.$$

Now, since we assumed $f(x)$ to be of bounded variation in $[-1, 1]$, it follows that $F(\gamma')$ is of bounded variation in $[\varrho, \pi - \varrho]$. Consequently, $F(\gamma') = F_1(\gamma') - F_2(\gamma')$ where $F_1(\gamma')$, $F_2(\gamma')$ are monotone in $[\varrho, \pi - \varrho]$. Now

$$\sqrt{n}\int_{\rho}^{\pi-\rho} \sin\left[(n+1)\gamma' - \frac{\pi}{4}\right] F_1(\gamma')\,d\gamma'$$

$$=\sqrt{n}\left[F_1(\varrho)\int_{\rho}^{\xi}\sin\left[(n+1)\gamma'-\frac{\pi}{4}\right]d\gamma'+F_1(\pi-\varrho)\int_{\xi}^{\pi-\rho}\sin\left[(n+1)\gamma'-\frac{\pi}{4}\right]d\gamma'\right]$$

where $\varrho < \xi < \pi - \varrho$. Clearly, the absolute value of this integral is less than $2\sqrt{n}\,[|F_1(\varrho)| + |F_1(\pi-\varrho)|]/(n+1)$, which tends to zero as $n \to \infty$. We get an analogous result by substituting $F_2(\gamma')$ for $F_1(\gamma')$, whence $\lim_{n\to\infty} I_{3,n} = 0$ and, therefore, for a given positive number σ, there exists a positive integer n_1 such that for $n > n_1$ we have

$$|I_{3,n}| < \frac{\sigma}{5} \text{ for } n > n_1. \tag{22}$$

Let us now consider the integrals $I_{1,n}$, $I_{2,n}$. By (18') of Sec. 5 we have

$$(n+1)\frac{P_n(x') - P_{n+1}(x')}{1-x'} = \frac{dP_n(x')}{dx'} + \frac{dP_{n+1}(x')}{dx'},$$

whence

$$I_{1,n} = \tfrac{1}{2}\int_{-1}^{-1+\varepsilon}\left[\frac{dP_n(x')}{dx'} + \frac{dP_{n+1}(x')}{dx'}\right] f(x')dx';$$

$$I_{2,n} = \tfrac{1}{2}\int_{1-\varepsilon}^{1}\left[\frac{dP_n(x')}{dx'} + \frac{dP_{n+1}(x')}{dx'}\right] f(x')dx'.$$

Now, if ε is a positive number such that $0 < \varepsilon < 1$, we shall suppose for simplicity that $f(x')$ is increasing in $[-1, -1+\varepsilon]$ and $[1-\varepsilon, 1]$. (Otherwise we could substitute for $f(x)$ the difference of two such functions). Then in $[-1, -1+\varepsilon]$, $f(x')$ increases from $f(-1+)$ to $f(-1+\varepsilon)$, and in $[1-\varepsilon, 1]$ from $f(1-\varepsilon)$ to $f(1-)$ and we have

$$\begin{aligned}2[I_{1,n} + I_{2,n}] &= f(-1+)[P_n(-1+\xi_1) + P_{n+1}(-1+\xi_1)]\\ &+ f(-1+\varepsilon)[P_n(-1+\varepsilon) + P_{n+1}(-1+\varepsilon)\\ &- P_n(-1+\xi_1) - P_{n+1}(-1+\xi_1)]\\ &+ f(1-\varepsilon)[P_n(1-\xi_2)+P_{n+1}(1-\xi_2)-P_n(1-\varepsilon)-P_{n+1}(1-\varepsilon)]\\ &+ 2f(1-) - f(1-)[P_n(1-\xi_2) + P_{n+1}(1-\xi_2)],\end{aligned}$$

or

$$
(23)\quad \begin{aligned}
I_{1,n}+I_{2,n} &= f(1-)+\tfrac{1}{2}[f(-1+)-f(-1+\varepsilon)][P_n(-1+\xi_1)\\
&\quad +P_{n+1}(-1+\xi_1)]+\tfrac{1}{2}f(-1+\varepsilon)[P_n(-1+\varepsilon)+P_{n+1}(-1+\varepsilon)]\\
&\quad +\tfrac{1}{2}[f(1-\varepsilon)-f(1-)][P_n(1-\xi_2)+P_{n+1}(1-\xi_2)]\\
&\quad -\tfrac{1}{2}f(1-\varepsilon)[P_n(1-\varepsilon)+P_{n+1}(1-\varepsilon)]
\end{aligned}
$$

where $-1<\xi_1<-1+\varepsilon$; $1-\varepsilon<\xi_2<1$.

For a given σ, determine ε so that

$$|f(-1+)-f(-1+\varepsilon)|<\sigma/5, \quad |f(1-\varepsilon)-f(1-)|<\sigma/5.$$

Now assume $|f(x)|<L$ in $[-1, 1]$. Then corresponding to the ε just determined, there exists a positive integer n_2 such that for $n>n_2$ we have

$$|P_n(1-\varepsilon)|<\sigma/5L, \quad [|P_n(-1+\varepsilon)|<\sigma/5L] \text{ for } n>n_2.$$

Finally, if n_0 is the greater of the two integers n_1, n_2 then by (21), (22) and (23) it follows that for $n>n_0$

$$|S_n(1)-f(1-)|<\sigma \text{ for } n>n_0$$

which proves the theorem.

Similar reasoning applies to the point -1.

8. The preceding results will be illustrated by the following example.

Expand in a series of Legendre polynomials the function $f(x)$ defined by:

$$f(0)=0;\ f(x)=1 \text{ for } 0<x\leqq 1;\ f(x)=-1 \text{ for } -1\leqq x<0.$$

$f(x)$ is odd and of bounded variation. Consequently, the coefficients

$$a_n=\frac{2n+1}{2}\int_{-1}^{1} f(x)P_n(x)dx$$

are zero for n even. For $n=2m+1$ we have

$$a_{2m+1}=\frac{4m+3}{2}\int_{-1}^{1} f(x)\,P_{2m+1}(x)dx=(4m+3)\int_{0}^{1} P_{2m+1}(x)dx.$$

However, by (12) of Sec. 4

$$(n+1)\int_0^1 P_n(x)dx = \int_0^1 P'_{n+1}(x)dx - \int_0^1 xP'_n(x)dx$$

$$= 1 - P_{n+1}(0) - [xP_n(x)]_0^1 + \int_0^1 P_n(x)dx,$$

therefore, for $n = 2m+1$

$$\int_0^1 P_{2m+1}(x)dx = -\frac{P_{2m+2}(0)}{2m+1} = -\frac{1}{2m+1}\binom{-\frac{1}{2}}{m+1},$$

whence

$$a_{2m+1} = -\frac{4m+3}{2m+1}\binom{-\frac{1}{2}}{m+1} = (-1)^m \frac{4m+3}{m+1}\,\frac{1\times 3\times 5\times \ldots \times (2m-1)}{2^{m+1}\times 1\times 2\times 3\times \ldots \times m},$$

and, therefore, for $0 < x \leqq 1$

$$1 = -\sum_{m=0}^{\infty} \frac{4m+3}{2m+1}\binom{-\frac{1}{2}}{m+1} P_{2m+1}(x).$$

The series is uniformly convergent in any interval interior to [0, 1].

15. Series of Stieltjes-Neumann

This article is based on results of Neumann and Stieltjes. (Cf. F. E. Neumann [81], T. J. Stieltjes [104a].) For the proof given in the text, see L. Fejér [33c].

1a. THEOREM. For $0 < \omega < 1$ we have

$$\boxed{\frac{2^{\omega}}{(1-x)^{\omega}} \sim c_0 + c_1 P_1(x) + c_2 P_2(x) + \ldots + c_n P_n(x) + \ldots} \tag{1}$$

where

$$c_0 = \frac{1}{1-\omega}, \quad c_1 = \frac{3\omega}{(1-\omega)(2-\omega)},$$

$$c_n = (2n+1)\frac{\omega(\omega+1)(\omega+2)\ldots(\omega+n-1)}{(1-\omega)(2-\omega)\ldots(n+1-\omega)}, \quad (n=2, 3, \ldots). \tag{2}$$

Proof. We have

$$c_n = \frac{2n+1}{2}\int_{-1}^{1} \frac{2^\omega}{(1-x)^\omega} P_n(x)dx,$$

and by the formula of Rodrigues [Sec. 1, (7)]

$$\frac{2 \cdot 2^n \cdot n!}{(2n+1)2^\omega} c_n = \int_{-1}^{1} \frac{1}{(1-x)^\omega} \frac{d^n(x^2-1)^n}{dx^n} dx. \tag{3}$$

Integrating by parts and observing that $d^{n-1}(x^2-1)^n/dx^{n-1}$ has zeros of the first order at $x = -1$, $x = 1$ we have

$$\int_{-1}^{1} \frac{1}{(1-x)^\omega} \frac{d^n(x^2-1)^n}{dx^n} dx = -\omega\int_{-1}^{1} (1-x)^{-\omega-1} \frac{d^{n-1}(x^2-1)^n}{dx^{n-1}} dx.$$

Integrating by parts again and observing that $d^{n-2}(x^2-1)^n/dx^{n-2}$ has zeros of the second order at -1 and 1 and continuing in this way we have

$$\int_{-1}^{1} \frac{1}{(1-x)^\omega} \frac{d^n(x^2-1)^n}{dx^n} dx$$

$$= (-1)^n\omega(\omega+1)\ldots(\omega+n-1)\int_{-1}^{1} (1-x)^{-\omega-n}(x^2-1)^n dx$$

$$= \omega(\omega+1)\ldots(\omega+n-1)\int_{-1}^{1} (1-x)^{-\omega}(1+x)^n dx.$$

Now if we set $x = 1 - 2t$ in the integral, then in view of the properties of the Gamma and Beta functions (Appendix, Th. 19 and Th. 23) we get

$$\int_{-1}^{1} \frac{1}{(1-x)^\omega} \frac{d^n(x^2-1)^n}{dx^n} dx = \omega(\omega+1)\ldots(\omega+n-1)2^{n+1-\omega}\int_{0}^{1} t^{-\omega}(1-t)^n dt$$

$$= \omega(\omega+1)\ldots(\omega+n-1)2^{n+1-\omega}\frac{\Gamma(1-\omega)\Gamma(n+1)}{\Gamma(n-\omega+2)}$$

$$= \omega(\omega+1)\ldots(\omega+n-1)2^{n+1-\omega}\frac{n!\Gamma(1-\omega)}{(n+1-\omega)(n-\omega)\ldots(1-\omega)\Gamma(1-\omega)}$$

$$= \frac{2^{n+1-\omega}n!\omega(\omega+1)\ldots(\omega+n-1)}{(1-\omega)(2-\omega)\ldots(n+1-\omega)},$$

and (2) follows at once from (3).

1b. We may evaluate asymptotically the order of magnitude of the coefficients c_n of the expansion (1). Indeed, we have by (2)

$$c_n = \frac{\Gamma(1-\omega)}{\Gamma(\omega)} \frac{(2n+1)\Gamma(n+\omega)}{\Gamma(n+2-\omega)},$$

and by Stirling's formula (cf. Appendix, Th. 20)

$$\boxed{c_n = \frac{2\Gamma(1-\omega)}{\Gamma(\omega)} n^{2\omega-1}[1+O(n^{-1})]}. \tag{4}$$

1c. Proceeding to the convergence of series (1) the theorem of Hobson guarantees that for $0 < \omega < \frac{3}{4}$, (1) is convergent to $2^\omega/(1-x)^\omega$ at every point x interior to $[-1, 1]$; in fact $2^\omega(1-x)^{-(\omega+\frac{1}{4})}(1+x)^{-\frac{1}{4}}$ is integrable in $[-1, 1]$. Then observing that $2^\omega(1-\cos\gamma)^{-\omega}\sin^{1/2}\gamma$ has a bounded derivative in every interval interior to $[0, \pi]$ and that therefore its Fourier series converges uniformly to the same function in such an interval (Ch. II, Theorem 17, Cor. 2^0) it follows that for $0 < \omega < \frac{3}{4}$, $-1 < x < 1$ we have

$$\boxed{\begin{gathered}\frac{2^\omega}{(1-x)^\omega} = c_0 + c_1 P_1(x) + \ldots + c_n P_n(x) + \ldots; \\ 0 < \omega < 3/4, -1 < x < 1\end{gathered}} \tag{5}$$

where the coefficients c_n are given by (2), and the series converges uniformly in every interval interior to $[-1, 1]$.

In particular for $\omega = \frac{1}{2}$, we have $2 = c_0 = c_1 = \ldots$, whence

$$\boxed{\frac{1}{\sqrt{2}\sqrt{1-x}} = P_0 + P_1(x) + P_2(x) + \ldots + P_n(x) + \ldots, -1 < x < 1}. \tag{6}$$

If $\omega > 3/4$ series (1) does not converge in any point interior to $[-1, 1]$. In fact we have by (4) and by Stieltjes formula (4) of Sec. 12. 1

$$c_n P_n(x) = \frac{2\Gamma(1-\omega)}{\Gamma(\omega)} n^{2\omega-1} \frac{2}{\sqrt{\pi}} n^{-\frac{1}{2}} \left[\frac{\cos\left[(n+\frac{1}{2})\gamma - \frac{\pi}{4}\right]}{(2 \sin \gamma)^{\frac{1}{2}}} + O\left(\frac{1}{n}\right) \right]$$

and since $2\omega - 3/2 \geqq 0$ the convergence of series (1) at a point $x = \cos \gamma$ interior to $[-1, 1]$ requires that

$$\lim_{n\to\infty} \left[\cos\ (n + \tfrac{1}{2})\gamma - \frac{\pi}{4} \right] = 0. \tag{7}$$

Now if the ratio of γ to π is irrational, expanding $\gamma/(2\pi)$ in a continued fraction and denoting by P_l/Q_l its lth convergent, we have

$$\left| \frac{\gamma}{2\pi} - \frac{P_l}{Q_l} \right| < \frac{1}{Q_l^2}, \qquad | Q_l\gamma - 2P_l\pi | < 2\pi/Q_l$$

and therefore for any positive number ε there exist infinitely many pairs of integers such that

$$r\gamma - 2s\pi = \theta\varepsilon, \quad [\,|\theta| < 1, \quad r > 0]$$

whence

$$\cos \left[(n + r + \tfrac{1}{2})\gamma - \frac{\pi}{4} \right] = \cos \left[(n + \tfrac{1}{2})\gamma - \frac{\pi}{4} + \theta\varepsilon \right],$$

and (7) then requires $(n + \frac{1}{2})\gamma - \pi/4 = (2m + 1)\,\pi/2$, ($m$ an integer) which contradicts the supposition that the ratio of γ to π is irrational.

If now $\gamma = r\,(s\pi)$, (r and s integers), we have

$$\cos \left[(n + 2s + \tfrac{1}{2})\gamma - \frac{\pi}{4} \right] = \cos \left[(n + \tfrac{1}{2})\gamma - \frac{\pi}{4} \right]$$

whence for any integer n, $(n + \frac{1}{2})\gamma - \pi/4 = (2m + 1)\pi/2$, ($m$ an integer). Moreover, since $(n + \frac{3}{2})\gamma - \pi/4 = (2m_1 + 1)\pi/2$, $(m_1 > m)$, it follows, by subtracting corresponding members of the preceding equalities, that $\gamma = (m_1 - m)\pi$, which is impossible since $0 < \gamma < \pi$.

1d. In view of the relation $P_n(1) = 1$, $P_n(-1) = (-1)^n$ [Sec. 6, (20)], formula (4) implies that series (1) is divergent at point 1 for any ω, and oscillating at point -1, for $\omega \geqq \frac{1}{2}$.

If now $0 < \omega < \frac{1}{2}$, series (1) at point $x = -1$ is an alternating series with every term less in absolute value than the preceding term and with the nth term approaching zero; therefore, it converges.

Note: The sum of the series at -1 is equal to 1 for $0 < \omega < \frac{1}{2}$. In fact the Poisson sum of (1) at the point -1 equals $\lim_{x \to -1+0} 2^\omega/(1-x)^\omega = 1$, [cf. Sec. 25. 1], and the Poisson sum coincides with the ordinary sum of (1) at the point $x = -1$, since the series is convergent there [Abel's theorem, cf. G. Sansone, *Lezioni sulla Teoria delle Funzioni di una variabile complessa*, I, (3rd edition, Padova, 1950), p. 20].

16. Series of Legendre Polynomials for a Finite Interval

If a function $f(t)$ is defined in a finite interval $[a, b]$, it is sometimes necessary in the applications to expand the function in a series of orthogonal polynomials in this interval. Clearly the substitution

$$(1) \qquad x = \frac{2}{b-a}\left[t - \frac{b+a}{2}\right], \quad a<b, \quad \left[t = \frac{b-a}{2}x + \frac{b+a}{2}\right]$$

transforms the interval $[a, b]$ of the t-axis into the interval $[-1, 1]$ of the x axis. It will therefore be sufficient to consider the expansion in series of Legendre polynomials of

$$f\left[\frac{b-a}{2}x + \frac{b+a}{2}\right] \sim \sum_{n=0}^{\infty} a_n P_n(x),$$

$$a_n = \frac{2n+1}{2}\int_{-1}^{1} f\left[\frac{b-a}{2}x + \frac{b+a}{2}\right] P_n(x)dx$$

and to make the substitution (1) in this expansion.

The substitutions can be avoided by the following considerations: Make the substitution (1) in

$$P_n(x) = \frac{1}{2^n n!} \frac{d^n (x+1)^n (x-1)^n}{dx^n} \qquad \text{[Sec. 1, (7)]}$$

and let

$$X_n(t) = P_n\left[\frac{2}{b-a}\left(t - \frac{b+a}{2}\right)\right].$$

Then we have

$$(2) \qquad \boxed{X_n(t) = \frac{1}{n!(b-a)^n} \frac{d^n (t-a)^n (t-b)^n}{dt^n}},$$

and equations (35_1), (35_2) of Sec. 9.1 become

$$(3_1) \qquad \int_a^b X_n(t) X_m(t)dt = 0 \text{ for } n \neq m;$$

$$(3_2) \qquad \int_a^b X_n^2(t)dt = \frac{b-a}{2n+1}.$$

Also

$$(4_1) \qquad X_0 = 1, \quad X_1(t) = \frac{2t}{b-a} - \frac{b+a}{b-a},$$

and the recurrence relation (13) of Sec. 4 becomes

$$(4_2) \quad (n+1)X_{n+1}(t) = (2n+1)X_1(t)X_n(t) - nX_{n-1}(t).$$

The polynomials $X_n(t)$ will be called the *Legendre polynomials* relative to the interval $[a, b]$.

Finally we conclude that if $f(t)$ is integrable in $[a, b]$ and if its expansion in series of Legendre polynomials relative to the interval $[a, b]$ is

$$(5) \qquad f(t) \sim \sum_{n=0}^{\infty} a_n X_n(t), \quad a_n = \frac{2n+1}{b-a} \int_a^b f(t) X_n(t)dt,$$

then the criteria for convergence given in Secs. 11 and 14 apply to this series. Obviously in theorem Sec. 14.6 the assumption must be made that $f(t)/\sqrt[4]{(t-a)(b-t)}$ is integrable in $[a, b]$.

(For this section see Picone [886], p. 265).

17. Ferrers' Functions associated with Legendre Functions

1a. If m is a positive integer, and $-1 \leqq x \leqq 1$, the functions

$$P_n^m(x) = (1-x^2)^{(\frac{1}{2})m} \frac{d^m P_n(x)}{dx^m}, \quad (m = 1, 2, \ldots, n), \tag{1}$$

where $(1-x^2)^{(\frac{1}{2})m}$ indicates the numerical value of the root are called the *Ferrers' functions associated with the Legendre functions of degree n and order m* [35].

REMARK: It should be observed that the symbol $P_n^{(m)}(x)$ denotes as usual the mth derivative of $P_n(x)$, that is $P_n^{(m)}(x) = d^m P_n/dx^m$ and, therefore $P_n^m(x) = (1-x^2)^{m/2} P_n^{(m)}(x)$.

In view of Rodrigues' formula [Sec. 1.3] we have:

$$P_n^m(x) = \frac{(1-x^2)^{(\frac{1}{2})m}}{2^n n!} \frac{d^{n+m}}{dx^{n+m}} (x^2-1)^n, \quad (m = 1, 2, \ldots, n). \tag{2}$$

1b. The differential equation which satisfies the functions $P_n^m(x)$ is easily established. From the differential equation of the Legendre polynomials [Sec. 3], differentiating m times with respect to x, we have

$$(1-x^2)\frac{d^{m+2} P_n}{dx^{m+2}} - 2mx \frac{d^{m+1} P_n}{dx^{m+1}} - m(m-1)\frac{d^m P_n}{dx^m}$$
$$- 2x \frac{d^{m+1} P_n}{dx^{m+1}} - 2m \frac{d^m P_n}{dx^m} + n(n+1)\frac{d^m P_n}{dx^m} = 0,$$

or

$$(1-x^2)\frac{d^{m+2} P_n}{dx^{m+2}} - 2(m+1)x \frac{d^{m+1} P_n}{dx^{m+1}} + (n-m)(n+m+1)\frac{d^m P_n}{dx^m} = 0,$$

and therefore

$$(1-x^2)\frac{d^2[(1-x^2)^{-(\frac{1}{2})m} P_n^m]}{dx^2} - 2(m+1)x \frac{d[(1-x^2)^{-(\frac{1}{2})m} P_n^m]}{dx}$$
$$+ (n-m)(n+m+1)(1-x^2)^{-(\frac{1}{2})m} P_n^m = 0,$$

which yields, after multiplying by $(1-x^2)^{(\frac{1}{2})m}$ and simplifying,

$$(3)\qquad \boxed{(1-x^2)\frac{d^2P_n^m}{dx^2}-2x\frac{dP_n^m}{dx}+\left[n(n+1)-\frac{m^2}{1-x^2}\right]P_n^m(x)=0},$$

or

$$(3')\qquad \frac{d}{dx}\left[(1-x^2)\frac{dP_n^m}{dx}\right]+\left[n(n+1)-\frac{m^2}{1-x^2}\right]P_n^m(x)=0.$$

2. The functions P_n^m, P_r^m, satisfy the relation

$$(4)\qquad \boxed{\begin{aligned}&\int_{-1}^{1}P_n^m(x)\,P_r^m(x)\,dx=0, \qquad (n\neq r),\\ &\int_{-1}^{1}[P_n^m(x)]^2\,dx=\frac{2}{2n+1}\,\frac{(n+m)!}{(n-m)!}\end{aligned}}.$$

Now if we multiply the differential equation of P_r^m

$$\frac{d}{dx}\left[(1-x^2)\frac{dP_r^m}{dx}\right]+\left[r(r+1)-\frac{m^2}{1-x^2}\right]P_r^m(x)=0$$

by P_n^m and subtract from (3′) multiplied by P_r^m we get

$$\frac{d}{dx}\left[(1-x^2)\left\{P_r^m(x)\frac{dP_n^m}{dx}-P_n^m(x)\frac{dP_r^m}{dx}\right\}\right]$$
$$+\,(n-r)(n+r+1)P_n^m(x)P_r^m(x)=0.$$

Then integrating with respect to x between -1 and 1 we get

$$(n-r)(n+r+1)\int_{-1}^{1}P_n^m(x)P_r^m(x)dx=0,$$

so that if $n\neq r$ we have the first of equations (4).

In order to obtain the second relation of (4) observe that

$$P_n^{m+1}(x)=(1-x^2)^{(\frac{1}{2})m}(1-x^2)^{\frac{1}{2}}\frac{d}{dx}[(1-x^2)^{-(\frac{1}{2})m}P_n^m(x)]$$

that is

$$P_n^{m+1}(x)=(1-x^2)^{\frac{1}{2}}\frac{dP_n^m}{dx}+m(1-x^2)^{-\frac{1}{2}}xP_n^m(x).$$

Now squaring and integrating with respect to x between -1 and 1 to get

$$\int_{-1}^{1}[P_n^{m+1}(x)]^2dx = \int_{-1}^{1}(1-x^2)\left[\frac{dP_n^m}{dx}\right]^2 dx + 2m\int_{-1}^{1} xP_n^m(x)\frac{dP_n^m}{dx}\,dx$$

$$+ m^2\int_{-1}^{1}\frac{x^2}{1-x^2}[P_n^m(x)]^2dx = -\int_{-1}^{1}P_n^m(x)\frac{d}{dx}\left\{(1-x^2)\frac{dP_n^m}{dx}\right\}dx$$

$$+ m[x\{P_n^m(x)\}^2]_{-1}^{1} - m\int_{-1}^{1}[P_n^m(x)]^2dx + \int_{-1}^{1}\frac{m^2x^2}{1-x^2}[P_n^m(x)]^2dx,$$

observing that $[x\{P_n^m(x)\}^2]_{-1}^{1} = 0$ and making use of (3′)

$$\int_{-1}^{1}[P_n^{m+1}(x)]^2dx = \int_{-1}^{1}[P_n^m(x)]^2\left\{n(n+1) - \frac{m^2}{1-x^2} - m + \frac{m^2x^2}{1-x^2}\right\}dx,$$

we obtain finally the recurrence formula

$$\int_{-1}^{1}[P_n^{m+1}(x)]^2dx = (n-m)(n+m+1)\int_{-1}^{1}[P_n^m(x)]^2dx.$$

From this formula with the use of (35_2) of Sec. 9, we get

$$\int_{-1}^{1}[P_n^m(x)]^2dx$$
$$= (n-m+1)(n-m+2)\ldots n(n+m)(n+m-1)\ldots(n+1)$$
$$\times\int_{-1}^{1}[P_n(x)]^2dx = \frac{(n+m)!}{(n-m)!}\,\frac{2}{2n+1},$$

which is the second relation of (4).

3. From (4) it follows that for any integer m the system of functions

$$\left\{\left[\frac{2n+1}{2}\,\frac{(n-m)!}{(n+m)!}\right]^{1/2}P_n^m(x)\right\}, \quad (n = m, m+1, \ldots) \tag{5}$$

represents an orthonormal system in $[-1, 1]$, and it is readily seen that it is complete there. In fact the terms of the sequence $\{x^n\}$ can be expressed as linear combinations of the sequence

$$P_m^{(m)}(x), \quad P_{m+1}^{(m)}(x), \quad P_{m+2}^{(m)}(x), \ldots; \tag{6}$$

and consequently there does not exist a square integrable function orthogonal $[-1, 1]$ to the functions of sequence (6) and therefore there is no integrable function orthogonal in $[-1, 1]$ to the functions of sequence (5) which differ from (6) by the factor $(1-x^2)^{m/2}$ multiplied by a constant.

We have in particular by Vitali's completeness theorem [Ch. I, Sec. 6, Th. 26]

$$\sum_{n=m}^{\infty} \frac{2n+1}{2} \frac{(n-m)!}{(n+m)!} \left[\int_x^1 P_n^m(x)\, dx \right]^2 = 1 - x. \tag{7}$$

4. In this and in the following section the integral formulas of Schläfli and of Laplace established in Secs. 2 and 6 will be extended to the Ferrer functions.

From Schläfli's formula differentiating m times under the integral sign with respect to the parameter x, we have

$$\frac{d^m P_n}{dx^m} = \frac{(n+m)!}{2\pi i n!} \oint_C \frac{(t^2-1)^n}{2^n (t-x)^{n+m+1}}\, dt,$$

and therefore the other integral formula of Schläfli

$$\boxed{P_n^m(x) = \frac{(n+m)!}{2\pi i n!} (1-x^2)^{m/2} \oint_C \frac{(t^2-1)^n}{2^n (t-x)^{n+m+1}}\, dt}, \tag{8}$$

where C is any regular closed curve surrounding the point x.

5. Suppose $-1 < x < 1$, and let the curve C in (8) be the circle with center at x and radius $|x^2 - 1|^{1/2}$. Then we get all the points C by letting

$$t = x + i\sqrt{1-x^2}\, e^{i\tau}$$

where $-\pi \leqq \tau \leqq \pi$ and the radical denotes the positive square root. However, (cf. Sec. 6)

$$\frac{1}{2\pi i}\oint_C \frac{(t^2-1)^n dt}{2^n(t-x)^{n+m+1}}$$

$$= -\frac{1}{2\pi i}\int_{-\pi}^{\pi}\frac{1}{2^n}\frac{i^n 2^n(1-x^2)^{n/2}[x+i\sqrt{1-x^2}\cos\tau]^n(1-x^2)^{1/2}}{i^{n+m+1}(1-x^2)^{(n+m+1)/2}\, e^{i(n+m+1)\tau}}e^{i(n+1)\tau}d\tau$$

$$= (-i)^m\frac{(1-x^2)^{-m/2}}{2\pi}\int_{-\pi}^{\pi}[x+i\sqrt{1-x^2}\cos\tau]^n\cos m\tau\, d\tau$$

$$-(-i)^{m+1}\frac{(1-x^2)^{-m/2}}{2\pi}\int_{-\pi}^{\pi}[x+i\sqrt{1-x^2}\cos\tau]^n\sin m\tau\, d\tau.$$

The last integral vanishes as is seen by changing τ to $-\tau$; the other integral can be expressed as the sum of two equal integrals between $-\pi$ and 0 and 0 and π. Therefore we get

$$\frac{1}{2\pi i}\oint_C \frac{(t^2-1)^n dt}{2^n(t-x)^{n+m+1}}$$

$$= (-i)^m\frac{(1-x^2)^{-m/2}}{\pi}\int_0^{\pi}[x+i\sqrt{1-x^2}\cos\tau]^n\cos m\tau\, d\tau,$$

and finally in view of (8) we have Laplace's formula

$$\boxed{P_n^m(x) = \frac{(n+m)!}{n!\,\pi}(-i)^m\int_0^{\pi}[x+i\sqrt{1-x^2}\cos\tau]^n\cos m\tau\, d\tau} \tag{9}$$

The last formula also holds for $x = \pm 1$.

6a. From (9) for $-1 \leqq x \leqq 1$ follows the inequality

$$|P_n^m(x)| \leqq \frac{(n+m)!}{n!}, \quad [n \geqq m,\ -1 \leqq x \leqq 1], \tag{10}$$

and from (1) we then get the other inequality:

$$\left|\frac{d^m P_n}{dx^m}\right| \leqq \frac{(n+m)!}{n!}(1-x^2)^{-m/2}, \quad [n \geqq m,\ -1 \leqq x \leqq 1]. \tag{10'}$$

Note that for the case $m = 1$, namely for $P_n'(x)$ we have already obtained the limit expressed by (18) of Sec. 10.5.

6b. A new inequality for $P_n^{(m)}(x)$ [and therefore for $P_n^m(x)$] is given by the following

THEOREM. For all $n \geqq m \geqq 1$ we have in the interval $[-1, 1]$

$$(11)\qquad \boxed{\,|\, P_n^{(m)}(x) \,|\, \leqq P_n^{(m)}(1) = \binom{n+m}{n-m}(2m-1)!! \leqq \frac{n^{2m}}{m!}\,}.$$

Proof. From the expansion in series (2) of Sec. 1.1

$$(12)\qquad (1-2\varrho x+\varrho^2)^{-1/2} = \sum_{n=0}^{\infty} \varrho^n P_n(x);$$

differentiating the two sides of the equation m times with respect to x we get

$$(12_m)\qquad \frac{(2m-1)!!}{(1-2\varrho x+\varrho^2)^{(2m+1)/2}} = \sum_{n=0}^{\infty} \frac{d^m P_{n+m}}{dx^m}\varrho^n.$$

In particular we have

$$\frac{d}{dx}[1-2\varrho x+\varrho^2]^{-1/2}$$
$$= -\tfrac{1}{2}[1-2\varrho x+\varrho^2]^{-3/2}(-2\varrho) = 1!!\,\varrho[1-2\varrho x+\varrho^2]^{-3/2},$$

$$\frac{d^2}{dx^2}[1-2\varrho x+\varrho^2]^{-1/2}$$
$$= 1!!(-\tfrac{3}{2})\varrho(-2\varrho)[1-2\varrho x+\varrho^2]^{-5/2} = 3!!\,\varrho^2[1-2\varrho x+\varrho^2]^{-5/2},$$

and in general

$$\frac{d^m}{dx^m}[1-2\varrho x+\varrho^2]^{-1/2} = (2m-1)!!\,\varrho^m[1-2\varrho x+\varrho^2]^{-(2m+1)/2}.$$

On the other hand, for a value of ϱ such that $|\,\varrho\,| < \frac{1}{4}$ and for x in $[-1, 1]$ the series

$$\sum_{n=0}^{\infty} \frac{d^m P_{n+m}(x)}{dx^m}\varrho^n$$

is uniformly convergent; in fact by (8_1), (8_2) of Sec. 1.4 the series of the absolute values is majorized by the series

$$2^{2m-2}\sum_{n=0}^{\infty}(n+m)^2(n+m-1)\dots(n+1)(2^2\,|\,\varrho\,|\,)^n,$$

and since

$$\lim_{n\to\infty}(n+m+1)^2(n+m)\dots(n+2)/(n+m)^2(n+m-1)\dots(n+1)=1,$$

the last series is convergent for $|\,\varrho\,|<\frac{1}{4}$, therefore (12) can be differentiated term by term m times with respect to x to obtain (12_m).

REMARK: After (11) has been established at the end of this section it will also follow that the expansion (12) holds for $|\,\rho\,|<1$ and $-1\leqq x\leqq 1$.

Letting $x=1$ in (12_m) we have

$$\text{(13)}\qquad \frac{(2m-1)!!}{(1-\varrho)^{2m+1}}=\sum_{n=0}^{\infty}\left[\frac{d^mP_{n+m}(x)}{dx^m}\right]_{x=1}\varrho^n,$$

also

$$\frac{(2m-1)!!}{(1-\varrho)^{2m+1}}=(2m-1)!!\sum_{n=0}^{\infty}\frac{(2m+1)(2m+2)\dots(2m+n)}{n!}\varrho^n$$

$$=(2m-1)!!\sum_{n=0}^{\infty}\binom{2m+n}{n}\varrho^n,$$

whence by (13)

$$\text{(14)}\qquad \left[\frac{d^m P_{n+m}(x)}{dx^m}\right]_{x=1}=\binom{2m+n}{n}(2m-1)!!,$$

and also changing $n+m$ to n

$$\left[\frac{d^mP_n(x)}{dx^m}\right]_{x=1}=\binom{n+m}{n-m}(2m-1)!!=\binom{n+m}{2m}(2m-1)!!$$

$$\text{(15)}\qquad =\frac{(n+m)(n+m-1)\dots(n-m+1)}{2^m m!}$$

$$=\frac{1}{m!}\,\frac{(n+m)(n-m+1)}{2}\,\frac{(n+m-1)(n-m+2)}{2}\dots\frac{(n+1)n}{2}\leqq\frac{n^{2m}}{m!}.$$

Now, obviously for $|z| < 1$

$$(1-z)^{-(2m+1)/2} = \sum_{l=0}^{\infty} \alpha_l^{(m)} z^l, \quad \alpha_l^{(m)} = \binom{-\frac{2m+1}{2}}{l} (-1)^l,$$

whence

$$(1-2\varrho\cos\gamma+\varrho^2)^{-(2m+1)/2} = (1-\varrho e^{i\gamma})^{-(2m+1)/2}(1-\varrho e^{-i\gamma})^{-(2m+1)/2}$$
$$= \sum_{l=0}^{\infty} \alpha_l^{(m)} e^{li\gamma} \varrho^l \sum_{l=0}^{\infty} \alpha_l^{(m)} e^{-li\gamma} \varrho^l,$$

and multiplying the two series by Cauchy's rule and using (12_m) we have

$$(2m-1)!!\, e^{-in\gamma} \sum_{l=0}^{n} \alpha_l^{(m)} \alpha_{n-l}^{(m)} e^{2il\gamma} = \left[\frac{d^m P_{n+m}(x)}{dx^m}\right]_{x=\cos\gamma}$$

and since $\alpha_l^{(m)} > 0$ it follows that [cf. (14)]

$$\left|\frac{d^m P_{n+m}(x)}{dx^m}\right| \leqq P_{n+m}^{(m)}(1) = \binom{2m+n}{n} (2m-1)!!.$$

Finally, changing $n+m$ to n and using (15) we get (11).

18. *Harmonic Polynomials and Spherical Harmonics.*

1a. A homogeneous polynomial V of the nth degree in the variables x, y, z, which satisfies the equation

$$\Delta_2 V = \frac{\partial^2 V}{\partial x^2} + \frac{\partial^2 V}{\partial y^2} + \frac{\partial^2 V}{\partial z^2} = 0, \tag{1}$$

is called a *Laplace polynomial of degree n.*

1b. THEOREM: There exist only $2n+1$ linearly independent Laplace polynomials of degree n. (M. Picone: *Appunti di Analisi Superiore* (Naples, 1940), p. 113, [88b].)

Let

$$V(x, y, z) = U_0(y, z) + \frac{x}{1!} U_1(y, z) + \frac{x^2}{2!} U_2(y, z) + \ldots + \frac{x^n}{n!} U_n(y, z); \tag{2}$$

where the polynomials $U_0, U_1, \ldots, U_n$ are homogeneous in the variables y, z and respectively of degrees $n, n-1, \ldots, 0$. Clearly,

$$U_k(y, z) = \left[\frac{\partial^k V}{\partial x^k}\right]_{x=0}, \qquad (k = 0, 1, \ldots, n).$$

From the expression (2) of V we get

$$\Delta_2 V = \Delta_2 \sum_{k=0}^{n} \frac{1}{k!} x^k U_k(y, z) = \sum_{k=0}^{n} \frac{1}{k!} \Delta_2 x^k U_k(y, z)$$

$$= \sum_{k=0}^{(n-2)} \frac{1}{k!} x^k U_{k+2}(y, z) + \sum_{k=0}^{n} \frac{x^k}{k!} \left[\frac{\partial^2 U_k}{\partial y^2} + \frac{\partial^2 U_k}{\partial z^2}\right],$$

and since $\Delta_2 U_{n-1} \equiv 0$, $\Delta_2 U_n \equiv 0$, it follows that a necessary and sufficient condition for the identical vanishing of $\Delta_2 V$ is

$$\text{(3)} \qquad U_{k+2}(y, z) = -\frac{\partial^2 U_k}{\partial y^2} - \frac{\partial^2 U_k}{\partial z^2}, \qquad (k = 0, 1, \ldots, n-2).$$

This relation determines uniquely the polynomials $U_2, \ldots, U_n$ from two arbitrary polynomials U_0 and U_1.

The $n + 1$ coefficients of U_0 and the n coefficients of U_1 are arbitrary. Consequently, the polynomials U_k depend on $2n + 1$ arbitrary constants, so that letting

$$U_0 = \sum_{h=0}^{n} c_h y^h z^{n-h}, \qquad U_1 = \sum_{k=0}^{(n-1)} c_{n+k+1} y^k z^{n-1-k},$$

we have

$$\text{(4)} \quad V = c_0 V_0 + c_1 V_1 + \ldots + c_n V_n + c_{n+1} V_{n+1} + \ldots + c_{2n} V_{2n}$$

where the polynomials $V_0, V_1, \ldots, V_n, V_{n+1}, \ldots, V_{2n}$ have the following significance. The polynomial V_n for $h = 0, 1, \ldots, n$ is the polynomial obtained from formula (2) when the polynomials $U_0, U_1, \ldots, U_n$ are calculated from (3) by taking $U_0 = y^h z^{n-h}$, $[h = 0, 1, \ldots, n]$ and $U_1 \equiv 0$. The sequence of polynomials U so obtained will be denoted by $U_0^{(h)}, U_1^{(h)}, \ldots, U_n^{(h)}$, $[h = 0, 1, \ldots, n]$.

The polynomial V_{n+1+k} for $k = 0, 1, \ldots, n-1$ is the polynomial obtained from formula (2) when the polynomials U_0,

$U_1, \ldots, U_n$ are calculated from (3) by taking $U_1 \equiv 0$, $U_1 = y^k z^{n-1-k}$, $[k = 0, 1, \ldots, n-1]$. The sequence of polynomials U so obtained will be denoted by $U_0^{(n+k+1)}, U_1^{(n+k+1)}, \ldots, U_n^{(n+k+1)}$; $[k = 0, 1, \ldots, n-1]$. Clearly, then

$$(5)\quad \begin{cases} V_h = y^h z^{n-h} + \dfrac{x^2}{2!} U_2^{(h)}(y, z) + \dfrac{x^3}{3!} U_3^{(h)}(y, z) + \ldots + \dfrac{x^n}{n!} U_n^{(h)}(y, z), \\ V_{n+1+k} = xy^k z^{n+1-k} + \dfrac{x^2}{2!} U_2^{(n+k+1)}(y, z) \\ \qquad + \dfrac{x^3}{3!} U_3^{(n+k+1)}(y, z) + \ldots + \dfrac{x^n}{n!} U_n^{(n+k+1)}(y, z) \\ \qquad (h = 0, 1, \ldots, n;\ k = 0, 1, \ldots, n-1). \end{cases}$$

It follows that the polynomials $V_0, V_1, \ldots, V_n, V_{n+1}, \ldots, V_{2n}$ are linearly independent and from (4) the theorem then follows at once.

2a. Let $V(x, y, z)$ be the most general Laplace polynomial of degree n; and consider x, y, z as the cartesian coordinates of a point M with respect to a set of orthogonal axes with origin at the point 0.

Let $\overline{OM} = \varrho$, denote by P the point of intersection of the radius $\overline{OM}$ with the unit sphere ω with center at 0, and denote by $\xi(P)$, $\eta(P)$, $\zeta(P)$, the coordinates of P. Then we have

$$V(x, y, z) = \varrho^n V(\xi(P), \eta(P), \zeta(P)).$$

On the sphere ω take the point $P_0 \equiv (1, 0, 0)$ as pole, the positive x-axis as polar axis, the half-plane determined by the x-axis and the positive y-axis as the half-plane of the prime meridian and denote by φ and θ the polar colatitude and the longitude of a point P on the sphere ω : φ is clearly the angle between 0 and π that the polar axis forms with the radius vector $\overline{OP}$, and θ is the angle between 0 and 2π that the semiplane through the x-axis and the point P forms with the half-plane of the prime meridian.

Then

$$\xi(P) = \cos\varphi; \quad \eta(P) = \sin\varphi\cos\theta; \quad \zeta(P) = \sin\varphi\sin\theta;$$
$$[0 \leqq \varphi \leqq \pi,\ 0 \leqq \theta \leqq 2\pi]$$

whence

$$V(x, y, z) = \varrho^n V(\cos\varphi,\ \sin\varphi\cos\theta,\ \sin\varphi\sin\theta).$$

The function

$$Y(P) = Y(\varphi, \theta) = V(\cos\varphi,\ \sin\varphi\cos\theta,\ \sin\varphi\sin\theta)$$

is called the Laplace function or *spherical harmonic of the nth order.*

$Y(\varphi, \theta)$ is a polynomial in $\cos\varphi$, $\sin\varphi\cos\theta$, $\sin\varphi\sin\theta$, and from the definition it follows that the spherical harmonics $Y(P)$ of order n are those functions of the point $P \equiv (\varphi, \theta)$ of the unit sphere ω such that the function $\varrho^n Y(P)$ of the point $M \equiv (\varrho, P)$, (that is the point $M \equiv (x, y, z)$ which is a distance ϱ from the origin, and such that the radius $\overline{OM}$ intersects the unit sphere ω at point P) is a Laplace polynomial of the nth degree.

2b. In particular denote by $Y_n^m(\varphi, \theta)$ the $2n + 1$ spherical harmonics

$$Y_n^m(\varphi, \theta) = V_m(\cos\varphi, \sin\varphi\cos\theta, \sin\varphi\sin\theta), \quad (m = 0, 1, \ldots, 2n),$$

where the $V_m(x, y, z)$ are the $2n + 1$ harmonic polynomials of the nth degree defined in Sec. 18.1b. We shall prove that they are linearly independent.

Suppose there were $2n + 1$ constants $c_0, c_1, \ldots, c_{2n}$ not all zero for which

$$\sum_{m=0}^{(2n)} c_m Y_n^m(\varphi, \theta) \equiv 0.$$

It would follow then that the harmonic function

$$\sum_{m=0}^{(2n)} \varrho^n c_m Y_n^m(\varphi, \theta) = \sum_{m=0}^{(2n)} c_m V_m(x, y, z)$$

vanishes identically on the surface of ω and therefore vanishes identically by a theorem of Picone. (cf. e.g. M. Picone [88b], p. 46, Theorem II). (We will give a direct proof in Sec. 18.4b that there are $2n + 1$ linearly independent spherical harmonics of order n). But $\sum_{m=0}^{(2n)} c_m V_m(x, y, z) \equiv 0$ is impossible since the $2n + 1$ polynomials V_m are linearly independent.

Conversely if $Y_n(\varphi, \theta)$ is an arbitrary spherical harmonic of order n, then for the corresponding Laplace polynomial $\varrho^n Y_n(\varphi, \theta)$ we have

$$\varrho^n Y_n(\varphi, \theta) = \sum_{m=0}^{(2n)} c_m V_m(x, y, z) = \sum_{m=0}^{(2n)} c_m \varrho^n Y_n^m(\varphi, \theta)$$

where $c_0, c_1, \ldots, c_{2n}$ are constants, and therefore $Y_n(\varphi, \theta) = \sum_{m=0}^{(2n)} c_m Y_n^m(\varphi, \theta)$.

We have therefore proved that there exist $2n + 1$ spherical harmonics of the nth order, and not more than $2n + 1$, which are linearly independent.

3. We now show that a necessary and sufficient condition for a polynomial $Y(\varphi, \theta)$ in $\cos\theta$, $\sin\theta$, $\cos\varphi$, $\sin\varphi$, to be a spherical harmonic of the nth order is that it satisfies the following partial differential equation of the second order

$$\text{(I)} \qquad \boxed{\frac{1}{\sin\varphi}\frac{\partial}{\partial\varphi}\left[\sin\varphi\frac{\partial Y}{\partial\varphi}\right] + \frac{1}{\sin^2\varphi}\frac{\partial^2 Y}{\partial\theta^2} + n(n+1)Y = 0}\,.$$

a) The condition is necessary. First, write the equation $\Delta_2 u = 0$ in terms of polar coordinates

$$\text{(6)} \qquad \frac{1}{\varrho^2}\left[\frac{\partial\left(\varrho^2\dfrac{\partial u}{\partial\varrho}\right)}{\partial\varrho} + \frac{1}{\sin\varphi}\frac{\partial\left(\sin\varphi\dfrac{\partial u}{\partial\varphi}\right)}{\partial\varphi} + \frac{1}{\sin^2\varphi}\frac{\partial^2 u}{\partial\theta^2}\right] = 0,$$

(Cf., e. g. G. Sansone [98a] *Lezioni di Analisi Matematica*; II (7th ed; Padova, 1948), p. 91). Then, if $u = \varrho^n Y(\varphi, \theta)$, we have

$$\frac{\partial u}{\partial\varrho} = n\varrho^{n-1}Y, \qquad \varrho^2\frac{\partial u}{\partial\varrho} = n\varrho^{n+1}Y, \qquad \frac{\partial}{\partial\varrho}\left(\varrho^2\frac{\partial u}{\partial\varrho}\right) = n(n+1)\varrho^n Y;$$

$$\sin\varphi\frac{\partial u}{\partial\varphi} = \varrho^n\sin\varphi\frac{\partial Y}{\partial\varphi}, \quad \frac{1}{\sin\varphi}\frac{\partial\left(\sin\varphi\dfrac{\partial u}{\partial\varphi}\right)}{\partial\varphi} = \varrho^n\frac{1}{\sin\varphi}\frac{\partial}{\partial\varphi}\left[\sin\varphi\frac{\partial Y}{\partial\varphi}\right];$$

$$\frac{1}{\sin^2\varphi}\frac{\partial^2 u}{\partial\theta^2} = \varrho^n\frac{1}{\sin^2\varphi}\frac{\partial^2 Y}{\partial\theta^2},$$

and substituting in (6) and canceling out ϱ^{n-2}, (I) follows.

b) The condition is sufficient.

Let $Y(\varphi, \theta)$ be a polynomial in $\cos\theta$, $\sin\theta$, $\cos\varphi$, $\sin\varphi$ which satisfies (I); if we express the powers of the sines and cosines of θ in terms of the sines and cosines of multiples of θ and express the products in terms of the usual reduction formulas, we get

$$Y(\varphi, \theta) = \sum_m [a_m \cos m\theta + b_m \sin m\theta],$$

where a_m, b_m are polynomials in $\sin\varphi$, $\cos\varphi$. Then, since $Y(\varphi, \theta)$ must satisfy (I), we have

$$\sum_m \left[\left\{\frac{d\left(\sin\varphi \dfrac{da_m}{d\varphi}\right)}{\sin\varphi\, d\varphi} + a_m\left[n(n+1) - \frac{m^2}{\sin^2\varphi}\right]\right\}\cos m\theta \right.$$
$$\left. + \left\{\frac{d\left(\sin\varphi \dfrac{db_m}{d\varphi}\right)}{\sin\varphi\, d\varphi} + b_m\left[n(n+1) - \frac{m^2}{\sin^2\varphi}\right]\right\}\sin m\theta\right] \equiv 0.$$

The sum in question has a finite number of terms and is identically zero. Consequently, the coefficients of $\cos m\theta$, $\sin m\theta$ are also zero. (If $\frac{1}{2}\alpha_0 + \sum_{m=1}^{n} (\alpha_m \cos m\theta + \beta_m \sin m\theta) \equiv 0$, α_m and β_m are the Fourier constants of the identically vanishing function and are therefore zero). Therefore, a_m and b_m for the same value of the index m are polynomials in $\sin\varphi$, $\cos\varphi$, which satisfy the same second order differential equation:

$$\frac{d\left(\sin\varphi \dfrac{dz}{d\varphi}\right)}{\sin\varphi\, d\varphi} + \left[n(n+1) - \frac{m^2}{\sin^2\varphi}\right] z = 0,$$

or letting $\cos\varphi = x$, $[\sin\varphi = \sqrt{1-x^2}]$, a_m and b_m are polynomials in x and $\sqrt{1-x^2}$ which satisfy the equation

$$(7) \qquad \frac{d}{dx}\left[(1-x^2)\frac{dz}{dx}\right] + \left[n(n+1) - \frac{m^2}{1-x^2}\right] z = 0.$$

We shall now prove that z has the form

$$z = (1-x^2)^{m/2} Z(x),$$

where $Z(x)$ is a polynomial in x and $\sqrt{1-x^2}$ and $m \leqq n$.

Since z is a polynomial in x and $\sqrt{1-x^2}$, the two products $(1-x^2)^{1/2}z'$, $(1-x^2)^{3/2}z''$ have a finite value for $x = \pm 1$, and by (7) it follows that z must vanish for $x = \pm 1$. Therefore

$$z = (1-x^2)^p Z(x), \quad Z(1) \neq 0, \quad p > 0,$$

where $Z(x)$ is a polynomial in x and $\sqrt{1-x^2}$.

Substituting this expression in (7) and clearing of fractions we have

$$(1-x^2)^2 Z'' - 2(2p+1)x(1-x^2)Z' \\ + [\{n(n+1) - 2p\}(1-x^2) + 4p^2x^2 - m^2]\, Z = 0.$$

Now, letting $x = \pm 1$, we get $p = m/2$ and the equation becomes

$$\Omega(Z) = (1-x^2)Z'' - (2m+1)xZ' \\ + [n(n+1) - m(m+1)]\, Z = 0. \tag{8}$$

It then remains to be proved that $m \leqq n$.

Clearly $Z(x) = Z_1(x) + Z_2(x)\sqrt{1-x^2}$ where $Z_1(x)$, $Z_2(x)$ are polynomials in x. Substituting in (8) we get

$$\Omega(Z_1) + \frac{T(x)}{(1-x^2)^{3/2}} = 0,$$

where $\Omega(Z_1)$ and $T(x)$ are polynomials in x. This implies that $\Omega(Z_1) = 0$ and if l is the highest degree of terms of Z_1 we have

$$-\,l(l-1) - 2(m+1)l + n(n+1) - m(m+1) = 0,$$
$$l^2 + (2m+1)l - n(n+1) + m(m+1) = 0,$$
$$l = [-\,(2m+1) \pm (2n+1)]/2$$

and since l must be a positive integer, $l = n - m$ with $n \geqq m$.

It follows at once that a_m and b_m differ by constant factors from the functions $P_n^m(x)$ defined in Sec. 17.1.

Indeed, it follows from (1) and (3′) of Sec. 17.1 that $P_n^m(x)$ is a polynomial in x and $\sqrt{1-x^2}$ that satisfies the equation

$$\frac{d}{dx}\left[(1-x^2)\frac{dP_n^m}{dx}\right] + \left[n(n+1) - \frac{m^2}{1-x^2}\right]P_n^m(x) = 0. \tag{9}$$

Multiplying (7) by $P_n^m(x)$ and (9) by z and subtracting we have

$$\frac{d}{dx}\left[(1-x^2)\left\{\frac{dz}{dx}P_n^m(x)-\frac{dP_n^m(x)}{dx}z\right\}\right]=0.$$

Integrating, we get

$$(1-x^2)\left[\frac{dz}{dx}P_n^m(x)-\frac{dP_n^m(x)}{dx}z\right]=c,\quad [c=\text{constant}].$$

Since z and $P_n^m(x)$ are polynomials in x and $\sqrt{1-x^2}$, their derivatives are at most divisible by $\sqrt{1-x^2}$ and therefore the first member of the preceding equation vanishes for $x=\pm 1$, whence $c=0$, and therefore

$$\frac{dz}{dx}P_n^m(x)-\frac{dP_n^m(x)}{dx}z=0.$$

Consequently, in an interval included between two consecutive zeros of $P_n^m(x)$ we have $z=cP_n^m(x)$, $[c=\text{constant}]$.

It is now easy to prove that the constant c remains the same for the whole interval $[-1, 1]$.

Clearly $P_n(x)$ has n simple zeros interior to $[-1, 1]$. Then by Rolle's theorem, $P_n^{(m)}(x)$ has $n-m$ simple zeros, $\beta_1, \beta_2, \ldots, \beta_{n-m}$, $\beta_1<\beta_2<\ldots<\beta_{n-m}$ interior to $[-1, 1]$, and P_n^m will admit besides these $n-m$ zeros, two zeros $\beta_0=-1$, $\beta_{n-m+1}=1$. Between β_r and β_{r+1}, β_{r+1} and β_{r+2} we have respectively

$$z=c_r P_n^m(x),\quad z=c_{r+1}P_n^m(x),$$

where c_r and c_{r+1} are constants; and since $[P_n^m(x)]'_{x=\beta_{r+1}}\neq 0$, it follows that $c_r=c_{r+1}$.

It follows that $Y(\varphi, \theta)$ is given by the expression

$$(10)\qquad Y(\varphi,\theta)=\sum_{m=0}^{n}[h_m\cos m\theta+k_m\sin m\theta]P_n^m(\cos\varphi)$$

where $P_n^m(\cos\varphi)$ is the value which $P_n^m(x)$ assumes when we substitute $\cos\varphi$ for x, namely by (1) of Sec. 17.1.

$$P_n^m(\cos\varphi) = \sin^m\varphi\left[\frac{d^m P_n(x)}{dx^m}\right]_{x=\cos\varphi}. \tag{11}$$

Finally, we get the following expression for $Y(\varphi, \theta)$

$$Y(\varphi,\theta) = \sum_{m=0}^{n}[h_m \cos m\theta + k_m \sin m\theta]\sin^m\varphi\left[\frac{d^m P_n(x)}{dx^m}\right]_{x=\cos\varphi}, \tag{10'}$$

where h_m and k_m are constants.

From (10′) it follows that $Y(\varphi, \theta)$ is a homogeneous polynomial of the nth degree in $\cos\varphi$, $\sin\varphi\cos\theta$, $\sin\varphi\sin\theta$.

In fact we have

$$\cos m\theta = \cos^m\theta - \binom{m}{2}\cos^{m-2}\theta\sin^2\theta + \ldots;$$

$$\sin m\theta = \binom{m}{1}\cos^{m-1}\theta\sin\theta - \binom{m}{3}\cos^{m-3}\theta\sin^3\theta + \ldots$$

and since every term of the polynomial $d^m P_n/dx^m$ has the form Ax^{n-m-2k} with k an integer not greater than $[(n-m)/2]$ it follows that the terms of the expansion (10′) are given by

$$\begin{aligned}\cos^{m-r}\theta\sin^r\theta\sin^m\varphi(\cos\varphi)^{n-m-2k} &\\ = (\cos\theta\sin\varphi)^{m-r}(\sin\varphi\sin\theta)^r(\cos\varphi)^{n-m-2k} &\\ = (\cos\theta\sin\varphi)^{m-r}(\sin\varphi\sin\theta)^r(\cos\varphi)^{n-m-2k}[\cos^2\varphi &\\ + \sin^2\varphi\cos^2\theta + \sin^2\varphi\sin^2\theta]^k &\end{aligned}$$

which is precisely a homogeneous polynomial of the nth degree in $\cos\varphi$, $\sin\varphi\cos\theta$, $\sin\varphi\sin\theta$.

Now if we let $x = \varrho\cos\varphi$, $y = \varrho\sin\varphi\cos\theta$, $z = \varrho\sin\varphi\sin\theta$, then $u = \varrho^n Y(\theta, \varphi)$, is a homogeneous polynomial of the nth degree in x, y, z and since Y satisfies (I), the polynomial u is clearly harmonic in view of the transformation given in a).

This proves that the condition is sufficient.

4a. Every system of $2n+1$ linearly independent spherical harmonics of the nth order is called *a fundamental system of spherical harmonics of the nth order.*

4b. The theorem proved in Sec. 18.3 can be completed as follows: If $P_n(x)$ is the nth Legendre polynomial [of the nth degree], the $2n + 1$ functions of the point $P \equiv (\varphi, \theta)$ of ω

$$\text{(II)} \quad \boxed{\begin{gathered} U_n^0(P) = P_n(\cos\varphi) \\ U_n^m(P) = P_n^{(m)}(\cos\varphi)\sin^m\varphi\cos m\theta, \\ V_n^m(P) = P_n^{(m)}(\cos\varphi)\sin^m\varphi\sin m\theta \\ (m = 1, 2, \ldots, n) \end{gathered}}$$

[where $P_n^{(m)}(x)$ denotes the derivative of the mth order of $P_n(x)$ with respect to x], form a fundamental system of spherical harmonics of the nth order.

By the theorem proved in Sec. 18.3, the $2n + 1$ functions defined by (II) are spherical harmonics of the nth order. It will therefore be sufficient to prove that they are linearly independent.

Suppose, then, that there exist constants $h_0, h_1, \ldots, h_n, k_1, \ldots, k_n$ such that we get identically

$$h_0 P_n(\cos\varphi) + \sum_{m=1}^{n} P_n^{(m)}(\cos\varphi)\sin^m\varphi\,[h_m\cos m\theta + k_m\sin m\theta] \equiv 0.$$

Then multiplying both sides by $\cos m\theta$, or by $\sin m\theta$, and integrating with respect to θ between 0 and 2π we get

$$h_m P_n^{(m)}(\cos\varphi)\sin^m\varphi \equiv 0, \quad k_m P_n^{(m)}(\cos\varphi)\sin^m\varphi \equiv 0,$$
$$[m = 0, 1, \ldots, n]$$

whence it follows that h_m and k_m are all zero.

4c. Finally, in view of the relation between the Ferrers functions $P_n^m(x)$ associated with the Legendre functions and the $P_n^{(m)}(x)$

$$P_n^m(x) = (1 - x^2)^{m/2} P_n^{(m)}(x)$$

and (11), we have that a fundamental system of $2n + 1$ spherical harmonics of the point $P \equiv (\varphi, \theta)$ of ω is given by

$$\text{(II')} \quad \boxed{\begin{gathered} U_n(P) = P_n(\cos\varphi) \\ U_n^m(P) = P_n^m(\cos\varphi)\cos m\theta, \quad V_n^m(P) = P_n^m(\cos\varphi)\sin m\theta, \\ (m = 1, 2, \ldots, n) \end{gathered}}\,.$$

19. Integral Properties of Spherical Harmonics and the Addition Theorem for Legendre Polynomials

1a. Two spherical harmonics of different order are orthogonal over the sphere ω [unit sphere with center at the origin]; namely

$$\text{(III)} \qquad \boxed{\int_\omega Y_n(P)\,Y_m(P)\,d\omega(P) = 0, \quad n \neq m},$$

where $d\omega(P)$ denotes the element of area of the surface ω.

In fact we have from (I) of Sec. 18.3

$$\frac{1}{\sin\varphi}\frac{\partial}{\partial\varphi}\left[\sin\varphi\,\frac{\partial Y_n}{\partial\varphi}\right] + \frac{1}{\sin^2\varphi}\frac{\partial^2 Y_n}{\partial\theta^2} = -\,n(n+1)\,Y_n,$$

$$\frac{1}{\sin\varphi}\frac{\partial}{\partial\varphi}\left[\sin\varphi\,\frac{\partial Y_m}{\partial\varphi}\right] + \frac{1}{\sin^2\varphi}\frac{\partial^2 Y_m}{\partial\theta^2} = -\,m(m+1)Y_m,$$

and multiplying the first equation by $Y_m \sin\varphi$ and the second by $Y_n \sin\varphi$ and subtracting we have

$$\frac{\partial}{\partial\varphi}\left[\sin\varphi\left(Y_m\frac{\partial Y_n}{\partial\varphi} - Y_n\frac{\partial Y_m}{\partial\varphi}\right)\right] + \frac{1}{\sin\varphi}\frac{\partial}{\partial\theta}\left[Y_m\frac{\partial Y_n}{\partial\theta} - Y_n\frac{\partial Y_m}{\partial\varphi}\right]$$
$$= [m(m+1) - n(n+1)]Y_m Y_n \sin\varphi$$

from which, integrating with respect to θ between 0 and 2π and with respect to φ between 0 and π and recalling that $d\omega(P) = \sin\varphi\; d\varphi d\theta$, we have

$$[m(m+1) - n(n+1)]\int_\omega Y_m(P)\,Y_n(P)\,d\omega(P)$$
$$= \int_0^{2\pi}\left[\sin\varphi\left(Y_m\frac{\partial Y_n}{\partial\varphi} - Y_n\frac{\partial Y_m}{\partial\varphi}\right)\right]_{\varphi=0}^{\varphi=\pi} d\theta$$
$$+ \int_0^{\pi}\left[Y_m\frac{\partial Y_n}{\partial\theta} - Y_n\frac{\partial Y_m}{\partial\theta}\right]_{\theta=0}^{\theta=2\pi}\frac{d\varphi}{\sin\varphi},$$

and since the right side is zero, (III) follows at once if $m \neq n$.

1b. The fundamental spherical harmonics of the nth order

$$\text{(II}'\text{)}\quad U_n^0(P); U_n^m(P) = P_n^m(\cos\varphi)\cos m\theta,\quad V_n^m(P) = P_n^m(\cos\varphi)\sin m\theta,$$
$$(m = 1, 2, \ldots, n)$$

are mutually orthogonal over the sphere ω.

In fact, if we take the integral of the product of two of these functions over the sphere ω and change the surface integral to a double integral by successive integrations with respect to θ and φ, the contribution of the integration with respect to θ is zero (the functions 1, $\sin\theta$, $\sin 2\theta$, . . ., $\sin n\theta$, $\cos\theta$, $\cos 2\theta$, . . ., $\cos n\theta$, are orthogonal in $[0, 2\pi]$) which establishes the property referred to.

1c. From 1a and 1b it follows that the fundamental system of spherical harmonics (II), as n varies, describes a denumerably infinite number of mutually orthogonal spherical harmonics over the sphere ω.

2a. The following relations are easily established:

$$\text{(IV)}\quad \boxed{\begin{gathered} \int_\omega [U_n^0(P)]^2\, d\omega(P) = \frac{4\pi}{2n+1}, \\ \int_\omega [U_n^m(P)]^2\, d\omega(P) = \int_\omega [V_n^m(P)]^2\, d\omega(P) = \frac{2\pi}{2n+1}\frac{(n+m)!}{(n-m)!} \\ (m = 1, \ldots, n) \end{gathered}}\,.$$

To establish the first formula of IV we note that $U_n^0 = P_n(\cos\varphi)$ and $d\omega = \sin\varphi\, d\varphi\, d\theta$ whence (cf. Sec. 9.1, (35_2))

$$\int_\omega [U_n^0(P)]^2 d\omega = \int_0^{2\pi} d\theta \int_0^\pi [P_n(\cos\varphi)]^2 \sin\varphi d\varphi = 2\pi \int_{-1}^1 [P_n(x)]^2 dx = \frac{4\pi}{2n+1}.$$

To establish the formula for $U_n^m(P)$ we note that

$$\begin{aligned} \int_\omega [U_n^m(P)]^2 d\omega &= \int_0^{2\pi} d\theta \int_0^\pi [P_n^m(\cos\varphi)\cos m\theta]^2 \sin\varphi\, d\varphi \\ &= \int_0^{2\pi} \cos^2 m\theta\, \mathrm{d}\theta \int_0^\pi [P_n^m(\cos\varphi)]^2 \sin\varphi\, d\varphi \\ &= \pi \int_{-1}^1 [P_n^m(x)]^2 dx = \frac{2\pi}{2n+1}\frac{(n+m)!}{(n-m)!}, \quad \text{(cf. Sec. 17.2, (4))}. \end{aligned}$$

The remaining formula for $V_n^m(P)$ can be established similarly.

2b. Let P and Q be two variable points on the sphere ω, and indicate by $\gamma(P, Q)$ the angle, between 0 and π, formed by the two vector radii $\overline{OP}$, $\overline{OQ}$ (O is the center of ω). Then we prove that, for P fixed and Q varying over ω, $P_n[\cos\gamma\,(P, Q)]$ is a spherical harmonic of the nth order of the spherical coordinates of Q.

In fact, if φ, θ and φ', θ' are the spherical coordinates of P and Q with respect to a stationary pole P_0 of ω, if we refer the points of ω to a system of orthogonal cartesian axes with the positive x-axis coincident with OP_0, and with the plane xy coincident with the first meridian, and let $P \equiv (x, y, z)$, $Q \equiv (x', y', z')$ then we have $x = \cos\varphi$, $y = \sin\varphi\cos\theta$, $z = \sin\varphi\ \sin\theta$; $x' = \cos\varphi'$, $y' = \sin\varphi'\cos\theta'$, $z' = \sin\varphi'\sin\theta'$, whence

$$\begin{aligned}\cos\gamma &= xx' + yy' + zz' \\ &= \cos\varphi\cos\varphi' + \sin\varphi\sin\varphi'[\cos\theta\cos\theta' + \sin\theta\sin\theta'],\end{aligned}$$

or

$$\cos\gamma = \cos\varphi\cos\varphi' + \sin\varphi\sin\varphi'\cos(\theta - \theta'). \tag{12}$$

Therefore, $P_n(\cos\gamma)$ is a polynomial in $\cos\varphi'$, $\sin\varphi'$, $\cos\theta'$, $\sin\theta'$, and to prove the proposition in question it will suffice to show that (I) of Sec. 18.3 is satisfied with respect to the variables φ', θ'.

It is easily seen that

$$\begin{aligned}\frac{1}{\sin\varphi'}\frac{\partial}{\partial\varphi'}&\left[\sin\varphi'\frac{\partial P_n(\cos\gamma)}{d\varphi'}\right] \\ &= [-\cos\varphi\sin\varphi' + \sin\varphi\cos\varphi'\cos(\theta - \theta')]^2 P_n''(\cos\gamma) \\ &\quad + \frac{\cos\varphi'}{\sin\varphi'}[-\cos\varphi\sin\varphi' + \sin\varphi\cos\varphi'\cos(\theta - \theta')]\,P_n'(\cos\gamma) \\ &\quad - (\cos\varphi\cos\varphi' + \sin\varphi\sin\varphi'\cos(\theta - \theta')]\,P_n'(\cos\gamma),\end{aligned}$$

$$\begin{aligned}\frac{1}{\sin^2\varphi'}\frac{\partial^2 P_n(\cos\gamma)}{\partial\theta'^2} &= P_n''(\cos\gamma)\sin^2\varphi\sin^2(\theta - \theta') \\ &\qquad - \frac{\sin\varphi}{\sin\varphi'}\cos(\theta - \theta')P_n'(\cos\gamma).\end{aligned}$$

If we now add these two equations we find that the coefficient of P_n'' (cos γ) on the right side of the resulting equation is

$$\cos^2\varphi(1-\cos^2\varphi')+\sin^2\varphi\cos^2\varphi'\cos^2(\theta-\theta')$$
$$-2\sin\varphi\cos\varphi\sin\varphi'\cos\varphi'\cos(\theta-\theta')+\sin^2\varphi[1-\cos^2(\theta-\theta')]$$
$$=1-[\cos\varphi\cos\varphi'+\sin\varphi\sin\varphi'\cos(\theta-\theta')]^2=1-\cos^2\gamma,$$

and the coefficient of $P_n'(\cos\gamma)$ is

$$\frac{1}{\sin\varphi'}(-\cos\varphi\cos\varphi'\sin\varphi'+\sin\varphi\cos^2\varphi'\cos(\theta-\theta')$$
$$-\cos\varphi\cos\varphi'\sin\varphi'-\sin\varphi\sin^2\varphi'\cos(\theta-\theta')-\sin\varphi\cos(\theta-\theta')]$$
$$=\frac{1}{\sin\varphi'}[-2\cos\varphi\cos\varphi'\sin\varphi'-2\sin\varphi\sin^2\varphi'\cos(\theta-\theta')]=-2\cos\gamma,$$

so that the left side of (I) (equation of the spherical harmonics of order n) is

$$(1-\cos^2\gamma)P_n''(\cos\gamma)-2(\cos\gamma)P_n'(\cos\gamma)+n(n+1)P_n(\cos\gamma)$$

and this vanishes identically (cf. Sec. 3).

2c. We now prove the following formulas for a spherical harmonic $Y_n(P)$ of the nth order.

$$(V_1)\qquad \frac{2n+1}{4\pi}\int_\omega Y_n(Q)P_m[\cos\gamma(P,Q)]d\omega(Q)=0\quad\text{if } m\neq n,$$

$$(V_2)\qquad \frac{2n+1}{4\pi}\int_\omega Y_n(Q)P_n[\cos\gamma\,(P,Q)]d\omega(Q)=Y_n(P).$$

The first of these formulas is a consequence of the orthogonality on the sphere ω of two spherical harmonics of different order.

Since the function $\varrho^n Y_n$ together with its partial derivatives of the second order are continuous functions of the coordinates x, y, z of the point (ϱ, P) both within and on the unit sphere ω, and $\varrho^n Y_n$ satisfies the equation $\Delta_2\,\varrho^n Y_n=0$, we have by Stokes formula at a point (ϱ', P) inside ω, $(0\leqq\varrho'<1)$

$$\varrho'^n Y_n(P) = \frac{1}{4\pi}\int_\omega \left[\varrho^n Y_n(Q)\,\frac{d\frac{1}{r}}{dn} - \frac{1}{r}\,\frac{d\varrho^n Y_n(Q)}{dn}\right] d\omega(Q), \tag{13}$$

where $1/r$ is the reciprocal of the distance r of the point (ϱ', P) (inside ω) from the variable point (ϱ, Q) on ω, and n is the inner normal to the surface ω. (Cf. Appendix, Th. 22.)

On ω we have

$$\varrho^n Y_n(Q) = Y_n(Q), \quad \frac{d\varrho^n Y_n(Q)}{dn} = -\,nY_n(Q), \tag{14}$$

and we must then evaluate $1/r$ and $d(1/r)/dn$ on ω.

Let $\varrho' = \overline{OP}$, $\varrho = \overline{OQ}$, $(\varrho' < \varrho)$ and $\gamma = \angle POQ$. Now by (2) of Sec. 1.1, we have

$$\frac{1}{r} = \frac{1}{(\varrho'^2 - 2\varrho\varrho'\cos\gamma + \varrho^2)^{1/2}} = \frac{1}{\varrho}\left[1 - 2\frac{\varrho'}{\varrho}\cos\gamma + \left(\frac{\varrho'}{\varrho}\right)^2\right]^{-1/2}$$
$$= \sum_{m=0}^{\infty} \frac{\varrho'^m}{\varrho^{m+1}} P_m(\cos\gamma)$$

and differentiating in the $\overline{QO}$ direction,

$$\frac{d\frac{1}{r}}{d\varrho} = \sum_{m=0}^{\infty} (m+1)\frac{\varrho'^m}{\varrho^{m+2}} P_m(\cos\gamma)$$

and for $\varrho = 1$, that is on ω, we have

$$\begin{aligned} \left(\frac{1}{r}\right)_\omega &= \sum_{m=0}^{\infty} \varrho'^m P_m[\cos\gamma(P,Q)], \\ \left(\frac{d\frac{1}{r}}{dn}\right)_\omega &= \sum_{m=0}^{\infty} (m+1)\varrho'^m P_m[\cos\gamma(P,Q)] \end{aligned} \tag{15}$$

and substituting (14) and (15) in (13) we obtain

$$\varrho'^n Y_n(P) = \frac{1}{4\pi}\int_\omega Y_n(Q)\left[\sum_{m=0}^{\infty}(m+1)\varrho'^m P_m[\cos\gamma(P,Q)]\right.$$
$$\left. + n\sum_{m=0}^{\infty}\varrho'^m P_m[\cos\gamma(P,Q)]\right] d\omega(Q)$$
$$= \frac{1}{4\pi}\int_\omega Y_n(Q)\left[\sum_{m=0}^{\infty}(n+m+1)\varrho'^m P_m[\cos\gamma(P,Q)]\right] d\omega(Q).$$

We may now integrate term by term on the right side since the series is uniformly convergent whence, by virtue of the orthogonality of Y_n and P_m for $n \neq m$, we get

$$\varrho'^n Y_n(P) = \varrho'^n \frac{2n+1}{4\pi} \int_\omega Y_n(Q) P_n[\cos\gamma(P, Q)] d\omega(Q),$$

from which (V_2) follows at once.

2d. From (V_2) changing P and Q to Q and Q', and letting $Y_n(Q') = P_n[\cos\gamma(P, Q')]$ we have, in particular,

$$\text{(VI)} \qquad \boxed{\begin{aligned} &\frac{2n+1}{4\pi}\int_\omega P_n[\cos\gamma(P, Q')]\, P_n[\cos\gamma(Q, Q')]\, d\omega(Q') \\ &\qquad = P_n[\cos\gamma(P, Q)] \end{aligned}}\,.$$

3a. The following formula gives the so-called addition theorem of spherical harmonics

$$\text{(VII)} \qquad \begin{aligned} P_n[\cos\gamma(P, Q)] &= U_n^0(P)U_n^0(Q) \\ &+ 2\sum_{m=1}^{n}\frac{(n-m)!}{(n+m)!}[U_n^m(P)U_n^m(Q) + V_n^m(P)V_n^m(Q)], \end{aligned}$$

or if $P \equiv (\varphi, \theta)$, $Q \equiv (\varphi', \theta')$ (cf. (12) and (II'))

$$\text{(VII')} \qquad \boxed{\begin{aligned} &P_n[\cos\varphi\cos\varphi' + \sin\varphi\sin\varphi'\cos(\theta-\theta')] \\ &= P_n(\cos\varphi)P_n(\cos\varphi') \\ &+ 2\sum_{m=1}^{n}\frac{(n-m)!}{(n+m)!}P_n^m(\cos\varphi)P_n^m(\cos\varphi')\cos m(\theta-\theta') \end{aligned}}\,.$$

In fact, for P fixed, the function $P_n[\cos\gamma\,(P, Q)]$ is a spherical harmonic of Q of the nth order (Sec. 19.2 b) which therefore admits of the following decomposition [Sec. 18.4]

$$P_n[\cos\gamma(P, Q)] = A_0 U_n^0(Q) + \sum_{m=1}^{n}[A_m U_n^m(Q) + B_m V_n^m(Q)],$$

where A_m, B_m are constants.

Multiplying both sides of the last equation by $U_n^0(Q)$, $U_n^m(Q)$,

$V_n^m(Q)$, integrating over ω, and using (IV), (V_1), (V_2) we have

$$\frac{4\pi}{2n+1} A_0 = \int_\omega U_n^0(Q) P_n[\cos \gamma(P, Q)] d\omega(Q) = \frac{4\pi}{2n+1} U_n^0(P),$$

$$\frac{2\pi}{2n+1} \frac{(n+m)!}{(n-m)!} A_m = \int_\omega P_n[\cos \gamma(P, Q)] U_n^m(Q) d\omega(Q) = \frac{4\pi}{2n+1} U_n^m(P);$$

and similarly for B_m, we then get (VII).

3b. Letting $\cos \varphi = x$, $\cos \varphi' = x'$, $-1 \leqq x, x' \leqq 1$, $\sin \varphi = \sqrt{1-x^2}$, $\sin \varphi' = \sqrt{1-x'^2}$, where the radicals denote absolute values, $\theta - \theta' = \tau$, we have from (VII′) the so-called addition formula for Legendre polynomials.

$$\text{(VIII)} \quad \boxed{\begin{aligned} &P_n(xx' + \sqrt{1-x^2}\sqrt{1-x'^2}\cos \tau) \\ &= P_n(x)P_n(x') + 2\sum_{m=1}^{m} \frac{(n-m)!}{(n+m)!} P_n^m(x) P_n^m(x') \cos m\tau \end{aligned}}.$$

Integrating this result with respect to τ between 0 and π, we obtain the following formula

$$\text{(VIII′)} \quad \boxed{\frac{1}{\pi}\int_0^\pi P_n(xx' + \sqrt{1-x^2}\sqrt{1-x'^2}\cos \tau) d\tau = P_n(x) P_n(x')}.$$

3c. Let $f(Q)$ be a function of a variable point Q on ω. If φ and θ denote as usual the polar colatitude and longitude of point Q; $0 \leqq \varphi \leqq \pi$, $0 \leqq \theta \leqq 2\pi$, $Q \equiv (\varphi, \theta)$, $f(Q)$ will be called integrable over ω if the function $f(\varphi, \theta) \sin \varphi$ is integrable in the rectangle

$$0 \leqq \varphi \leqq \pi, \quad 0 \leqq \theta \leqq 2\pi.$$

Denoting the integral, as in the preceding sections, by

$$\int_\omega f(Q) d\omega(Q)$$

we have by Fubini's theorem (Cf. G. Fubini [40], or S. Saks [97a], p. 69).

$$\int_\omega f(Q)\,d\omega(Q) = \int_0^{2\pi} d\theta \int_0^\pi f(\varphi, \theta) \sin\varphi\, d\varphi = \int_0^\pi \sin\varphi\, d\varphi \int_0^{2\pi} f(\varphi, \theta)\,d\theta,$$

$$\int_\omega |f(Q)|\,d\omega(Q) = \int_0^{2\pi} d\theta \int_0^\pi |f(\varphi, \theta)| \sin\varphi\, d\varphi = \int_0^\pi \sin\varphi\, d\varphi \int_0^{2\pi} |f(\varphi, \theta)|\,d\theta$$

whence

$$\left|\int_\omega f(Q)\,d\omega(Q)\right| \leqq \int_\omega |f(Q)|\,d\omega(Q).$$

An important consequence of the foregoing is the fact that if $f(Q)$ is a function of the variable point Q on ω, then the function

$$Y_n(P) = \int_\omega f(Q) P_n[\cos\gamma(P, Q)]\,d\omega(Q)$$

is a spherical harmonic of the nth order of P.

If $P \equiv (\varphi, \theta)$, $Q \equiv (\varphi', \theta')$ then by (VII′)

$$Y_n(P) = Y_n(\varphi, \theta) = P_n(\cos\varphi) \int_0^{2\pi} d\theta' \int_0^\pi f(\varphi', \theta') P_n(\cos\varphi') \sin\varphi' d\varphi'$$

$$+ 2 \sum_{m=1}^{n} \frac{(n-m)!}{(n+m)!} P_n^m(\cos\varphi) \cos m\theta$$

$$\times \int_0^{2\pi} \cos m\theta'\, d\theta' \times \int_0^\pi f(\varphi', \theta') P_n^m(\cos\varphi') \sin\varphi' d\varphi'$$

$$+ 2 \sum_{m=1}^{n} \frac{(n-m)!}{(n+m)!} P_n^m(\cos\varphi) \sin m\theta \int_0^{2\pi} \sin m\theta' d\theta' \int_0^\pi f(\varphi', \theta') P_n^m(\cos\varphi') \sin\varphi' d\varphi'$$

which shows that $Y_n(P)$ is a linear combination of the fundamental spherical harmonics of order n: $P_n(\cos\varphi)$, $P_n^m(\cos\varphi) \cos m\theta$, $P_n^m(\cos\varphi) \sin m\theta$.

20. Completeness of Spherical Harmonics with Respect to Square Integrable Functions

1. On the unit sphere ω, let (φ, θ) denote the usual system of spherical coordinates [φ the polar colatitude, θ the longitude, $0 \leqq \varphi \leqq \pi$, $0 \leqq \theta \leqq 2\pi$] and furthermore let $\{\Phi_n(P)\}$, $P \equiv (\varphi, \theta)$, denote a sequence of orthonormal functions defined over ω:

$$\int_\omega \Phi_n(P)\Phi_m(P)\,d\omega(P) = 0, \quad (n \neq m); \qquad \int_\omega [\Phi_n(P)]^2\,d\omega(P) = 1.$$

Applying Vitali's reasoning for functions of only one variable

[Ch. I, Sec. 6.6] we obtain the following necessary and sufficient condition for the system $\{\Phi_n(P)\}$ to be a complete system with respect to square integrable functions defined over ω. Let T denote the spherical isosceles triangle bounded by the two meridians $\theta = 0, \theta$, and of length φ. Then the condition is:

$$\text{(1)} \quad \begin{aligned} \text{area } T &= \sum_{n=1}^{\infty} \left[\int_T \Phi_n(P)\, d\omega(P) \right]^2 \\ &= \sum_{n=1}^{\infty} \left[\int_0^{\theta} d\theta' \int_0^{\varphi} \Phi_n(\varphi', \theta') \sin \varphi' d\varphi' \right]^2 \end{aligned}$$

where the area $T = (1 - \cos \varphi)\theta$.

REMARK: It should be observed that given the orthonormal system $\{\Phi_n(P)\}$ over ω, even without the assumption of completeness, Bessel's inequality holds [Ch. I, Sec. 4.3],

$$\sum_{n=1}^{\infty} \left[\int_T \Phi_n(P)\, d\omega(P) \right]^2 \leqq \text{area } T.$$

2. Let $\{\Phi_n(P)\}$ be the sequence of spherical harmonics

$$\left[\frac{2n+1}{4\pi}\right]^{1/2} P_n(\cos\varphi), \qquad \left[\frac{2n+1}{2\pi}\frac{(n-m)!}{(n+m)!}\right]^{1/2} P_n^m(\cos\varphi)\cos m\theta,$$

$$\left[\frac{2n+1}{2\pi}\frac{(n-m)!}{(n+m)!}\right]^{1/2} P_n^m(\cos\varphi)\sin m\theta,$$

then the equation of completeness (1) becomes

$$(1-\cos\varphi)\theta = \frac{1}{4\pi}\sum_{n=1}^{\infty}(2n+1)\theta^2\left[\int_0^{\varphi} P_n(\cos\varphi')\sin\varphi' d\varphi'\right]^2$$

$$+\frac{1}{\pi}\sum_{n=1}^{\infty}(2n+1)\sum_{m=1}^{n}\frac{(n-m)!}{(n+m)!}\,\frac{1-\cos m\theta}{m^2}\left[\int_0^{\varphi} P_n^m(\cos\varphi')\sin\varphi' d\varphi'\right]^2,$$

and also, letting $\cos\varphi' = x'$, $\cos\varphi = x$, and inverting the order of integration on the right side, which is permissible, we have

$$(1 - x)\theta = \frac{1}{4\pi} \sum_{n=1}^{\infty} (2n + 1)\theta^2 \left[\int_x^1 P_n(x')dx' \right]^2$$

$$+ \frac{2}{\pi} \sum_{m=1}^{\infty} \frac{1 - \cos m\theta}{m^2} \sum_{n=m}^{\infty} \frac{2n+1}{2} \frac{(n-m)!}{(n+m)!} \left[\int_x^1 P_n^m(x')\, dx' \right]^2 .$$

Now, referring to (7) of Sec. 17.3, the equation of completeness becomes

$$\theta = \frac{\theta^2}{2\pi} + \frac{2}{\pi} \sum_{m=1}^{\infty} \frac{1 - \cos m\theta}{m^2},$$

and finally

$$\theta(2\pi - \theta) = 4 \sum_{m=1}^{\infty} \frac{1 - \cos m\theta}{m^2}.$$

Letting, $\theta = \pi + \alpha$, the last formula becomes

$$\begin{aligned} \pi^2 - \alpha^2 &= 4 \sum_{m=1}^{\infty} \frac{1 - (-1)^m \cos m\alpha}{m^2} \\ &= 4 \sum_{m=1}^{\infty} \frac{1}{m^2} + 4 \sum_{m=1}^{\infty} (-1)^{m-1} \frac{\cos m\alpha}{m^2} \end{aligned} \tag{2}$$

and since

$$\sum_{m=1}^{\infty} \frac{1}{m^2} = \frac{\pi^2}{6}, \quad \sum_{m=1}^{\infty} (-1)^{m-1} \frac{\cos m\alpha}{m^2} = \tfrac{1}{12}\pi^2 - \tfrac{1}{4}\alpha^2 \text{ for } -\pi \leqq \alpha \leqq \pi,$$

[Ch. II: Sec. 2.3; Sec. 3.1] equation (2) is identically satisfied.

21. Laplace Series for an Integrable Function

a. DEFINITION. Let $f(P)$ be a function defined over ω, and integrable there. If we define $Y_n(P)$ by

$$Y_n(P) = \frac{2n+1}{4\pi} \int_{\omega} f(Q) P_n[\cos \gamma(P, Q)] d\omega(Q), \quad (n=0, 1, \ldots), \tag{1}$$

then $Y_n(P)$ is a spherical harmonic of order n of the coordinates of the point P [Sec. 19.3 c], and the series $\sum_{n=0}^{\infty} Y_n(P)$ is called *the Laplace series of $f(P)$ over ω*, and we write

$$f(P) \sim \sum_{n=0}^{\infty} Y_n(P).$$

b. THEOREM. If $f(P)$ is a continuous function over ω and its Laplace series is uniformly convergent over ω, then

$$f(P) = \sum_{n=0}^{\infty} Y_n(P).$$

Proof: It is sufficient to follow the reasoning of Theorem 8 of Ch. II [Sec. 3.1]. Letting

$$\varphi(Q) = \sum_{m=0}^{\infty} Y_m(Q) \tag{2}$$

$\varphi(Q)$ is continuous over ω, since the functions $Y_m(Q)$ are also. Then multiplying both sides of equation (2) by $(2n+1/4\pi) P_n[\cos\gamma(P, Q)]$ and integrating over ω, since the series

$$\sum_{m=0}^{\infty} P_n[\cos\gamma(P, Q)] Y_m(Q)$$

is uniformly convergent ($|P_n| \leqq 1$), it is, therefore, possible to integrate term by term. We get [cf. (V_1), (V_2)]

$$\frac{2n+1}{4\pi}\int_\omega \varphi(Q) P_n[\cos\gamma(P, Q)] d\omega(Q) = Y_n(P),$$

and comparing with (1) for any point P of ω we have

$$\int_\omega [f(Q) - \varphi(Q)] P_n[\cos\gamma(P, Q)] d\omega(Q) = 0,$$

and in view of (VII), and the linear independence of $U_n^0(P)$, $U_n^m(P)$, $V_n^m(P)$ it follows that the difference $f(Q) - \varphi(Q)$ is orthogonal to the sequence of spherical harmonics $\{U_n^0(Q), U_n^m(Q), V_n^m(Q)\}$, and therefore, vanishes almost everywhere over ω [Sec. 20.2].

Finally, since $f(Q) - \varphi(Q)$ is continuous over ω, it follows that $f(Q) - \varphi(Q) \equiv 0$.

22. *Criterion for Pointwise Convergence of Laplace Series*

a. The first results on the pointwise convergence of the series of spherical harmonics relative to a function $f(\theta, \varphi)$ fixed on a unit sphere ω are due to Poisson [90c, d], Dirichlet [29a], and Bonnet [11], but the first exhaustive study of the problem was made by U. Dini [28c, e, f]. We shall limit ourselves to giving a simple criterion of pointwise convergence which will be derived from the theorem of Sec. 14.7, while in the following paragraphs we shall consider (C, k) summability and Poisson summability of the same series.

b. Let $f(P)$ be integrable over the sphere ω. Then

$$f(P) \sim \sum_{n=0}^{\infty} Y_n(P).$$

Taking the pole of the spherical coordinates at the point P, and denoting by $f(P; \varphi, \theta)$ the values of f referred to such a system of coordinates we have

$$f(P) \sim \sum_{n=0}^{\infty} \frac{2n+1}{4\pi} \int_0^{\pi} P_n(\cos \varphi) \sin \varphi \, d\varphi \int_0^{2\pi} f(P; \varphi, \theta) d\theta$$

or letting

$$\frac{1}{2\pi} \int_0^{2\pi} f(P; \varphi, \theta) d\theta = \Phi(P; \varphi),$$

that is indicating by $\Phi(P; \varphi)$ the mean of values of $f(P; \varphi, \theta)$ along the parallel, all of whose points have colatitude φ, we have

$$f(P) \sim \sum_{n=0}^{\infty} \frac{2n+1}{2} \int_0^{\pi} \Phi(P; \varphi) P_n(\cos \varphi) \sin \varphi \, d\varphi.$$

This series represents the Legendre series

$$\sum_{n=0}^{\infty} \frac{2n+1}{2} P_n(\cos \bar{\varphi}) \int_0^{\pi} \Phi(P; \varphi) P_n(\cos \varphi) \sin \varphi \, d\varphi,$$

of the function $\Phi(P; \varphi)$ at the point $\bar{\varphi} = 0$.

From the theorem of Sec. 14.7, it follows that if $\Phi(P; \varphi)$ is a function of φ of bounded variation in $[0, \pi]$, then the series $\sum_{n=0}^{\infty} Y_n(P)$ converges at P to

$$\lim_{\varphi \to +0} \Phi(P; \varphi).$$

In particular if $\Phi(P; \varphi)$ is of bounded variation with respect to φ in the interval $[0, \pi]$, and if $f(P)$ is continuous at P, then $f(P) = \sum_{n=0}^{\infty} Y_n(P)$.

23. (C, k) *Summation of Laplace Series*

1a. (C, k) summation of Laplace series was considered for the first time by L. Fejér for the case $k = 2$, and subsequently by others for the case $k > 0$. In this section we shall use the results of Sec. 7 regarding (C, k) summation of Fourier series to derive the results of a recent memorandum of L. Fejér that will lead very simply to the properties of (C, k) summation of Laplace series for $k > \frac{1}{2}$. [Cf. also for the Bibliography: L. Fejér, [33d]; E. W. Hobson, [48b], pp. 349—359].

1b. We have seen that for $|\varrho| < 1$ we have [Sec. 1.1]

$$(1_1)\quad \frac{1}{(1-2\varrho\cos\varphi+\varrho^2)^{1/2}} = 1+P_1(\cos\varphi)\varrho+\ldots+P_n(\cos\varphi)\varrho^n+\ldots$$

$$(1_2)\quad \frac{1-\varrho^2}{(1-2\varrho\cos\varphi+\varrho^2)^{3/2}} = 1+3P_1(\cos\varphi)\varrho+\ldots+(2n+1)P_n(\cos\varphi)\varrho^n+\ldots.$$

For $|\varrho| < 1$ we have also

$$(2)\quad \frac{1}{(1-\varrho)^{k+1}} = (1-\varrho)^{-(k+1)} = 1+\binom{k+1}{1}\varrho+\ldots+\binom{k+n}{n}\varrho^n+\ldots$$

whence multiplying series (1_2) and (2) by Cauchy's rule

$$\frac{1}{(1-\varrho)^{k+1}}\frac{1-\varrho^2}{(1-2\varrho\cos\varphi+\varrho^2)^{3/2}} = \sum_{n=0}^{\infty}\varrho^n\left[\binom{k+n}{n}\right.$$

$$(3)\quad +\binom{k+n-1}{n-1}3P_1(\cos\varphi)+\binom{k+n-2}{n-2}5P_2(\cos\varphi)+\ldots$$

$$\left.+\binom{k}{0}(2n+1)P_n(\cos\varphi)\right] = \sum_{n=0}^{\infty} S_n^{(k)}(\varphi)\varrho^n,$$

where we have set

$$S_n^{(k)}(\varphi) = \binom{k+n}{n} + \binom{k+n-1}{n-1} 3P_1(\cos\varphi) + \dots + \binom{k}{0}(2n+1)P_n(\cos\varphi) \tag{4}$$

It follows at once that the mean $\Sigma_n^{(k)}(\varphi)$ of the first $n+1$ terms of the series

$$\sum_{n=0}^{\infty} (2n+1)P_n(\cos\varphi) \tag{5}$$

is given by [cf. Ch. II, Sec. 6.2, (4)]

$$\Sigma_n^{(k)}(\varphi) = \frac{S_n^{(k)}(\varphi)}{A_n^{(k)}}, \qquad A_n^{(k)} = \binom{n+k}{n}. \tag{6}$$

But we have

$$\frac{1}{(1-\varrho)^{k+1}} \frac{1-\varrho^2}{(1-2\varrho\cos\varphi+\varrho^2)^{3/2}} = \frac{1}{(1-2\varrho\cos\varphi+\varrho^2)^{1/2}} \left\{ \frac{1}{(1-\varrho)^{k+1}} \frac{1-\varrho^2}{1-2\varrho\cos\varphi+\varrho^2} \right\}$$

and therefore

$$\sum_{n=0}^{\infty} S_n^{(k)}(\varphi)\varrho^n = \sum_{n=0}^{\infty} P_n(\cos\varphi)\varrho^n \times \left\{ \frac{1}{(1-\varrho)^{k+1}} \left[1 + \sum_{n=1}^{\infty} 2\varrho^n \cos n\varphi \right] \right\}. \tag{7}$$

If now we denote by $\mathfrak{S}_n^{(k)}/A_n^{(k)}$ the mean of order k of the first $n+1$ terms of the series

$$1 + 2\cos\varphi + 2\cos 2\varphi + \dots + 2\cos n\varphi + \dots \tag{8}$$

that is [cf. Ch. II, Sec. 7.2, (8)]

$$\boxed{\begin{aligned}\mathfrak{S}_n^{(k)}(\varphi) &= \binom{k+n}{n} + \binom{k+n-1}{n-1} 2\cos\varphi \\ &+ \binom{k+n-2}{n-2} 2\cos 2\varphi + \ldots + \binom{k}{0} 2\cos n\varphi\end{aligned}}, \tag{9}$$

then from (7), in view of the remarks of Ch. II, Sec. 6.3e, we get

$$\boxed{\begin{aligned}S_n^{(k)}(\varphi) &= P_0(\cos\varphi)\mathfrak{S}_n^{(k)}(\varphi) \\ &+ P_1(\cos\varphi)\mathfrak{S}_{n-1}^{(k)}(\varphi) + \ldots + P_n(\cos\varphi)\mathfrak{S}_0^{(k)}(\varphi)\end{aligned}}. \tag{10}$$

2. Some bounds for the sums $\Sigma_n^{(1)}(\varphi)$ will now be derived. We have seen that [cf. Ch. II, Sec. 4.4]

$$1 + 2\cos\varphi + 2\cos 2\varphi + \ldots + 2\cos n\varphi = \frac{\sin(n+\frac{1}{2})\varphi}{\sin\dfrac{\varphi}{2}},$$

therefore [cf. Ch. II, Sec. 6.5]

$$\mathfrak{S}_n^{(1)}(\varphi) = \frac{1}{\sin\dfrac{\varphi}{2}}\left[\sin\frac{\varphi}{2} + \sin\frac{3\varphi}{2} + \ldots + \sin(n+\tfrac{1}{2})\varphi\right] = \frac{\sin^2(n+\frac{1}{2})\varphi}{\sin^2\dfrac{\varphi}{2}},$$

whence by (10)

$$\begin{aligned}S_n^{(1)}(\varphi) &= \left\{\frac{\sin\frac{1}{2}(n+1)\varphi}{\sin\frac{1}{2}\varphi}\right\}^2 P_0(\cos\varphi) \\ &+ \left\{\frac{\sin\frac{1}{2}n\varphi}{\sin\frac{1}{2}\varphi}\right\}^2 P_1(\cos\varphi) + \ldots + \left\{\frac{\sin\frac{1}{2}\varphi}{\sin\frac{1}{2}\varphi}\right\}^2 P_n(\cos\varphi).\end{aligned} \tag{11}$$

We have for $0 < \varphi < \pi$, [Sec. 10.3]

$$|\,P_n(\cos\varphi)\,| < \frac{k_1}{(n\sin\varphi)^{1/2}}, \quad [n > 0,\ 0 < \varphi < \pi],$$

where k_1 is a positive constant independent of n and φ, whence by (11)

$$| S_n^{(1)}(\varphi) | < \frac{1}{\sin^2 \frac{1}{2}\varphi} \frac{k_1}{\sin^{1/2}\varphi} \left\{ \frac{1}{k_1} + \frac{1}{1^{1/2}} + \frac{1}{2^{1/2}} + \ldots + \frac{1}{n^{1/2}} \right\}$$

$$< \frac{k_1}{(\sin^2 \frac{1}{2}\varphi) \sin^{1/2} \varphi} \left\{ \frac{1}{k_1} + \int_0^n \frac{dx}{x^{1/2}} \right\} = \frac{k_1}{(\sin^2 \frac{1}{2}\varphi) \sin^{1/2} \varphi} \left\{ \frac{1}{k_1} + 2n^{1/2} \right\},$$

and therefore, since $A_n^{(1)} = n + 1$,

$$(12_1) \qquad \boxed{|\Sigma_n^{(1)}(\varphi)| < \frac{k_2}{n^{1/2}(\sin^2 \frac{1}{2}\varphi) \sin^{1/2}\varphi}}, \qquad [0<\varphi<\pi,\ n=1,2,\ldots],$$

where k_2 is a constant independent of n and φ.

For $\pi/2 \leqq \varphi \leqq \pi$ and $n \geqq 0$, we have $| P_n (\cos \varphi) | \leqq 1$, and $1/\sin^2 (\varphi/2) \leqq 2$, whence by (11)

$$| S_n^{(1)}(\varphi) | \leqq 2 \left[| \sin \tfrac{1}{2} (n + 1)\varphi |^2 + | \sin \tfrac{1}{2} n\varphi |^2 + \ldots + | \sin \tfrac{1}{2}\varphi |^2 \right] \leqq 2(n + 1),$$

$$(12_2) \qquad \boxed{| \Sigma_n^{(1)}(\varphi) | \leqq 2, \text{ for } \pi/2 \leqq \varphi \leqq \pi}.$$

Moreover, since for arbitrary φ, $| \sin 2\varphi | \leqq 2 | \sin \varphi |$ and by induction $| \sin (n + 1)\varphi | = | \sin n\varphi \cos \varphi + \cos n\varphi \sin \varphi | \leqq n | \sin \varphi | + | \sin \varphi | = (n + 1) | \sin \varphi |$, we have

$$| \sin \tfrac{1}{2}(n + 1)\varphi/\sin \tfrac{1}{2}\varphi |^2 \leqq (n + 1)^2,$$

and from (11) for $0 \leqq \varphi \leqq \pi$

$$| S_n^{(1)}(\varphi) | \leqq (n+1)^2+n^2+\ldots+1^2 = \frac{(n+1)(n+2)(2n+3)}{6} < k_3 n^3$$

where k_3 is a constant independent of n and φ and therefore

$$(12_3) \qquad \boxed{| \Sigma_n^{(1)}(\varphi) | < k_3 n^2, \text{ for } 0 \leqq \varphi \leqq \pi,\ n = 1, 2, \ldots}.$$

REMARK: The bounds (12_1), (12_2), (12_3) have been obtained by entirely elementary reasoning. In the next section we shall see that (12_1) and (12_3) also exist for the mean $\Sigma_n^{(k)}(\varphi)$ where $0 \leqq k \leqq 1$.

3a. An upper bound will now be derived for the means $\Sigma_n^{(k)}(\varphi)$ for $0 \leqq k \leqq 1$, namely

$$(13_1) \quad \boxed{\left|\Sigma_n^{(k)}(\varphi)\right| \leqq \frac{B}{n^{k-\frac{1}{2}}} \frac{1}{\varphi^{1+k} \sin^{1/2}\varphi}; \ (0 \leqq k \leqq 1, 0<\varphi<\pi, n=1, 2, \ldots)},$$

where B is a constant independent of n and φ.

Indeed from (6) and (10) we have

$$|\Sigma_n^{(k)}(\varphi)| = |S_n^{(k)}(\varphi)/A_n^{(k)}| \leqq \sum_{r=0}^{n} |P_r(\cos\varphi)\mathfrak{S}_{n-r}^{(k)}(\varphi)/A_n^{(k)}|,$$

but

$$|P_r(\cos\varphi)| < \frac{k_1}{(r\sin\varphi)^{1/2}}, \quad [k_1 > 0, \text{Sec. } 10.3],$$

$$|\mathfrak{S}_{n-r}^{(k)}(\varphi)| < H'/\varphi^{k+1}, \quad (\text{Ch. II, Sec. } 7.2, (11)],$$

$$A_n^{(k)} > B_2 n^k, \quad [B_2 > 0,\ n = 1, 2, \ldots, \text{Ch. II, Sec. } 7.1, (3)],$$

therefore

$$|\Sigma_n^{(k)}(\varphi)| \leqq \frac{1}{B_2 n^k} \frac{H'}{\varphi^{k+1}} \frac{k_1}{\sin^{1/2}\varphi}\left[\frac{\sin^{1/2}\varphi}{k_1} + \frac{1}{1^{1/2}} + \frac{1}{2^{1/2}} + \ldots + \frac{1}{n^{1/2}}\right]$$

$$\leqq \frac{B_1}{n^k \varphi^{k+1}\sin^{1/2}\varphi}\left[\frac{1}{k_1} + 2n^{1/2}\right] \leqq \frac{B}{n^{k-\frac{1}{2}}\varphi^{k+1}\sin^{1/2}\varphi}.$$

3b. We shall generalize (12_3) by showing that

$$(13_2) \quad \boxed{|\Sigma_n^{(k)}(\varphi)| < 2n^2, \text{ for } 0 \leqq \varphi \leqq \pi,\ n = 3, 4, \ldots}.$$

Indeed from (6) and (4) we have

$$\Sigma_n^{(k)}(\varphi) = \frac{S_n^{(k)}(\varphi)}{A_n^{(k)}} = 1 + \frac{n}{k+n} 3P_1(\cos\varphi)$$

$$+ \frac{n(n-1)}{(k+n)(k+n-1)} 5P_2(\cos\varphi) + \ldots$$

$$+ \frac{n(n-1)\ldots 1}{(k+n)(k+n-1)\ldots(k+1)} (2n+1)P_n(\cos\varphi),$$

and therefore for $k \geqq 0$

$$| \Sigma_n^{(k)}(\varphi) | \leqq 1 + 3 + 5 + \ldots + (2n + 1) = (n + 1)^2,$$

and since for $n \geqq 3$ we have $(n + 1)^2 \leqq 2n^2$, (13_2) follows.

4a. The bounds obtained in Sec. 23.2 will now be used to study $(C, 1)$ summability of Laplace series.

Let $f(P)$ be integrable over sphere ω and let

$$f(P) \sim \sum_{n=0}^{\infty} Y_n(P). \tag{14}$$

Then

$$Y_n(P) = \frac{2n+1}{4\pi} \int_\omega f(Q) P_n[\cos \gamma(P, Q)] \, d\omega(Q) \tag{15}$$

and assuming as in Sec. 22 that the pole of spherical coordinates is at point P and denoting by $f(P; \varphi, \theta)$ the values of f referred to such a system of coordinates, we have

$$Y_n(P) = \frac{1}{4\pi} \int_0^\pi (2n + 1) P_n(\cos \varphi) \sin \varphi \, d\varphi \int_0^{2\pi} f(P; \varphi, \theta) \, d\theta.$$

Now, if we denote by $S_n^*(P)$ the $(C, 1)$ sum of the first $(n + 1)$ terms of the series (14) we have

$$S_n^*(P) = \frac{1}{4\pi} \int_0^\pi \Sigma_n^{(1)}(\varphi) \sin \varphi \, d\varphi \int_0^{2\pi} f(P; \varphi, \theta) \, d\theta, \tag{16}$$

or, letting

$$S_{1,n}^*(P) = \frac{1}{4\pi} \int_\varepsilon^{\pi-\varepsilon} \Sigma_n^{(1)}(\varphi) \sin \varphi \, d\varphi \int_0^{2\pi} f(P; \varphi, \theta) \, d\theta, \tag{17_1}$$

$$S_{2,n}^*(P) = \frac{1}{4\pi} \int_{\pi-\varepsilon}^{\pi} \Sigma_n^{(1)}(\varphi) \sin \varphi \, d\varphi \int_0^{2\pi} f(P; \varphi, \theta) \, d\theta, \tag{17_2}$$

$$S_{3,n}^*(P) = \frac{1}{4\pi} \int_0^{\varepsilon} \Sigma_n^{(1)}(\varphi) \sin \varphi \, d\varphi \int_0^{2\pi} f(P; \varphi, \theta) \, d\theta, \tag{17_3}$$

$$\boxed{S_n^*(P) = S_{1,n}^*(P) + S_{2,n}^*(P) + S_{3,n}^*(P)}. \tag{18}$$

From (17_1) and (12_1) we have

$$(19_1)\qquad \begin{aligned} |S^*_{1,n}(P)| &\leqq k_2 \frac{1}{4\pi}\int_\varepsilon^{\pi-\varepsilon} \frac{\sin\varphi}{n^{1/2}\sin^2\frac{1}{2}\varphi\sin^{1/2}\varphi}\,d\varphi\int_0^{2\pi}|f(P;\varphi,\theta)|\,d\theta \\ &\leqq \frac{k_2}{4\pi n^{1/2}}\frac{1}{\sin^2\frac{1}{2}\varepsilon\sin^{1/2}\varepsilon}\int_\omega |f(Q)|\,d\omega(Q) = \frac{C(\varepsilon)}{n^{1/2}}, \end{aligned}$$

where $C(\varepsilon)$ is independent of n and of the position of the point P on ω.

We also have from (17_2) and (12_2) for $0 < \varepsilon \leqq \pi/2$,

$$\begin{aligned} |S^*_{2,n}(P)| &\leqq \frac{1}{4\pi}\int_{\pi-\varepsilon}^{\pi} |\Sigma_n^{(1)}(\varphi)|\sin\varphi\,d\varphi\int_0^{2\pi}|f(P;\varphi,\theta)|\,d\theta \\ &\leqq \frac{1}{2\pi}\int_{\pi-\varepsilon}^{\pi}\sin\varphi\,d\varphi\int_0^{2\pi}|f(P;\varphi,\theta)|\,d\theta. \end{aligned}$$

Now if ε is fixed, then for any value of the index n and for any point P on ω, we have

$$(19_2)\qquad |S^*_{2,n}(P)| \leqq \eta(\varepsilon),$$

where $\lim_{\varepsilon\to 0}\eta(\varepsilon) = 0$.

Let $S(P)$ be a constant [depending on the point P of ω]; we have from (17_3)

$$\begin{aligned} S^*_{3,n}(P) = {} & \frac{S(P)}{4\pi}\int_0^\varepsilon \Sigma_n^{(1)}(\varphi)\sin\varphi\,d\varphi\int_0^{2\pi}d\theta \\ &+ \frac{1}{4\pi}\int_0^\varepsilon \Sigma_n^{(1)}(\varphi)\sin\varphi\,d\varphi\int_0^{2\pi}[f(P;\varphi,\theta) - S(P)]\,d\theta, \end{aligned}$$

so that letting

$$(20)\qquad \boxed{T_{n,\varepsilon}(P) = \frac{1}{4\pi}\int_0^\varepsilon \Sigma_n^{(1)}(\varphi)\sin\varphi\,d\varphi\int_0^{2\pi}[f(P;\varphi,\theta) - S(P)]\,d\theta},$$

we have

$$(21)\qquad \boxed{S^*_{3,n}(P) = \frac{S(P)}{4\pi}\int_0^\varepsilon \Sigma_n^{(1)}(\varphi)\sin\varphi\,d\varphi\int_0^{2\pi}d\theta + T_{n,\varepsilon}(P)}.$$

Now if $f(P; \theta, \varphi) = 1$ on ω, then for any n we have

$$\frac{1}{4\pi}\int_0^\pi \Sigma_n^{(1)}(\varphi) \sin \varphi \, d\varphi \int_0^{2\pi} d\theta = 1,$$

therefore

$$\frac{1}{4\pi}\int_0^{2\pi} d\theta \int_0^\varepsilon \Sigma_n^{(1)}(\varphi) \sin \varphi \, d\varphi = 1 + \frac{\bar{\theta}_1 \bar{C}(\varepsilon)}{n^{1/2}} + \theta_2 \bar{\eta}(\varepsilon)$$

where $0 < |\theta_1|, |\theta_2| \leqq 1$; and $\bar{C}(\varepsilon)$, $\bar{\eta}(\varepsilon)$ are independent of n and of the point P on ω. Also $\lim_{\varepsilon\to 0} \bar{\eta}(\varepsilon) = 0$. Therefore we get

$$S_n^*(P) - S(P) = \frac{\theta_1 C(\varepsilon) + \bar{\theta}_1 \bar{C}(\varepsilon) S(P)}{n^{1/2}} + \theta_2 \eta(\varepsilon) + \bar{\theta}_2 \bar{\eta}(\varepsilon) S(P) + T_{n,\varepsilon}(P),$$

where $0 < |\theta_1|, |\theta_2| < 1$.

It follows that a necessary and sufficient condition in order to have

$$\lim_{n\to\infty} S_n^*(P) = S(P),$$

is that for any $\sigma > 0$ there exists a positive $\varepsilon_0 < \pi/2$ such that to every positive number $\varepsilon < \varepsilon_0$ there corresponds a number n' such that for all $n > n'$ we get

$$|T_{n,\varepsilon}(P)| < \sigma. \tag{22}$$

If as P varies over a spherical domain I of ω, $S(P)$ remains bounded, and (22) is valid uniformly with respect to P, then $S_n^*(P)$ converges uniformly to $S(P)$ over I as $n \to \infty$.

As in Sec. 22 let $\Phi(P, \varphi)$ denote the mean value of $f(P; \varphi, \theta)$ along the parallel with spherical center P and spherical radius φ, namely,

$$\Phi(P; \varphi) = \frac{1}{2\pi}\int_0^{2\pi} f(P; \varphi, \theta)\, d\theta, \tag{23}$$

and let

$$F(P, \varphi) = F(\varphi) = \frac{1}{\varphi}\int_0^\varphi |\Phi(P; \varphi) - S(P)|\, d\varphi. \tag{24_1}$$

We now show that if

$$\boxed{\lim_{\varphi\to+0} F(P; \varphi) = 0}, \tag{24_2}$$

equation (22) holds.

For $\delta > 0$ there exists an $\varepsilon_0 > 0$ such that for $0 < \varphi < \varepsilon_0$,

$$0 \leqq F(P; \varphi) < \delta. \tag{25}$$

It then follows from (20) for values of $\varepsilon < \varepsilon_0$ that

$$T_{n,\varepsilon}(P) = \tfrac{1}{2}\int_0^\varepsilon [\Phi(P;\varphi) - S(P)]\, \Sigma_n^{(1)}(\varphi) \sin\varphi\, d\varphi.$$

Now taking n so that $1/n < \varepsilon$, and letting

$$U_{n,\varepsilon}(P) = \tfrac{1}{2}\int_{1/n}^\varepsilon [\Phi(P;\varphi) - S(P)]\, \Sigma_n^{(1)}(\varphi) \sin\varphi\, d\varphi, \tag{26_1}$$

$$V_n(P) = \tfrac{1}{2}\int_0^{1/n} [\Phi(P;\varphi) - S(P)]\, \Sigma_n^{(1)}(\varphi) \sin\varphi\, d\varphi, \tag{26_2}$$

we also get

$$T_{n,\varepsilon}(P) = U_{n,\varepsilon}(P) + V_n(P). \tag{27}$$

By (12_1)

$$|U_{n,\varepsilon}(P)| \leqq \tfrac{1}{2}\int_{1/n}^\varepsilon |\Phi(P;\varphi) - S(P)| \frac{k_2}{n^{1/2}\sin^2\frac{1}{2}\varphi}\sqrt{2\sin\tfrac{1}{2}\varphi\cos\tfrac{1}{2}\varphi}\, d\varphi$$

$$\leqq \frac{k_2}{\sqrt{2}n^{1/2}}\int_{1/n}^\varepsilon |\Phi(P;\varphi) - S(P)| \frac{1}{\sin^{3/2}\frac{1}{2}\varphi}\, d\varphi,$$

but for $0 < \varphi < \pi$ we have $\pi^{3/2}/\varphi^{3/2} \geqq 1/\sin^{3/2}\frac{1}{2}\varphi$, so that letting $L = k_2\pi^{3/2}/\sqrt{2}$ and integrating by parts we get, in view of (25),

$$|U_{n,\varepsilon}(P)| \leqq \frac{L}{n^{1/2}}\int_{1/n}^\varepsilon |\Phi(P;\varphi) - S(P)| \frac{1}{\varphi^{3/2}}\, d\varphi$$

$$= \frac{L}{n^{1/2}}\left[\frac{1}{\varphi^{3/2}}\int_0^\varphi |\Phi(P;\varphi) - S(P)|\, d\varphi\right]_{1/n}^\varepsilon$$

$$+ \frac{3}{2}\frac{L}{n^{1/2}}\int_{1/n}^\varepsilon \frac{1}{\varphi^{5/2}}\left\{\int_0^\varphi |\Phi(P;\varphi) - S(P)|\, d\varphi\right\} d\varphi,$$

$$|U_{n,\varepsilon}(P)| \leqq \frac{L}{n^{1/2}}\left\{\left[\frac{F(\varphi)}{\varphi^{1/2}}\right]_{1/n}^\varepsilon + \frac{3}{2}\int_{1/n}^\varepsilon \frac{F(\varphi)}{\varphi^{3/2}}\, d\varphi\right\}$$

$$\leqq \frac{L}{n^{1/2}}\left[\frac{\delta}{\varepsilon^{1/2}} + 3\delta\left(n^{1/2} - \frac{1}{\varepsilon^{1/2}}\right)\right] \leqq L\delta\left[1 + 3\left(1 - \frac{1}{(n\varepsilon)^{1/2}}\right)\right] < 4L\delta.$$

Finally by (26_2) and (12_3), $[0 \leqq \sin\varphi \leqq \varphi \leqq 1/n]$,

$$|V_n(P)| \leqq \tfrac{1}{2}\int_0^{1/n} |\Phi(P;\varphi) - S(P)|\, k_3 n^2 \frac{1}{n}\, d\varphi = \frac{k_3}{2} F\left(P; \frac{1}{n}\right),$$

whence by (27)

$$|T_{n,\varepsilon}(P)| < 4L\delta + \frac{k_3}{2} F\left(P; \frac{1}{n}\right), \tag{28}$$

and in view of (24_2) it follows that $T_{n,\varepsilon}(P) \to 0$ as $\delta \to 0$ and $n \to \infty$.

Thus we have proved the theorem: A sufficient condition for the $(C, 1)$ sum of the Laplace series of $f(P)$ to have the value $S(P)$ at a point P, is

$$\lim_{\varphi \to +0} \frac{1}{\varphi} \int_0^{\varphi} |\Phi(P;\varphi) - S(P)|\, d\varphi = 0, \tag{24}$$

where

$$\frac{1}{2\pi}\int_0^{2\pi} f(P;\varphi,\theta)\, d\theta = \Phi(P;\varphi). \tag{23}$$

4b. In particular: if $f(P)$ is continuous at a point P_0 of ω, then the $(C, 1)$ sum of the Laplace series of $f(P)$ is equal to $f(P_0)$.

Indeed for $\sigma > 0$, and for φ sufficiently small we have

$$f(P_0; \varphi, \theta) = f(P_0) + \vartheta\sigma \text{ where } 0 \leqq |\vartheta| < 1;$$
$$\Phi(P_0; \varphi) = f(P_0) + \vartheta_1\sigma \text{ where } 0 \leqq |\vartheta_1| < 1,$$

and

$$\frac{1}{\varphi}\int_0^{\varphi} |\Phi(P_0;\varphi) - f(P_0)|\, d\varphi < \sigma.$$

4c. We also have if $f(P)$ is continuous in a domain I (over the entire sphere ω) and I' is a domain of ω entirely interior to I (i.e., there exists a $\varrho > 0$ such that every spherical zone with center at any point of I' and with spherical radius ϱ has all its points interior to I) the $(C, 1)$ sum of the Laplace series of $f(P)$ converges uniformly to $f(P)$ in I'.

Indeed, under this hypothesis it is sufficient to take $S(P) = f(P)$

and observe that (28) holds uniformly for all P in I' (or over the entire sphere ω).

5a. In this paragraph we shall study the question of summing of Laplace series for

$$\boxed{1 > k > \tfrac{1}{2}},$$

under the hypothesis (24_2).

Using all the notations of the preceding paragraph and using (13_1) instead of (12_1) we have

$$| S^*_{1,n}(P) | \leqq \frac{C(\varepsilon)}{n^{k-\frac{1}{2}}},$$

where $C(\varepsilon)$ is independent of the position of point P on ω. We cannot apply (12_2) to $S^*_{2,n}(P)$, since (12_2) was established for the case $k = 1$; however using (13_1) we have

$$| S^*_{2,n}(P) | \leqq \frac{B}{4\pi n^{k-\frac{1}{2}}} \int_{\pi-\varepsilon}^{\pi} \frac{\sin\varphi\, d\varphi}{\varphi^{1+k} \sin^{1/2}\varphi} \int_0^{2\pi} | f(P; \varphi, \theta) | \, d\theta$$

which provides a majorant of $S^*_{2,n}(P)$ under the hypothesis that the right side is convergent, that is under the hypothesis that, if we denote by $C(P'; \varepsilon)$ the spherical zone that has for center the point P' on ω diametrically opposite to P and of spherical radius ε, the following integral has a finite value

$$\iint_{C(P';\varepsilon)} \frac{| f(P; \varphi, \theta) |}{| \pi - \varphi |^{1/2}} \sin\varphi\, d\varphi\, d\theta = \int_{\pi-\varepsilon}^{\pi} \frac{\sin\varphi}{| \pi - \varphi |^{1/2}} d\varphi \int_0^{2\pi} | f(P; \varphi, \theta) | \, d\theta. \tag{29}$$

Using (13_2) instead of (12_3), $[k_3 = 2]$ we also have

$$| V_n(P) | \leqq F\left(P; \frac{1}{n}\right),$$

and it remains to consider only the term $U_{n,\varepsilon}(P)$.

We have from (26_1) and (13_1)

$$| U_{n,\varepsilon}(P) | \leqq \frac{B}{2n^{k-\frac{1}{2}}} \int_{1/n}^{\varepsilon} | \Phi(P; \varphi) - S(P) | \frac{1}{\varphi^{k+1} \sin^{1/2} \varphi} \sin \varphi \, d\varphi$$

$$\leqq \frac{B}{2n^{k-\frac{1}{2}}} \int_{1/n}^{\varepsilon} | \Phi(P; \varphi) - S(P) | \frac{1}{\varphi^{k+\frac{1}{2}}} d\varphi$$

$$\leqq \frac{B}{2n^{k-\frac{1}{2}}} \left\{ \left[\frac{F(P; \varphi)}{\varphi^{k-\frac{1}{2}}} \right]_{\varphi=1/n}^{\varphi=\varepsilon} + (k + \tfrac{1}{2}) \int_{1/n}^{\varepsilon} \frac{F(P; \varphi)}{\varphi^{k+\frac{1}{2}}} d\varphi \right\},$$

and then, if for arbitrary $\delta > 0$ we determine ε_0 such that for $0 < \varphi \leqq \varepsilon_0$, $0 \leqq F(P; \varphi) < \delta$, and for every $\varepsilon < \varepsilon_0$ we take values of n such that $1/n < \varepsilon$, then we get for $k > \frac{1}{2}$

$$|U_{n,\varepsilon}(P)| \leqq \frac{B\delta}{2\varepsilon^{k-\frac{1}{2}} n^{k-\frac{1}{2}}} + \frac{B\delta(2k+1)}{4n^{k-\frac{1}{2}}} \int_{1/n}^{\varepsilon} \frac{1}{\varphi^{k+\frac{1}{2}}} d\varphi$$

$$\leqq \frac{B\delta}{2\varepsilon^{k-\frac{1}{2}} n^{k-\frac{1}{2}}} + \frac{B(2k+1)\delta}{2(2k-1)} \left[1 - \frac{1}{(n\varepsilon)^{k-\frac{1}{2}}} \right] \leqq \frac{B\delta}{2} \left[1 + \frac{2k+1}{2k-1} \right],$$

and the conclusions of Sec. 23.4 remain unchanged.

Therefore, we have proved: if $f(P; \varphi, \theta)$ is summable over ω, and if we let

$$\frac{1}{2\pi} \int_0^{2\pi} f(P; \varphi, \theta) \, d\theta = \Phi(P; \varphi)$$

then for a constant $S(P)$ we have

$$\lim_{\varphi \to +0} \frac{1}{\varphi} \int_0^{\varphi} | \Phi(P; \varphi) - S(P) | \, d\varphi = 0,$$

and if the integral

$$\iint_{C(P';\varepsilon)} \frac{| f(P; \varphi, \theta) |}{| \pi - \varphi |^{1/2}} \sin \varphi \, d\varphi \, d\theta = \int_{\pi-\varepsilon}^{\pi} \frac{\sin \varphi}{| \pi - \varphi |^{1/2}} d\varphi \int_0^{2\pi} | f(P; \varphi, \theta) | \, d\theta, \tag{29}$$

extended to the spherical zone $C(P'; \varepsilon)$ that has for center the point P' diametrically opposite to P, and a spherical radius ε, is convergent, then the Laplace series of the function $f(P)$ is (C, k) summable for $\frac{1}{2} < k < 1$, at point P, and its sum has the value $S(P)$.

It should be observed that (C, k) summability of Laplace series at a point P, for $1 > k > \frac{1}{2}$ requires that the integral (29) relative to a spherical zone with center at the point P' diametrically opposite to P be finite, a condition which was not necessary for $(C, 1)$ summability in Sec. 23.4.

It should also be observed that the integral (29) is finite if there exists a constant $M > 0$ and a nonnegative number $s < \frac{1}{2}$ such that

$$|f(P; \varphi, \theta)| < \frac{M}{|\pi - \varphi|^s}, \qquad [0 \leqq s < \tfrac{1}{2};\ \pi - \varepsilon \leqq \varphi \leqq \pi,\ 0 \leqq \theta \leqq 2\pi].$$

5b. Corresponding to Theorem 44 mentioned in Sec. 7.4 of Ch. II, we have for Laplace series the following theorem: The (C, k) sum of the Laplace series of a function $f(P)$ integrable over ω, converges almost everywhere to $f(P)$ for $k > \frac{1}{2}$. [Cf. G. Sansone, [98 f].]

24. Poisson Summation of Laplace Series

1a. Let $f(P)$ be a function integrable over ω and let us consider its Laplace series

$$(1) \qquad f(P) \sim \sum_{n=0}^{\infty} Y_n(P),$$

$$(2) \qquad Y_n(P) = \frac{2n+1}{4\pi} \int_{\omega} f(Q) P_n[\cos \gamma(P, Q)] d\omega(Q), \quad (n = 0, 1, \ldots).$$

In order to perform the Poisson summation of (1), we must consider the series [Ch. II, Sec. 8.1]

$$(3) \qquad \sum_{n=0}^{\infty} \varrho^n Y_n(P), \qquad (0 < \varrho < 1),$$

and since $| P_n \cos \gamma(P, Q) | \leqq 1$, from

$$| Y_n(P) | \leqq \frac{2n+1}{4\pi} \int_{\omega} | f(Q) | \, d\omega(Q)$$

it follows that the series (3) is majorized, apart from a constant factor, by the series $\sum_{n=0}^{\infty} (2n+1)\varrho^n$, which has radius of convergence equal to unity, and therefore series (3) is uniformly convergent for ϱ varying in an interval interior to $[-1, 1]$ and P varying over ω.

Then letting

$$u(\varrho; P) = \sum_{n=0}^{\infty} \varrho^n Y_n(P) \tag{4}$$

we have by (2)

$$u(\varrho; P) = \sum_{n=0}^{\infty} \frac{2n+1}{4\pi} \varrho^n \int_\omega f(Q) P_n[\cos \gamma(P, Q)] \, d\omega(Q),$$

and using the same reasoning as in Ch. II, Sec. 8.2 we also have

$$u(\varrho; P) = \frac{1}{4\pi} \int_\omega f(Q) \left[\sum_{n=0}^{\infty} (2n+1) \varrho^n P_n[\cos \gamma(P, Q)] \right] d\omega$$

or by (4) of Sec. 1.1

$$\boxed{u(\varrho; P) = \frac{1}{4\pi} \int_\omega f(Q) \frac{1-\varrho^2}{[1 - 2\varrho \cos \gamma(P, Q) + \varrho^2]^{3/2}} \, d\omega(Q)}, \tag{5}$$

(The Poisson Integral).

Recalling the theorem proved for $(C, 1)$ summability of Laplace series [Sec. 23.4] and Frobenius' theorem [Ch. II, Sec. 8.1] we have: the Poisson sum of the Laplace series of $f(P)$ has the value $S(P)$ at a point P, namely

$$\lim_{\rho \to 1-0} u(\varrho; P) = S(P),$$

and if we let

$$\frac{1}{2\pi} \int_0^{2\pi} f(P; \varphi, \theta) \, d\theta = \Phi(P; \varphi),$$

then

$$\lim_{\rho \to +0} \frac{1}{\varphi} \int_0^{\varphi} | \Phi(P; \varphi) - S(P) | \, dP = 0.$$

1b. If $f(P)$ is continuous at a point P_0 of ω, then the Poisson sum of the Laplace series of $f(P)$ has the value $f(P_0)$, namely

$$\lim_{\rho \to 1-0} u(\varrho;\ P_0) = f(P_0).$$

1c. If $f(P)$ is continuous in a domain I, (or over the whole sphere ω) and I' is a domain of ω entirely interior to I, then the Poisson sum of the Laplace series of $f(P)$ converges uniformly to $f(P)$ in I' (or over the whole sphere ω).

2. The preceding results will be used to solve the so-called Dirichlet problem for a sphere.

The function $\varrho^n Y_n(\varphi, \theta)$, $[P \equiv (\varphi, \theta)]$, is a spherical harmonic [Sec. 18.3a], and consequently satisfies the equation

$$\Delta_2 u = \frac{1}{\varrho^2}\left[\frac{\partial\left(\varrho^2 \dfrac{\partial u}{\partial \varrho}\right)}{\partial \varrho} + \frac{1}{\sin\varphi}\frac{\partial\left(\sin\varphi\dfrac{\partial u}{\partial\varphi}\right)}{\partial\varphi} + \frac{1}{\sin^2\varphi}\frac{\partial^2 u}{\partial\theta^2}\right] = 0,$$

and to prove that the function

$$u(\varrho;\varphi,\theta) = \sum_{n=0}^{\infty} \varrho^n Y_n(\varphi,\theta)$$

is also harmonic in the interior of ω $(\varrho < 1)$ it is sufficient to prove that the following derivatives of u

$$\frac{\partial u}{\partial \varrho},\ \frac{\partial^2 u}{\partial \varrho^2},\ \frac{\partial u}{\partial \varphi},\ \frac{\partial^2 u}{\partial \varphi^2},\ \frac{\partial^2 u}{\partial \theta^2}$$

can be obtained by term by term differentiation of the series on the right side of (4).

It is immediately clear that this is so for $\partial u/\partial\varrho$, $\partial^2 u/\partial\varrho^2$ because, except for a constant factor, the resulting series are majorized by the series $\Sigma n(2n+1)\varrho^{n-1}$, $\Sigma n(n-1)(2n+1)\varrho^{n-2}$ both having unit radius of convergence.

In view of (12) of Sec. 19.2 we have

$$\frac{\partial Y_n}{\partial\varphi} = \frac{2n+1}{4\pi}\int_\omega f(Q)\, P_n'[\cos\gamma(P,Q)]$$
$$\times [-\sin\varphi\cos\varphi' + \cos\varphi\sin\varphi'\cos(\theta-\theta')]\, d\omega(Q),$$

$$\frac{\partial^2 Y_n}{\partial \varphi^2} = \frac{2n+1}{4\pi} \int_\omega f(Q) P_n''[\cos \gamma(P, Q)]$$

$$\times [-\sin \varphi \cos \varphi' + \cos \varphi \sin \varphi' \cos (\theta - \theta')]^2 d\omega(Q)$$

$$- \frac{2n+1}{4\pi} \int_\omega f(Q) P_n'[\cos \gamma(P, Q)]$$

$$\times [\cos \varphi \cos \varphi' + \sin \varphi \sin \varphi' \cos (\theta - \theta')] d\omega(Q);$$

where

$$d\omega(Q) = \sin \varphi' d\varphi' d\theta'.$$

However,

$$| \cos \varphi \cos \varphi' + \sin \varphi \sin \varphi' \cos (\theta - \theta') | \leqq 1,$$

$$| -\sin \varphi \cos \varphi' + \cos \varphi \sin \varphi' \cos (\theta - \theta') | \leqq 1,$$

and by (11) of Sec. 17 we have

$$| P_n' | \leqq \binom{n+1}{n-1} = \frac{n(n+1)}{2};$$

$$| P_n'' | \leqq 3 \binom{n+2}{n-2} = \frac{(n-1)n(n+1)(n+2)}{8}.$$

Therefore, it is sufficient to observe that the series

$$\Sigma n(n+1)(2n+1)\varrho^n, \quad \Sigma (n-1)n(n+1)(n+2)(2n+1)\varrho^n$$

both have unit radius of convergence in order to justify term by term differentiation of series (4) with respect to φ.

Analogous considerations hold for differentiation with respect to θ, and consequently by Sec. 24.1, it follows that the function $u(\varrho; P)$ given by (5) is a harmonic function of the spherical coordinates ϱ, θ, φ of points in the interior of ω. Moreover, integral (5) or the equivalent series (4) solves the Dirichlet problem to construct a function continuous inside and on the unit sphere, harmonic in the interior, and which coincides on ω with a prescribed continuous function $f(P)$ $[= f(\varphi, \theta)]$.

REMARK: For other results on the Dirichlet problem, see M.

Picone [88b], p. 211, *et. seq.* and for the case of a space of n dimensions, see, for example, G. Sansone, *Lezioni sulle Funzioni di una variabile complessa,* (2nd ed. Padova, 1949), p. 577.

25. The Poisson Sum of Legendre Series

1. Let an integrable function $f(x)$ be defined in $[-1, 1]$ and consider on the surface of the sphere ω which has the segment $(-1, 1)$ for diameter a system of spherical coordinates (φ, θ) with origin at the center of the sphere. Now consider parallels on the surface of ω with respect to a pole at the point 1. On every such parallel assign to $f(\varphi, \theta)$ the value $f(\cos\varphi) = f(x)$ where x is the distance of the plane of the parallel from the center of the sphere. It follows [cf. Sec. 19.3, (VII')]

$$\begin{aligned}
Y_n(\varphi, \theta) &= \frac{2n+1}{4\pi}\int_0^{\pi} f(\cos\varphi')\sin\varphi' d\varphi' \\
&\quad\times \int_0^{2\pi} P_n[\cos\varphi\cos\varphi' + \sin\varphi\sin\varphi'\cos(\theta-\theta')]\,d\theta' \\
&= \frac{2n+1}{4\pi}\int_0^{\pi} f(\cos\varphi')\sin\varphi' d\varphi' \int_0^{2\pi}\Big[P_n(\cos\varphi)P_n(\cos\varphi') \\
&\quad + 2\sum_{m=1}^{n}\frac{(n-m)!}{(n+m)!}P_n^m(\cos\varphi)\,P_n^m(\cos\varphi')\cos m(\theta-\theta')\Big]\,d\theta' \\
&= \frac{2n+1}{2}P_n(\cos\varphi)\int_0^{\pi} f(\cos\varphi')\,P_n(\cos\varphi')\sin\varphi' d\varphi' \\
&= \frac{2n+1}{2}P_n(x)\int_{-1}^{1} f(x)\,P_n(x)\,dx,
\end{aligned}$$

whence

$$(1)\qquad f(x) \sim \sum_{n=0}^{\infty} a_n P_n(x), \qquad a_n = \frac{2n+1}{2}\int_{-1}^{1} f(x)\,P_n(x)\,dx.$$

We have proved, therefore, that the Legendre series of a function $f(x)$ defined in $[-1, 1]$ may be considered as a Laplace series of a function defined over the unit sphere ω, which assumes on every parallel with pole at the point 1 a constant value $f(\cos\varphi) = f(x)$.

We get at once the following theorem: Given a function $f(x)$ integrable in $[-1, 1]$ and its Legendre series (1), then the series

$$\sum_{n=0}^{\infty} a_n \varrho^n P_n(x) \tag{2}$$

converges uniformly and absolutely for all x in $[-1, 1]$ and for all ϱ such that $0 \leqq \varrho \leqq \delta < 1$ and at the point $x = \cos\varphi$, $0 \leqq \varphi \leqq \pi$, the series (2) has for its sum the Poisson integral

$$\frac{1}{4\pi}\int_0^{2\pi} d\theta' \int_0^{\pi} f(\cos\varphi') \frac{(1-\varrho^2)\sin\varphi'}{(1-2\varrho\cos\gamma+\varrho^2)^{3/2}}\, d\varphi', \tag{3}$$

$$\cos\gamma = \cos\varphi\cos\varphi' + \sin\varphi\sin\varphi'\cos(\theta-\theta'),$$

which is therefore independent of the longitude θ.

Clearly, then, for any fixed point P of ω at a spherical distance φ from the point 1 as pole, the average of the values that f assumes at the points of a parallel of ω with pole at P and spherical radius φ', which we will denote by $\Phi(\varphi, \varphi')$, is given by

$$\Phi(\varphi, \varphi') = \frac{1}{2\pi}\int_0^{2\pi} f(\cos\varphi\cos\varphi' + \sin\varphi\sin\varphi'\cos\theta')\, d\theta'$$

$$= \frac{1}{\pi}\int_0^{\pi} f(\cos\varphi\cos\varphi' + \sin\varphi\sin\varphi'\cos\theta')\, d\theta',$$

so that if

$$\lim_{\varphi'\to+0} \frac{1}{\varphi'}\int_0^{\varphi'} |\, \Phi(\varphi, \varphi') - S(\cos\varphi)\, |\, d\varphi' = 0, \tag{4}$$

we have

$$\lim_{\rho\to 1-0} \sum_{n=0}^{\infty} a_n \varrho^n P_n(\cos\varphi) = S(\cos\varphi).$$

It follows that if x is an interior point of $[-1, 1]$ at which $f(x)$ is continuous, and we put $S(\cos\varphi) = f(\cos\varphi) = f(x)$ or, if x is an interior point of $[-1, 1]$ at which $f(x)$ has a finite discontinuity, we put

$$2S(\cos\varphi) = \lim_{x\to(\cos\varphi)-0} f(x) + \lim_{x\to(\cos\varphi)+0} f(x)$$

then (4) is satisfied, and, consequently, if $f(x)$ is integrable in $[-1, 1]$ and x is a point interior to $[-1, 1]$ at which $f(x)$ is continuous, then the Poisson sum of the Legendre series of $f(x)$ converges to $f(x)$, namely

$$\lim_{\rho\to 1-0} \sum_{n=0}^{\infty} a_n \varrho^n P_n(x) = f(x).$$

If, on the other hand, x is a point interior to $[-1, 1]$ at which $f(x)$ has a finite discontinuity, then the Poisson sum of the Legendre series converges to $[f(x+) + f(x-)]/2$, namely

$$\lim_{\rho\to 1-0} \sum_{n=0}^{\infty} a_n \varrho^n P_n(x) = \frac{f(x+) + f(x-)}{2}.$$

Moreover, if the $\lim_{x\to 1-} f(x)$ or $\lim_{x\to -1+} f(x)$ exists, we have

$$\lim_{\rho\to 1-} \sum_{n=0}^{\infty} a_n \varrho^n = f(1-) \quad \text{or} \quad \lim_{\rho\to -1+} \sum_{n=0}^{\infty} a_n (-1)^n \varrho^n = f(-1+).$$

2a. We shall now apply a theorem of Littlewood to deduce from the Poisson summability of the Legendre series of $f(x)$ a criterion for ordinary convergence of the series.

Let $f(x)$ be integrable in $[-1, 1]$ and let (1) represent the corresponding Legendre series. Then the corresponding Poisson series is $[x = \cos\varphi,\ 0 \leqq \varphi \leqq \pi]$

$$a_0 + a_1 \varrho P_1(\cos\varphi) + a_2 \varrho^2 P_2(\cos\varphi) + \ldots + a_n \varrho^n P_n(\cos\varphi) + \ldots$$

and by a theorem of Littlewood [cf., e. g., E. C. Titchmarsh, [109] pp. 233—235] we have that if $|a_n P_n(\cos\varphi)| = O(n^{-1})$, then the Legendre series (1) converges at the point $x = \cos\varphi$ if the Poisson sum exists there and the two sums are equal.

Let x be interior to $[-1, 1]$, then $|P_n(\cos\varphi)| < kn^{-\frac{1}{2}}$, k constant, [Sec. 10.3], so that the preceding condition is satisfied if $a_n = O(n^{-\frac{1}{2}})$. Therefore, we conclude: If in series (1)

$$a_n = O(n^{-\frac{1}{2}}), \tag{5}$$

then the series $a_0 + a_1 P_1(\cos\varphi) + \ldots + a_n P_n(\cos\varphi) + \ldots,$

$0 < \varphi < \pi$, converges wherever the limit

$$\lim_{\rho \to 1-0} \sum_{n=0}^{\infty} a_n \varrho^n P_n(\cos \varphi)$$

exists. This last condition is satisfied if $f(x)$ is continuous at point $x = \cos \varphi$, or has a finite discontinuity there.

2b. In view of (2_1) of Sec. 11.1 we have that (5) is satisfied if $f(x)$ is of bounded variation in $[-1, 1]$. We, therefore, have a new proof that the Legendre series of a function of bounded variation converges at every interior point of $[-1, 1]$, [cf. Sec. 14.6]. The convergence at points -1 and $+1$ is assured by the results of Sec. 14.7.

2c. Finally we observe that the condition imposed in Hobson's theorem (Sec. 14.6) that $f(x)(1-x^2)^{-1/4}$ be summable in $[-1, 1]$ implies $a_n = O(n^{1/2})$.

Indeed, by Sec. 10.3,

$$|P_n(x)| \leqq \sqrt{2} \frac{4}{\sqrt{\pi}} \frac{1}{\sqrt{n}} \frac{1}{\sqrt[4]{1-x^2}},$$

whence

$$|a_n| = \frac{2n+1}{2} \left| \int_{-1}^{1} f(x) P_n(x)\, dx \right|$$

$$\leqq \frac{2n+1}{2} \frac{4\sqrt{2}}{\sqrt{\pi}\sqrt{n}} \int_{-1}^{1} \frac{|f(x)|}{\sqrt[4]{1-x^2}}\, dx = O(n^{1/2}).$$

CHAPTER IV

Expansions in Laguerre and Hermite Series

1. Laguerre Polynomials

1. Consider the function $(1-z)^{-(\alpha+1)}\,e^{-\frac{xz}{1-z}}$, and its expansion in a power series in z for $|z|<1$.

Then

$$(1-z)^{-(\alpha+1)}\,e^{-\frac{xz}{1-z}} = \sum_{k=0}^{\infty}(-1)^k \frac{x^k}{k!}\,\frac{z^k}{(1-z)^{\alpha+k+1}}$$

$$= \sum_{k=0}^{\infty}(-1)^k \frac{x^k}{k!}\left[\sum_{m=0}^{\infty}(-1)^m \binom{-\alpha-k-1}{m} z^{m+k}\right].$$

Since the resulting double series is absolutely convergent, we may collect the terms in z^n and obtain

$$(1)\qquad \boxed{(1-z)^{-(\alpha+1)}\,e^{-\frac{xz}{1-z}} = 1 + \sum_{n=1}^{\infty} z^n L_n^{(\alpha)}(x)}$$

where

$$L_n^{(\alpha)}(x) = \sum_{m=0}^{n}(-1)^{n-m}\frac{1}{(n-m)!}(-1)^m\binom{-\alpha-n+m-1}{m}x^{n-m},$$

or

$$(2)\qquad (-1)^n L_n^{(\alpha}(x)$$
$$= \frac{x^n}{n!} + \sum_{m=1}^{n}(-1)^m \frac{(\alpha+n)(\alpha+n-1)\ldots(\alpha+n-m+1)}{(n-m)!\,m!}x^{n-m},$$

$$(3_1)\qquad \boxed{(-1)^n L_n^{(\alpha)}(x) = \sum_{m=0}^{n}(-1)^m \frac{\Gamma(\alpha+n+1)}{m!(n-m)!\,\Gamma(\alpha+n-m+1)}x^{n-m}}$$

$$(n = 1, 2, 3, \ldots),$$

(See Appendix, Th. 19 for the definition and properties of $\Gamma(a)$). For convenience of notation we also define

$$L_0^{(\alpha)}(x) = 1. \tag{3_2}$$

(a) For $x = 0$ we have

$$L_n^{(\alpha)}(0) = \frac{(\alpha+n)(\alpha+n-1)(\alpha+n-2)\ldots(\alpha+1)}{n!}. \tag{3_3}$$

(b) For the case $\alpha = 0$, the polynomials $L_n^{(0)}(x)$ will be redesignated $L_n(x)$. These polynomials are clearly given by the following

$$L_n(x) = \sum_{m=0}^{n} (-1)^{n-m} \frac{n!}{m![(n-m)!]^2} x^{n-m}$$

$$\boxed{L_n(x) = \sum_{k=0}^{n} (-1)^k \frac{1}{k!} \binom{n}{k} x^k}, \quad (n=1, 2, \ldots). \tag{4}$$

(c) The polynomials $L_n^{(\alpha)}(x)$ are called *Laguerre polynomials* (Tchebychef [108a], pp. 499—508; [108d]; Laguerre [63a]), These polynomials also occur in an unedited manuscript of Abel [1].

(d) For convenience of reference the expressions for the first few polynomials $L_n^{(\alpha)}(x)$ are given below

$$L_0^{(\alpha)}(x) = 1,$$

$$-L_1^{(\alpha)}(x) = \frac{x}{1!} - \frac{\alpha+1}{1!},$$

$$L_2^{(\alpha)}(x) = \frac{x^2}{2!} - \frac{\alpha+2}{1!\,1!}x + \frac{(\alpha+2)(\alpha+1)}{2!},$$

$$-L_3^{(\alpha)}(x) = \frac{x^3}{3!} - \frac{\alpha+3}{2!\,1!}x^2 + \frac{(\alpha+3)(\alpha+2)}{1!\,2!}x - \frac{(\alpha+3)(\alpha+2)(\alpha+1)}{3!},$$

$$L_4^{(\alpha)}(x) = \frac{x^4}{4!} - \frac{\alpha+4}{3!\,1!}x^3 + \frac{(\alpha+4)(\alpha+3)}{2!\,2!}x^2$$

$$- \frac{(\alpha+4)(\alpha+3)(\alpha+2)}{1!\,3!}x + \frac{(\alpha+4)(\alpha+3)(\alpha+2)(\alpha+1)}{4!}.$$

2. From (3_1) follows

$$x^\alpha e^{-x} L_n^{(\alpha)}(x) = \frac{1}{n!}\sum_{m=0}^{n}(-1)^{n-m}\frac{n!\Gamma(\alpha+n+1)}{m!(n-m)!\Gamma(\alpha+n-m+1)}x^{n+\alpha-m}e^{-x}$$

$$= \frac{1}{n!}\sum_{m=0}^{n}\binom{n}{m}(\alpha+n)\ldots(\alpha+n-m+1)x^{n+\alpha-m}[(-1)^{n-m}e^{-x}]$$

$$= \frac{1}{n!}\sum_{m=0}^{n}\binom{n}{m}\frac{d^m}{dx^m}x^{n+\alpha}\frac{d^{n-m}}{dx^{n-m}}e^{-x} = \frac{1}{n!}\frac{d^n}{dx^n}[x^{n+\alpha}e^{-x}],$$

whence

$$\boxed{L_n^{(\alpha)}(x) = \frac{x^{-\alpha}e^x}{n!}\frac{d^n}{dx^n}[x^{n+\alpha}e^{-x}]}, \quad (n = 1, 2, \ldots). \tag{5}$$

3. If we let

$$\varphi(x, z) = (1-z)^{-(\alpha+1)}e^{-xz/(1-z)} \tag{6}$$

then

$$(1-z)^2\frac{\partial\varphi}{\partial z} = [(\alpha+1)(1-z)-x]\varphi$$

and by (1)

$$(1-z)^2\sum_{n=1}^{\infty} nz^{n-1}L_n^{(\alpha)}(x) = [(\alpha+1-x)-(\alpha+1)z]\sum_{n=0}^{\infty} z^n L_n^{(\alpha)}(x).$$

Performing the indicated multiplication on the right side and identifying coefficients of z^n on the two sides, we get

$$\boxed{(n+1)L_{n+1}^{(\alpha)}(x)-[2n+\alpha+1-x]L_n^{(\alpha)}(x)+(n+\alpha)L_{n-1}^{(\alpha)}(x) = 0}, \tag{7}$$

$$(n = 0, 1, 2, \ldots)$$

where, for convenience, we define

$$\boxed{L_{-1}^{(\alpha)}(x) = 0}. \tag{7'}$$

4. Multiplying (7) by $L_n^{(\alpha)}(y)/\Gamma(\alpha+1)\binom{n+\alpha}{n}$, we get

$$(2n+\alpha+1-x)\frac{L_n^{(\alpha)}(x)L_n^{(\alpha)}(y)}{\Gamma(\alpha+1)\binom{n+\alpha}{n}}$$

$$=\frac{n+1}{\Gamma(\alpha+1)\binom{n+\alpha}{n}}L_{n+1}^{(\alpha)}(x)L_n^{(\alpha)}(y)+\frac{n+\alpha}{\Gamma(\alpha+1)\binom{n+\alpha}{n}}L_{n-1}^{(\alpha)}(x)L_n^{(\alpha)}(y).$$

Interchanging x and y and subtracting and noting that

$$(n+1)/\Gamma(\alpha+1)\binom{n+\alpha}{n}=\Gamma(n+2)/\Gamma(n+\alpha+1),$$

$$(n+\alpha)/\Gamma(\alpha+1)\binom{n+\alpha}{n}=\Gamma(n+1)/\Gamma(n+\alpha),$$

we get

$$(y-x)\frac{L_n^{(\alpha)}(x)L_n^{(\alpha)}(y)}{\Gamma(\alpha+1)\binom{n+\alpha}{n}}$$

$$=\frac{\Gamma(n+2)}{\Gamma(n+\alpha+1)}[L_{n+1}^{(\alpha)}(x)L_n^{(\alpha)}(y)-L_{n+1}^{(\alpha)}(y)L_n^{(\alpha)}(x)]$$

$$+\frac{\Gamma(n+1)}{\Gamma(n+\alpha)}[L_{n-1}^{(\alpha)}(x)L_n^{(\alpha)}(y)-L_{n-1}^{(\alpha)}(y)L_n^{(\alpha)}(x)].$$

Now letting $n=0, 1, 2, \ldots, n$ and adding we get

$$\boxed{\sum_{k=0}^{n}\frac{L_k^{(\alpha)}(x)L_k^{(\alpha)}(y)}{\Gamma(\alpha+1)\binom{k+\alpha}{k}}=\frac{\Gamma(n+2)}{\Gamma(n+\alpha+1)}\frac{L_{n+1}^{(\alpha)}(x)L_n^{(\alpha)}(y)-L_{n+1}^{(\alpha)}(y)L_n^{(\alpha)}(x)}{y-x}}. \tag{8}$$

5a. By (3_1)

$$(-1)^n\frac{d}{dx}L_n^{(\alpha)}(x)=\sum_{m=0}^{(n-1)}(-1)^m\frac{\Gamma(\alpha+n+1)}{m!(n-m-1)!\Gamma(\alpha+n-m+1)}x^{n-m-1}, \tag{9}$$

$$(-1)^{n-1}\frac{d}{dx}L_{n-1}^{(\alpha)}(x)=\sum_{m=0}^{(n-2)}(-1)^m\frac{\Gamma(\alpha+n)}{m!(n-m-2)!\Gamma(\alpha+n-m)}x^{n-m-2};$$

Moreover

$$\frac{\Gamma(\alpha+n+1)}{m!(n-m-1)!\Gamma(\alpha+n-m+1)}$$

$$-\frac{\Gamma(\alpha+n)}{(m-1)!(n-m-1)!\Gamma(\alpha+n-m+1)}=\frac{\Gamma(\alpha+n)}{m!(n-m-1)!\Gamma(\alpha+n-m)}.$$

Therefore, adding the two preceding relations (9) we have

$$(-1)^n\frac{d}{dx}[L_n^{(\alpha)}(x)-L_{n-1}^{(\alpha)}(x)]$$

$$=\sum_{m=0}^{n-1}(-1)^m\frac{\Gamma(\alpha+n)}{m!(n-m-1)!\Gamma(\alpha+n-m)}x^{n-m-1}$$

and finally

$$(10)\qquad \boxed{L_{n-1}^{(\alpha)}(x)=\frac{d}{dx}[L_{n-1}^{(\alpha)}(x)-L_n^{(\alpha)}(x)]},\quad (n=0, 1, 2,\ldots).$$

5b. If we note that the right side of the first of the relations (9) represents $(-1)^{n-1}L_{n-1}^{(\alpha+1)}$ we get

$$(11)\qquad \boxed{\frac{d}{dx}L_n^{(\alpha)}(x)=-L_{n-1}^{(\alpha+1)}(x)},\quad (n=0, 1, 2, \ldots),$$

and consequently by changing $n-1$ to n in (10), we have

$$(10')\qquad \boxed{L_n^{(\alpha)}=L_n^{(\alpha+1)}-L_{n-1}^{(\alpha+1)}}.$$

5c. Differentiating (7) and noting that by (10)

$$L'_{n+1}=L'_n-L_n,\qquad L'_{n-1}=L'_n+L_{n-1}$$

we get

$$(12)\qquad \boxed{x\frac{d}{dx}L_n^{(\alpha)}(x)=nL_n^{(\alpha)}(x)-(n+\alpha)L_{n-1}^{(\alpha)}(x)}.$$

5d. From (10) and (12) follows

$$(n+\alpha)\frac{d}{dx}[L_{n-1}^{(\alpha)}(x)-L_n^{(\alpha)}(x)]=(n+\alpha)L_{n-1}^{(\alpha)}(x)=nL_n^{(\alpha)}(x)-x\frac{d}{dx}L_n^{(\alpha)}(x)$$

or

$$(13)\quad (n+\alpha)\frac{d}{dx}L_{n-1}^{(\alpha)} = nL_n^{(\alpha)}(x) + (n+\alpha-x)\frac{d}{dx}L_n^{(\alpha)}(x).$$

Now differentiating (12) we get

$$x\frac{d^2}{dx^2}L_n^{(\alpha)}(x) + \frac{d}{dx}L_n^{(\alpha)}(x) = n\frac{d}{dx}L_n^{(\alpha)}(x) - (n+\alpha)\frac{dL_{n-1}^{(\alpha)}}{dx}$$

and, finally, by (13) we obtain the differential equation of the polynomials $L_n^{(\alpha)}(x)$

$$(14_1)\quad \boxed{x\frac{d^2L_n^{(\alpha)}(x)}{dx^2}+(\alpha-x+1)\frac{dL_n^{(\alpha)}(x)}{dx}+nL_n^{(\alpha)}(x)=0},$$

$$(n=0, 1, 2, \ldots),$$

which can also be written in the form

$$(14_2)\quad \boxed{\frac{d}{dx}\left[x^{\alpha+1}e^{-x}\frac{dL_n^{(\alpha)}(x)}{dx}\right]+ne^{-x}x^{\alpha}L_n^{(\alpha)}(x)=0}.$$

6. We now prove the following result due to Kogbetliantz ([58 b], p. 156) which gives an integral relation connecting $L_n^{(\alpha)}(x)$ and $L_n^{(\beta)}(x)$ for $\alpha > \beta$:

$$(15)\quad \boxed{L_n^{(\alpha)}(x)=\frac{\Gamma(n+\alpha+1)}{\Gamma(n+\beta+1)\Gamma(\alpha-\beta)}x^{-\alpha}\int_0^x(x-u)^{\alpha-\beta-1}u^{\beta}L_n^{(\beta)}(u)du}$$

$$\text{for } \alpha > \beta.$$

The relation follows at once by making use of (3_1), changing the variable in the integral with $u = xt$ and then making use of the Beta function (Appendix, Th. 23) thus:

$$\frac{\Gamma(n+\alpha+1)}{\Gamma(n+\beta+1)\Gamma(\alpha-\beta)}x^{-\alpha}\int_0^x (x-u)^{\alpha-\beta-1}u^\beta L_n^{(\beta)}(u)du$$

$$=\frac{1}{\Gamma(\alpha-\beta)}\sum_{m=0}^{n}(-1)^{n-m}\frac{\Gamma(n+\alpha+1)x^{-\alpha}}{m!(n-m)!\Gamma(\beta+n-m+1)} \times\int_0^x (x-u)^{\alpha-\beta-1}u^{\beta+n-m}du$$

$$=\frac{1}{\Gamma(\alpha-\beta)}\sum_{m=0}^{n}(-1)^{n-m}\frac{\Gamma(n+\alpha+1)x^{n-m}}{m!(n-m)!\Gamma(\beta+n-m+1)} \times\int_0^1 (1-t)^{\alpha-\beta-1}t^{\beta+n-m}dt$$

$$=\frac{1}{\Gamma(\alpha-\beta)}\sum_{m=0}^{n}(-1)^{n-m}\frac{\Gamma(n+\alpha+1)}{m!(n-m)!\Gamma(\beta+n-m+1)} \times\frac{\Gamma(\alpha-\beta)\Gamma(\beta+n-m+1)}{\Gamma(\alpha+n-m+1)}x^{n-m}$$

Q.E.D.

7a. The orthogonality of the polynomials $L_n^{(\alpha)}(x)$ in $[0, \infty)$ will now be established.

Multiplying (14_2) by $L_m^{(\alpha)}(x)$, we get

$$L_m^{(\alpha)}\frac{d}{dx}\left[x^{\alpha+1}e^{-x}\frac{dL_n^{(\alpha)}}{dx}\right]+e^{-x}x^\alpha nL_n^{(\alpha)}(x)L_m^{(\alpha)}(x)=0,$$

interchanging n and m and subtracting

$$\frac{d}{dx}\left[x^{\alpha+1}e^{-x}\left\{L_m^{(\alpha)}\frac{dL_n^{(\alpha)}}{dx}-L_n^{(\alpha)}\frac{dL_m^{(\alpha)}}{dx}\right\}\right]+e^{-x}x^\alpha(n-m)L_n^{(\alpha)}L_m^{(\alpha)}=0,$$

integrating between 0 and ∞ for $\alpha > -1$ we get

$$(n-m)\int_0^{+\infty}e^{-x}x^\alpha L_n^{(\alpha)}(x)L_m^{(\alpha)}(x)\,dx=0,$$

whence the orthogonality relation follows

$$(16_1)\qquad \boxed{\int_0^{+\infty}e^{-x}x^\alpha L_n^{(\alpha)}(x)L_m^{(\alpha)}(x)\,dx=0},\quad (n\neq m, \alpha>-1).$$

7b. We next show that

$$(16_2)\qquad \boxed{\int_0^{+\infty} e^{-x} x^{\alpha} [L_n^{(\alpha)}(x)]^2 dx = \frac{\Gamma(n+\alpha+1)}{n!}},$$

$$(n = 0, 1, 2, \ldots; \alpha > -1),$$

which provides the factor for normalizing the polynomials $L_n^{(\alpha)}(x)$.

Multiplying (7) by $e^{-x} x^{\alpha} L_{n-1}^{\alpha}(x)\,dx$, integrating between 0 and $+\infty$, and making use of (16_1) we get

$$(17_1)\qquad \int_0^{+\infty} e^{-x} x^{\alpha+1} L_n^{(\alpha)} L_{n-1}^{(\alpha)}\,dx + (n+\alpha)\int_0^{+\infty} e^{-x} x^{\alpha} [L_{n-1}^{(\alpha)}(x)]^2 dx = 0;$$

changing n to $n-1$ in (7), multiplying by $e^{-x} x^{\alpha} L_n^{(\alpha)}(x)\,dx$, and integrating between 0 and $+\infty$, we get

$$(17_2)\qquad \int_0^{+\infty} e^{-x} x^{\alpha+1} L_n^{(\alpha)} L_{n-1}^{(\alpha)}\,dx + n\int_0^{+\infty} e^{-x} x^{\alpha} [L_n^{(\alpha)}(x)]^2 dx = 0.$$

From (17_1) and (17_2) follows the recurrence relation

$$\int_0^{+\infty} e^{-x} x^{\alpha} [L_n^{(\alpha)}(x)]^2\,dx = \frac{n+\alpha}{n}\int_0^{+\infty} e^{-x} x^{\alpha} [L_{n-1}^{(\alpha)}(x)]^2\,dx.$$

Letting $n = 1, 2, \ldots, n$ in the preceding, using

$$\int_0^{+\infty} e^{-x} x^{\alpha} [L_0^{(\alpha)}(x)]^2 dx = \int_0^{+\infty} e^{-x} x^{\alpha}\,dx = \Gamma(\alpha+1),$$

and multiplying the resulting equations, we get

$$(18)\qquad \int_0^{+\infty} e^{-x} x^{\alpha} [L_n^{(\alpha)}(x)]^2 dx = \frac{(n+\alpha)(n+\alpha-1)\ldots(\alpha+1)\Gamma(\alpha+1)}{n!}$$

which is precisely (16_2).

7c. From the preceding, the system

$$\{\Gamma^{\frac{1}{2}}(n+1)\Gamma^{-\frac{1}{2}}(n+\alpha+1)\, e^{-x/2} x^{\alpha/2} L_n^{(\alpha)}(x)\}$$

is clearly orthonormal in $[0, +\infty)$. In Sec. 7.2 we shall establish the completeness of this system in $[0, +\infty)$ with respect to all square integrable functions.

2. *Hermite Polynomials and Tchebychef Orthogonal Polynomials*

1. If we expand e^{-z^2-2xz} in a power series in z, we have

$$e^{-z^2-2xz} = e^{x^2}e^{-(z+x)^2} = e^{x^2}\sum_{n=0}^{\infty}\left[\frac{d^n e^{-(z+x)^2}}{dz^n}\right]_{z=0}\frac{z^n}{n!}$$

whence

$$e^{-z^2-2xz} = e^{x^2}\sum_{n=0}^{\infty}\frac{d^n e^{-x^2}}{dx^n}\,\frac{z^n}{n!}.$$

The polynomial $H_n(x)$ is now defined by

$$(1)\qquad \boxed{H_n(x) = e^{x^2}\frac{d^n e^{-x^2}}{dx^n}},\qquad (n = 1, 2, \ldots);\; H_0(x) = 1.$$

Then

$$(2)\qquad \psi(x, z) = e^{-z^2-2xz} = \sum_{n=0}^{\infty}\frac{H_n(x)}{n!}z^n,$$

and the series obtained converges for all values of x and z.

From (1) follows at once

$$(3)\qquad \boxed{H_n(-x) = (-1)^n H_n(x)}.$$

Thus, the polynomials $H_n(x)$ are even or odd functions according as the index n is even or odd.

The polynomials (1) are called *Hermite polynomials* (Tchebychef [108a], pp. 49—508, [108d], (1859), Hermite [46a], (1864). However, these polynomials already occurred in the two works of Laplace: Treatise on Celestial Mechanics (1805) and Analytical Theory of Probability (1820), [65a, d]. These polynomials will play a role in Secs. 9 and 10 in the study of the expansion of a function in orthogonal polynomials in the interval $(-\infty, +\infty)$.

2a. Consider the series

$$(3')\qquad \sum_{n=1}^{\infty}\frac{H'_n(x)}{n!}z^n$$

obtained from (2) by term by term differentiation with respect to x. The series of absolute values of this series is the expansion in series of the function $2\,|\,z\,|\,e^{|z|^2+2|x|\,|z|}$, and, consequently, the series (3′), considered as a series of functions of x, is uniformly convergent. Therefore,

$$\frac{\partial\psi}{\partial x} = -\,2ze^{-z^2-2xz} = \sum_{n=1}^{\infty} \frac{H'_n(x)}{n!}\,z^n$$

or

$$\sum_{n=1}^{\infty} \frac{H'_n(x)}{n!}\,z^n = \sum_{n=0}^{\infty} \left[-\,2\,\frac{H_n(x)}{n!}\right] z^{n+1}$$

whence

$$\text{(4)} \qquad \boxed{H'_{n+1}(x) = -\,2(n+1)\,H_n(x)}\,.$$

2b. Differentiating (2) with respect to z, we have

$$\frac{\partial\psi}{\partial z} = -\,2(x+z)\,\psi,$$

whence

$$\sum_{n=1}^{\infty} \frac{H_n(x)}{(n-1)!}\,z^{n-1} = -\,2x\sum_{n=0}^{\infty} \frac{H_n(x)}{n!}\,z^n - 2\sum_{n=0}^{\infty} \frac{H_n(x)}{n!}\,z^{n+1}$$

which implies

$$H_{n+1}/n! + 2xH_n/n! + 2H_{n-1}/(n-1)! = 0$$

and, therefore, we get the recurrence formula

$$\text{(5)} \qquad \boxed{H_{n+1}(x) + 2xH_n(x) + 2nH_{n-1}(x) = 0}\,.$$

2c. Differentiating (5) and using (4) we have

$$H'_{n+1} + 2H_n + 2xH'_n + 2nH'_{n-1} = 0,$$
$$-\,2(n+1)H_n + 2H_n + 2xH'_n - H''_n = 0,$$

and, finally, the differential equation of the polynomials H_n

$$\text{(6)} \qquad \boxed{H''_n(x) - 2xH'_n(x) + 2nH_n(x) = 0}\,.$$

3. The first six Hermite polynomials can be easily calculated from (1) and are the following:

$$(7)\quad \begin{cases} H_0 = 1,\ H_1 = -2x,\ H_2 = 4x^2 - 2,\ H_3 = -8x^3 + 12x, \\ H_4 = 16x^4 - 48x^2 + 12,\ H_5 = -32x^5 + 160x^3 - 120x. \end{cases}$$

In order to determine explicitly the coefficients of the polynomials $H_n(x)$ we proceed as follows:

Let

$$H_n(x) = \sum_{k=0}^{n} a_k x^{n-k},$$

then

$$H'_n = \sum_{k=0}^{(n-1)} (n-k)\, a_k x^{n-k-1}, \quad H''_n = \sum_{k=0}^{(n-2)} (n-k)(n-k-1)\, a_k x^{n-k-2},$$

whence, substituting in (6), we get

$$\sum_{k=0}^{n-2} (n-k)(n-k-1)\, a_k x^{n-k-2} - 2\sum_{k=0}^{n-1} (n-k)\, a_k x^{n-k} + 2\sum_{k=0}^{n} n a_k x^{n-k} = 0.$$

Consequently

$$(n-k+2)(n-k+1)\, a_{k-2} - 2(n-k)a_k + 2na_k = 0$$

and therefore

$$(8)\qquad a_k = -(n-k+2)(n-k+1)\, a_{k-2}/2k.$$

From (1) follows $a_0 = (-1)^n 2^n$, whence by (8) we get

$$a_2 = \frac{n(n-1)}{1!}(-1)^{n-1} 2^{n-2},$$

$$a_4 = \frac{n(n-1)(n-2)(n-3)}{2!}(-1)^{n-2} 2^{n-4},$$

$$a_6 = \frac{n(n-1)(n-2)(n-3)(n-4)(n-5)}{3!}(-1)^{n-3} 2^{n-6};$$

. . . .

$$a_{2k} = \frac{n(n-1)\ldots(n-2k+1)}{k!}(-1)^{n-k} 2^{n-2k},$$

and since by (3) the a_k of odd index are zero, we have

$$(9_1)\quad \boxed{H_n(x) = (-1)^n \sum_{k=0}^{[n/2]} (-1)^k \frac{n(n-1)(n-2)\dots(n-2k+1)}{k!}(2x)^{n-2k}}$$

or

$$(9_2)\quad \boxed{H_n(x) = (-1)^n \sum_{k=0}^{[n/2]} (-1)^k \frac{\Gamma(n+1)}{\Gamma(k+1)\,\Gamma(n-2k+1)}(2x)^{n-2k}}$$

where $[n/2]$ denotes the largest integer $\leqq n/2$.

We note that for $x = 0$ we have [cf. (4)]

$$(10)\quad \boxed{\begin{aligned} &H_{2n}(0) = (-1)^n \frac{(2n)!}{n!}; \quad H_{2n+1}(0) = 0; \\ &H'_{2n}(0) = 0; \quad H'_{2n+1}(0) = 2(-1)^{n+1}\frac{(2n)!}{n!}(2n+1) \end{aligned}}.$$

4. We now establish Christoffel's summation formula which will be required in Sec. 10 in the consideration of expansions in series of polynomials $H_n(x)$.

Multiplying (5) by $H_n(y)/2^{n+1}n!$, we get

$$(10_1)\quad \frac{H_{n+1}(x)\,H_n(y)}{2^{n+1}n!} + x\,\frac{H_n(x)\,H_n(y)}{2^n n!} + \frac{H_{n-1}(x)\,H_n(y)}{2^n(n-1)!} = 0$$

whence, interchanging x and y and subtracting the result from (10_1) we have

$$(10_2)\quad \begin{aligned} &(y-x)\frac{H_n(x)\,H_n(y)}{2^n n!} \\ &= \frac{H_{n+1}(x)H_n(y)-H_{n+1}(y)H_n(x)}{2^{n+1}n!} + \frac{H_{n-1}(x)H_n(y)-H_{n-1}(y)H_n(x)}{2^n(n-1)!}. \end{aligned}$$

Finally, taking $n = 0, 1, 2, \ldots, n$ in (10_2) and adding the results, we obtain the formula in question

$$(11) \qquad \boxed{\sum_{k=0}^{n} \frac{H_k(x)H_k(y)}{2^k k!} = \frac{H_{n+1}(x)H_n(y) - H_{n+1}(y)H_n(x)}{2^{n+1} n!\,(y-x)}}.$$

5. An addition formula due to Runge [96] for the polynomials $H_n(x)$ is easily obtained. Clearly

$$e^{-(z\sqrt{2})^2 - 2\left(\frac{x+y}{\sqrt{2}}\right)(z\sqrt{2})} = e^{-z^2-2xz}\,e^{-z^2-2yz}$$

whence by (2)

$$\sum_{n=0}^{\infty} \frac{2^{n/2}}{n!} H_n\left(\frac{x+y}{\sqrt{2}}\right) z^n = \sum_{r=0}^{\infty} \frac{H_r(x)}{r!} z^r \sum_{s=0}^{\infty} \frac{H_s(y)}{s!} \cdot z^s.$$

Multiplying the two series on the right side by Cauchy's rule and indentifying coefficients of like powers of z we get

$$\frac{1}{n!}\, 2^{n/2} H_n\left(\frac{x+y}{\sqrt{2}}\right) = \sum_{r=0}^{n} \frac{1}{r!\,(n-r)!} H_r(x)\, H_{n-r}(y),$$

$$(12) \qquad \boxed{2^{n/2} H_n\left(\frac{x+y}{\sqrt{2}}\right) = \sum_{r=0}^{n} \binom{n}{r} H_r(x)\, H_{n-r}(y)}.$$

For other addition formulas for Laguerre and Hermite polynomials, cf. G. Palamá [84b]; L. Toscano [113c].

6a. The orthogonality relations are obtained in the usual way. Multiplying (6) by e^{-x^2}, we have

$$\frac{d}{dx}[e^{-x^2} H'_n(x)] + 2n\, e^{-x^2} H_n(x) = 0,$$

$$\frac{d}{dx}[e^{-x^2} H'_m(x)] + 2m\, e^{-x^2} H_m(x) = 0.$$

Then, multiplying the first of these equations by H_m and the second by H_n and subtracting, we get

$$\frac{d}{dx}[e^{-x^2}(H_m H'_n - H_n H'_m)] + 2(n-m)\, e^{-x^2} H_n(x)\, H_m(x) = 0.$$

Now integrating between $-\infty$ and $+\infty$ we have

$$2(n-m)\int_{-\infty}^{+\infty} e^{-x^2} H_n(x)\, H_m(x)\, dx = 0$$

whence for $n \neq m$,

$$\boxed{\int_{-\infty}^{+\infty} e^{-x^2} H_n(x)\, H_n(x)\, dx = 0} \,. \qquad (n \neq m). \tag{13}$$

6b. Next we show that

$$\boxed{\int_{-\infty}^{+\infty} e^{-x^2} H_n^2(x)\, dx = 2^n\, n!\, \sqrt{\pi}}\,, \quad (n = 0, 1, 2, \ldots).$$

Changing n to $n-1$ in (5) we have

$$H_n + 2xH_{n-1} + 2(n-1)H_{n-2} = 0.$$

Then multiplying by $e^{-x^2} H_n(x)\, dx$, integrating between $-\infty$ and $+\infty$ and using (13), we get

$$\int_{-\infty}^{+\infty} e^{-x^2} H_n^2(x)\, dx + 2\int_{-\infty}^{+\infty} e^{-x^2} xH_n(x) H_{n-1}(x)\, dx = 0.$$

Similarly, multiplying (5) by $e^{-x^2}\, H_{n-1} dx$ and integrating between $-\infty$ and $+\infty$, we get

$$2n\int_{-\infty}^{+\infty} e^{-x^2}\, H_{n-1}^2(x)\, dx + 2\int_{-\infty}^{+\infty} e^{-x^2} xH_n(x)\, H_{n-1}(x)\, dx = 0$$

which gives the recurrence relation

$$\int_{-\infty}^{+\infty} e^{-x^2} H_n^2(x)\, dx = 2n\int_{-\infty}^{+\infty} e^{-x^2} H_{n-1}^2(x)\, dx.$$

Finally, letting $n = 1, 2, \ldots, n$, noting that

$$\int_{-\infty}^{+\infty} e^{-x^2} H_0^2\, dx = \int_{-\infty}^{+\infty} e^{-x^2}\, dx = \sqrt{\pi},$$

(14) follows at once.

6c. From 6a and 6b it follows that the system $\{e^{-x^2/2} H_n(x)/\sqrt{2^n n!\, \sqrt{\pi}}\}$ is orthonormal in $(-\infty, +\infty)$. In Sec. 7, it will be proved that this system is complete with respect to square integrable functions.

7. The Legendre polynomials $P_n(x)$, as well as the polynomials $L_n^{(\alpha)}(x)$ and $H_n(x)$, are particular cases of polynomials which arise in the general problem of approximation in the mean of a function $f(x)$ defined in an interval $[a, b]$, finite or infinite, by means of a series of orthogonal polynomials. The existence of these polynomials is assured by the following

THEOREM. Suppose that $p(x)$ is a function (called the weight) defined in $[a, b]$, that the following integrals (called the moments) exist and are finite

$$\alpha_n = \int_a^b p(x)x^n dx, \qquad (n = 0, 1, 2, \ldots)$$

and that $\alpha_0 > 0$. Then it is possible to construct in a unique way, apart from the sign, a sequence of *Tchebychef polynomials* $\{\varphi_n(x)\}$ where $\varphi_n(x)$ is a polynomial in x of degree n, orthonormal with respect to $p(x)$ in $[a, b]$, namely

$$\int_a^b p(x)\varphi_r(x)\varphi_s(x)\,dx = \varepsilon_{r,s}, \quad \varepsilon_{r,s} = 0 \quad \text{for} \quad r \neq s, \quad \varepsilon_{r,r} = 1.$$

Proof. If we let

$$\Phi_n(x) = \varphi_n(x)/a_n = x^n - S_n x^{n-1} + \ldots,$$

then

$$\Phi_0(x) = 1; \qquad 1 = \int_a^b p(x)\,a_0^2\,dx, \qquad 1/a_0^2 = \alpha_0;$$

$$\Phi_1(x) = x - S_1; \quad \int_a^b p(x)[x - S_1]\,dx = 0,$$

$$S_1 = \alpha_1/\alpha_0; \qquad 1/a_1^2 = \int_a^b p(x)\Phi_1^2(x)\,dx.$$

To construct the successive polynomials $\Phi_2, \Phi_3, \ldots,$ we notice that for any constants c_n, λ_n the polynomial

$$\Phi_n(x) = (x - c_n)\Phi_{n-1} - \lambda_n\Phi_{n-2}, \qquad (n \geqq 2) \tag{15}$$

has its first coefficient equal to one and is orthogonal to Φ_0, $\Phi_1, \ldots, \Phi_{n-3}$; the orthogonality to Φ_{n-2} with respect to $p(x)$ implies

$$\int_a^b xp(x)\Phi_{n-1}(x)\Phi_{n-2}(x)\,dx = \lambda_n \int_a^b p(x)\Phi_{n-2}^2(x)\,dx, \tag{16$_1$}$$

the orthogonality to Φ_{n-1} with respect to $p(x)$ implies

$$\int_a^b xp(x)\,\Phi_{n-1}^2(x)\,dx = c_n\int_a^b p(x)\,\Phi_{n-1}^2(x)\,dx, \tag{16_2}$$

and the normalization of $a_n p^{1/2}(x)\Phi_n(x)$ in $[a, b]$ implies

$$1 = a_n^2\int_a^b p(x)\,\Phi_n^2(x)\,dx. \tag{16_3}$$

(16_1), (16_2), (16_3) determine respectively λ_n, c_n, a_n. The uniqueness follows from the fact that if $\omega(x)$ is a polynomial of degree n, with the first coefficient equal to unity, orthogonal to Φ_0, $\Phi_1, \ldots, \Phi_{n-1}$, then the difference $\omega - \Phi_n$ is a polynomial of degree $n - 1$, which may therefore be expressed as a linear combination with constant coefficients of $\Phi_0, \Phi_1, \ldots, \Phi_{n-1}$, namely,

$$\omega(x) - \Phi_n(x) = l_0\Phi_0(x) + l_1\Phi_1(x) + \ldots + l_{n-1}\Phi_{n-1}(x),$$

orthogonal to $\Phi_0, \Phi_1, \ldots, \Phi_{n-1}$ and this implies $l_0 = l_1 = \ldots = l_{n-1} = 0$ or $\omega(x) = \Phi_n(x)$.

By specifying a, b, $p(x)$ we obtain different classes of polynomials; for example, if a and b are finite and $p(x) = (x - a)^{\alpha-1}(b - x)^{\beta-1}$ where $\alpha > 0$, $\beta > 0$ we have the Jacobi polynomials [53]; for $a = -1$, $b = 1$, $\alpha = \beta = 1$ we have the Legendre polynomials [Ch. III] and for $a = -1$, $b = 1$, $\alpha = \beta = \frac{1}{2}$, if we let $x = \cos\varphi$, we get the system $1, \cos\varphi, \cos 2\varphi, \ldots, \cos n\varphi, \ldots$ (Cf. Ch. V; Sec. 6.1); and for $a = 0$, $b = +\infty$, $p(x) = x^\alpha e^{-x}$, $\alpha > -1$ we have the polynomials $L_n^{(\alpha)}(x)$; for $a = -\infty$, $b = +\infty$, $p(x) = e^{-x^2}$ the polynomials $H_n(x)$.

The results of Chapters II, III are applicable to the most general problem of expansions of a function $f(x)$ in series of Tchebychef polynomials. (For further study of the theory of the Tchebychef polynomials cf. G. Szegö, [107 a]. For the bibliography cf. J. A. Shohat, [100 b], E. Hille, [47 b], J. L. Walsh [118]).

8. For the zeros of the polynomials $\Phi_n(x)$ we have the

THEOREM. Let $\{\Phi_n(x)\}$ be a sequence of orthonormal polynomials with respect to the weight function $p(x)$; $p(x) \geqq 0$ for $a \leqq x \leqq b$; then every polynomial $\Phi_n(x)$, $(n \geqq 1)$ has n real simple zeros all interior to $[a, b]$.

Proof. First observe that if x_0 is a real zero of $\Phi_n(x)$ it is interior to $[a, b]$, because otherwise $\Phi_n(x)/(x - x_0)$ would be a polynomial of degree $n - 1$ with real coefficients, and since

$$0=\int_a^b p(x)\Phi_n(x)[\Phi_n(x)/(x-x_0)]dx=\int_a^b (x-x_0)p(x)[\Phi_n(x)/(x-x_0)]^2dx,$$

and the product $(x - x_0)p(x)$, where it is not zero, is always of the same sign, it would follow that the function $p(x)$ vanishes almost everywhere in $[a, b]$ whence

$$\int_a^b p(x)\,dx = 0,$$

which is impossible by hypothesis.

The relation

$$\int_a^b p(x)\,\Phi_n(x)\,dx = 0$$

implies that $\Phi_n(x)$ has at least one zero x_0 interior to $[a, b]$, and it is readily seen that x_0 is a simple zero. For, otherwise $\Phi_n(x)/(x-x_0)^2$ would be a polynomial with real coefficients of degree $n - 2$ and then we would have

$$0 = \int_a^b p(x)\Phi_n(x)[\Phi_n(x)/(x-x_0)^2]dx = \int_a^b [\sqrt{p(x)}\Phi_n(x)(x-x_0)^{-1}]^2dx,$$

and therefore $p(x)$ would vanish almost everywhere in $[a, b]$ which is impossible. Now, let $x_1, x_2, \ldots, x_l$, $(l \geqq 1)$ be the real zeros of $\Phi_n(x)$, (interior to $[a, b]$). We prove that it is impossible to have $l < n$. As x varies in $[a, b]$ the product $p(x)\,\Phi_n(x)(x - x_1)(x - x_2) \ldots (x - x_l)$ for the values of x for which it is not zero has a constant sign and therefore

$$\int_a^b p(x)\Phi_n(x)(x - x_1)(x - x_2) \ldots (x - x_l)dx \neq 0,$$

but since $\Phi_n(x)$ is orthogonal to the product $(x - x_1)(x - x_2) \ldots (x - x_l)$ with respect to the weight function $p(x)$, the last integral should be zero. On the other hand obviously we cannot have $l > n$, and therefore $l = n$ and the theorem is proved.

3. *Zeros of the Hermite and Laguerre Polynomials*

1a. For the bound of the maximum modulus of the zeros of the Hermite polynomials, E. Laguerre [63 b] has developed the following general theorem for the zeros of the polynomials with all roots real and distinct.

THEOREM. If $F(x)$ is a polynomial of degree n with real coefficients and with all roots real and distinct, then if x_0 is a root of $F(x)$ we have the inequality

$$3(n-2)F''^2(x_0) - 4(n-1)F'(x_0)F'''(x_0) \geqq 0. \tag{1}$$

Proof. (Cf. G. Szegö [107 a] p. 116 for the following proof.)

If $x_1, x_2, \ldots, x_{n-1}$ are the other $n-1$ roots of $F(x)$, let $F(x) = (x-x_0)f(x)$, then $f(x) = c(x-x_1) \ldots (x-x_{n-1})$ and by logarithmic differentiation, we get

$$\frac{1}{x-x_1} + \frac{1}{x-x_2} + \ldots + \frac{1}{x-x_{n-1}} = \frac{f'(x)}{f(x)}$$

whence differentiating again with respect to x and letting $x = x_0$

$$\frac{1}{(x_0-x_1)^2} + \frac{1}{(x_0-x_2)^2} + \ldots + \frac{1}{(x_0-x_{n-1})^2} = \frac{f'^2(x_0) - f''(x_0)f(x_0)}{f^2(x_0)}. \tag{2}$$

Now $F'(x) = f(x) + (x-x_0)f'(x)$, $F''(x) = 2f'(x) + (x-x_0)f''(x)$, $F'''(x) = 3f''(x) + (x-x_0)f'''(x)$ whence $F'(x_0) = f(x_0)$, $F''(x_0) = 2f'(x_0)$, $F'''(x_0) = 3f''(x_0)$, so that (2) may be written

$$\frac{1}{(x_0-x_1)^2} + \frac{1}{(x_0-x_2)^2} + \ldots + \frac{1}{(x_0-x_{n-1})^2} = \frac{3F''^2(x_0) - 4F'(x_0)F'''(x_0)}{12F'^2(x_0)}.$$

Now by the Lagrange-Cauchy inequality, (Appendix, Th. 24), we have

$$\frac{F''^2(x_0)}{4F'^2(x_0)} = \frac{f'^2(x_0)}{f^2(x_0)} = \left[\sum_{\nu=1}^{n-1} \frac{1}{x_0 - x_\nu}\right]^2 \leqq (n-1) \sum_{\nu=1}^{n-1} \frac{1}{(x_0-x_\nu)^2}$$

$$= (n-1)\frac{3F''^2(x_0) - 4F'(x_0)F'''(x_0)}{12F'^2(x_0)},$$

whence (1).

1b. Let $F(x) = H_n(x)$; then by (6) of 2, we have, $H_n''(x_0) = 2x_0 H_n'(x_0)$ and differentiating (6) we have

$$H_n'''(x_0) = 2H_n'(x_0) + 4x_0^2 H_n'(x_0) - 2nH_n'(x_0) = 2(2x_0^2 - n + 1) H_n'(x_0),$$

and (1) becomes

$$12(n-2)x_0^2 H_n'^2(x_0) - 8(n-1)(2x_0^2 - n + 1)H_n'^2(x_0) \geqq 0.$$

Now, dividing by $4H_n'^2(x_0) > 0$, $(n+2)x_0^2 \leqq 2(n-1)^2$ and therefore if x_0 is any zero of the Hermite polynomial, we have the bound

$$\boxed{|x_0| \leqq (n-1)\sqrt{2}/\sqrt{n+2}}. \tag{3}$$

2a. We now derive some bounds for the n zeros of $H_n(x)$. Let

$$f_n(x) = \pi^{-1/4}(2^n n!)^{-1/2} e^{-x^2/2} H_n(x). \tag{4}$$

then the entire transcendental function $f_n(x)$ has its zeros coinciding with the n zeros of $H_n(x)$, and from equation (6) of Sec. 2.2 we get for $f_n(x)$ the equation

$$f_n'' + [2n + 1 - x^2]f_n = 0, \tag{5}$$

and the zeros of $f_n(x)$ are by (3) in absolute value less than $\sqrt{2n+1}$.

This property also follows at once from (4) and (5), for, since by (4) f_n and f_n' are never zero simultaneously for finite values of x, for $x > \sqrt{2n+1}$ we have

$$[f_n f_n']' = f_n f_n'' + f_n'^2 = [x^2 - (2n+1)]f_n^2 + f_n'^2 > 0,$$

whence $f_n f_n'$ is increasing in the interval $(\sqrt{2n+1}, \infty)$, and since $\lim_{n\to\infty} f_n f_n' = 0$ we have in the same interval $f_n f_n' < 0$ and therefore $f_n(x) \neq 0$.

By (3) of Sec. 2.1, the number of positive zeros of $H_n(x)$ is $[n/2]$. If these zeros are arranged in increasing order and are denoted by $x_{1,n}\, x_{2,n}, \ldots, x_{[n/2],n}$ we have

$$0 < x_{1,n} < x_{2,n} < \ldots < x_{[n/2],n} < \sqrt{2n+1}. \tag{6}$$

2b. Now consider the equation

$$z'' + (2n+1)z = 0, \tag{7}$$

as x varies in $[0, \sqrt{2n+1}]$. By the theorem of Sec. 8.1b of Ch. III, between two consecutive zeros of a solution $f_n(x)$ of (5) lies at least one zero of $z(x)$, and if we assume n odd, $f_n(x)$ vanishes at 0, and the integral $z = \sin\sqrt{2n+1}\,x$ of (7), which vanishes at 0 has its νth zero at $\nu\pi/\sqrt{2n+1}$, whence

$$(8_1) \quad \nu\pi/\sqrt{2n+1} < x_{\nu,n}, \quad [\nu = 1, 2, \ldots, [n/2],\ n \equiv 1 \pmod{2}].$$

Now let n be even and note that $f_n(x) \neq 0$ in $(0, \pi/\sqrt{2n+1})$. In fact (7) has the solution $z = \cos\sqrt{2n+1}x$, and since we have $f_n'(0) = z'(0) = 0$, by Sturm's theorem of Sec. 8.1a of Ch. III, if $f_n(x)$ had a zero interior to $(0, \pi/\sqrt{2n+1})$, then $z(x)$ would necessarily vanish in the interior of such an interval, which is impossible.

Consequently, since $(\nu - \frac{1}{2})\pi/\sqrt{2n+1}$ is the νth positive zero of $z(x)$, then for n even, we have

$$(8_2) \quad (\nu - 1/2)\pi/\sqrt{2n+1} < x_{\nu,n},$$
$$[\nu = 1, 2, \ldots, [n/2];\ n \equiv 0, \pmod{2}].$$

2c. If in (5) we change the independent variable $x = t/\sqrt{2n+1}$ and let $f_n(t/\sqrt{2n+1}) = Z(t)$; then by (5) we get

$$\frac{d^2Z}{dt^2} + \left[1 - \frac{t^2}{(2n+1)^2}\right]Z = 0, \tag{9}$$

and between the zeros $x_{\nu,n}$ of $H_n(x)$ and the positive zeros $t_{\nu,n}$ of $Z(t)$ exists the relation $\sqrt{2n+1}\,x_{\nu,n} = t_{\nu,n}$.

Now if the parameter n in (9) increases, then the coefficient of Z increases, so that for fixed ν, by the theorem of Sec. 8.1b of Ch. III, we have that if n is odd, $n \geqq 2\nu + 1$, between two consecutive terms of the sequence

$$0,\ t_{1,2\nu+1},\ t_{2,2\nu+1},\ \ldots,\ t_{\nu,2\nu+1}$$

lies a positive zero of the solution $Z(t)$ of (9), whence $t_{\nu,n} \leqq t_{\nu,2\nu+1}$ and by (6)

$$(10_1) \quad \sqrt{2n+1}\, x_{\nu,n} \leqq \sqrt{4\nu+3}\, x_{\nu,2\nu+1} < 4\nu + 3, \quad (n \equiv 1, \text{ mod. } 2).$$

If n is even, we shall consider (9) for $n \geqq 2\nu$. By Sturm's theorem at least one zero of $Z(t)$ lies in the interval $(-t_{1,2\nu}, t_{1,2\nu})$ and since n is even, at least two zeros of which one is positive and one is negative, and therefore between two consecutive terms of the sequence $0, t_{1,2\nu}, t_{2,2\nu}, \ldots, t_{\nu,2\nu}$ there is at least one zero of $Z(t)$, whence $t_{\nu,n} \leqq t_{\nu,2\nu}$, or

$$(10_2) \quad \sqrt{2n+1}\, x_{\nu,n} \leqq \sqrt{4\nu+1}\, x_{\nu,2\nu} < 4\nu + 1, \quad [n \equiv 0, \text{ (mod. 2)}].$$

By (8_1), (8_2), (10_1), (10_2) we then have the inequalities

$$(11_1) \quad \nu\pi/\sqrt{2n+1} < x_{\nu,n} < (4\nu+3)/\sqrt{2n+1}, \quad [n=1, \text{ (mod. 2)}];$$

$$(11_2) \quad (\nu-1/2)\pi/(\sqrt{2n+1}) < x_{\nu,n} < (4\nu+1)/\sqrt{2n+1}, \quad (n \equiv 0, \text{ (mod. 2)}],$$

$$(\nu = 1, 2, \ldots, [n/2]).$$

3. In this section we shall derive bounds for the zeros of the polynomials $L_n^{(\alpha)}(x)$ for the case $\alpha > -1$. Under this hypothesis, the polynomials $L_n^{(\alpha)}(x)$ form an orthogonal system in $[0, \infty)$ with respect to the weight function $e^{-x}x^{\alpha}$, [Sec. 1.7a] and therefore [Sec. 2.8] the polynomial $L_n^{(\alpha)}(x)$ has exactly n zeros all real and positive. (For the determination of the number of real and complex zeros of the polynomials $L_n^{(\alpha)}(x)$ in the case $\alpha \leqq -1$, Cf., e.g., G. Sansone [98c].

3a. To obtain the first bound for the zeros of the polynomials $L_n^{(\alpha)}(x)$ set $F(x) = L_n^{(\alpha)}(x)$ in Laguerre's inequality of Sec. 3. 1a. Then from (14_1) of Sec. 1 and the fact that $[L_n^{(\alpha)}(x)]'_{x=x_0} \neq 0$ follows

$$(12) \quad (n+2)x_0^2 - 2[2n(n-1) + (n+2)(\alpha+1)]x_0 + (\alpha+1)[(n+2)(\alpha+1) + 4(n-1)] \leqq 0$$

whence, if

$$x_{1,n}, \quad x_{2,n}, \quad \ldots, \quad x_{n,n}$$

denote the n zeros of $L_n^{(\alpha)}(x)$, arranged in increasing order, we get the inequality

$$(13) \qquad 0<\alpha+1+2\frac{n-1}{n+2}\left[n-\sqrt{n^2+(n+2)(\alpha+1)}\right] \leqq x_{\nu,n}$$

$$\leqq \alpha+1+2\frac{n-1}{n+2}\left[n+\sqrt{n^2+(n+2)(\alpha+1)}\right], \quad (\nu=1, 2, \ldots, n).$$

3b. A second upper bound on the zeros of the polynomials $L_n^{(\alpha)}(x)$ can be obtained as follows:

If we let

$$(14) \qquad v(t) = e^{-t^2/2}t^{\alpha+1/2}L_n^{(\alpha)}(t^2)$$

then, by (14_1) of Sec. 1.5, the equation for $v(t)$ is

$$(15) \qquad \frac{d^2v}{dt^2}+\left[4n+2\alpha+2-t^2+\frac{1/4-\alpha^2}{t^2}\right]v=0,$$

whence for

$$t^2-[1/4-\alpha^2]/t^2-(4n+2\alpha+2)>0$$

and, therefore, for

$$x=t^2>2n+\alpha+1+\sqrt{(2n+\alpha+1)^2+1/4-\alpha^2}$$

the coefficient of v in (15) is positive, and arguing as in Sec. 3. 2a we get

$$(16) \qquad x_{\nu,n}<2\left[n+\frac{\alpha+1}{2}+\sqrt{\left[n+\frac{\alpha+1}{2}\right]^2+\frac{1/4-\alpha^2}{4}}\right],$$

$$(\nu=1, 2, \ldots, n).$$

3c. If we let

$$u(x)=e^{-x/2}x^{(\alpha+1)/2}L_n^{(\alpha)}(x)$$

then by (14_1), Sec. 1.5, we get

$$(17) \qquad u''+\left[\frac{n+(\alpha+1)/2}{x}+\frac{1-\alpha^2}{4x^2}-\frac{1}{4}\right]u=0.$$

Bessel's function $y=J_\alpha(x)$ of order $\alpha>-1$, namely

$$J_\alpha(x)=\sum_{\nu=0}^{\infty}(-1)^\nu(x/2)^{\alpha+2\nu}/\Gamma(\alpha+\nu+1)\Gamma(\nu+1)$$

satisfies the differential equation

$$\frac{d^2y}{dx^2} + \frac{1}{x}\frac{dy}{dx} + \left(1 - \frac{\alpha^2}{x^2}\right) y = 0$$

and has an infinite number of positive zeros $j_1, j_2, \ldots, j_\nu, \ldots$.

$$j_1 < j_2 < \ldots < j_\nu < \ldots, \quad \lim_{\nu\to\infty} j_\nu = \infty.$$

The function

$$U(x) = x^{1/2} J_\alpha \left[2x^{1/2}\left(n + \frac{\alpha+1}{2}\right)^{1/2}\right] \tag{18}$$

satisfies the differential equation

$$U'' + \left[\frac{n + (\alpha+1)/2}{x} + \frac{1-\alpha^2}{4x^2}\right] U = 0, \tag{19}$$

and has the sequence of zeros

$$(j_\nu/2)^2/[n + (\alpha+1)/2], \qquad (\nu = 1, 2, \ldots).$$

The $u(x)$ cannot vanish in the interior of $(0, (j_1/2)^2/[n + (\alpha+1)/2])$ since $\lim_{x\to 0}[U'u - Uu'] = 0$ and by virtue of (17) and (19) it would then follow [Ch. III, Sec. 8.1a] that $U(x)$ would have one positive zero in the interior of $(0, (j_1/2)^2/[n + (\alpha+1)/2])$ which is impossible. It therefore follows from (17) and (19) that

$$\boxed{\frac{(j_\nu/2)^2}{n + (\alpha+1)/2} < x_{\nu,n}}, \qquad [\nu = 1, 2, \ldots, n;\ n = 1, 2, \ldots]. \tag{20_1}$$

3d. Changing the independent variable in (17) by $x = \xi/[n + (\alpha+1)/2]$ we get

$$\frac{d^2u}{d\xi^2} + \left[\frac{1}{\xi} + \frac{1-\alpha^2}{4\xi^2} - \frac{1}{4[n + (\alpha+1)/2]^2}\right] u = 0.$$

Now, the coefficient of u decreases if we increase the parameter n, whence by a repetition of the argument of Sec. 3.2c, we deduce that for fixed ν, $[n + (\alpha+1)/2]x_{\nu,n}$ decreases as n increases. Consequently

$$[n + (\alpha + 1)/2] x_{\nu,n} < [\nu + (\alpha + 1)/2] x_{\nu,\nu},$$

and by (16)

$$(20_2) \quad \boxed{x_{\nu,n} < 2 \frac{\nu + (\alpha+1)/2}{n + (\alpha+1)/2} \left[\nu + \frac{\alpha+1}{2} + \sqrt{\left(\nu + \frac{\alpha+1}{2}\right)^2 + \frac{1/4 - \alpha^2}{4}} \right]},$$

$$(\nu = 1, 2, \ldots, n).$$

(20_1) and (20_2) thus give lower and upper bounds on the roots $x_{\nu,n}$.

In the case $\nu = 1$, the upper bound (20_2) can be improved: Indeed, from the expression for $L_n^{(\alpha)}(x)$ of Sec. 1.1 d, $x_{1,1} = \alpha + 1$, whence

$$(20_2) \qquad 0 < x_{1,n} < \frac{(\alpha + 1)(\alpha + 3)}{2n + \alpha + 1}.$$

4. Relations between the Polynomials $L_n^{(\alpha)}(x)$ and $H^n(x)$

1. If in formula (3_1) of Sec. 1.1

$$(-1)^n L_n^{(\alpha)}(u) = \sum_{k=0}^{n} (-1)^k \frac{\Gamma(\alpha + n + 1)}{k!(n - k)!\Gamma(n - k + \alpha + 1)} u^{n-k}$$

we let $\alpha = -1/2$, $u = x^2$ and observe that

$$\frac{n!\Gamma(n + 1/2)2^{2k}}{k!(n - k)!\Gamma(n - k + 1/2)}$$

$$= \frac{n!2^{2k}(n-1/2)\ldots(n-k+1/2)}{k!(n-k)!} = \frac{n!2^k(2n-1)(2n-3)\ldots(2n-2k+1)}{k!(n-k)!}$$

$$= \frac{2n(2n - 1)(2n - 2)\ldots(2n - 2k + 1)}{k!} = \frac{(2n)!}{k!(2n - 2k)!}$$

we get

$$(-1)^n 2^{2n} n! L_n^{(-1/2)}(x^2) =$$

$$= \sum_{k=0}^{n} (-1)^k \frac{n!\Gamma(n+1/2)2^{2k}}{k!(n-k)!\Gamma(n-k+1/2)} (2x)^{2(n-k)}$$

$$= \sum_{k=0}^{n} (-1)^k \frac{(2n)!}{k!(2n-2k)!} (2x)^{2(n-k)} = H_{2n}(x)$$

whence

$$(1_1) \qquad \boxed{H_{2n}(x) = (-1)^n 2^{2n} n! L_n^{(-1/2)}(x^2)},$$

and similarly

$$(1_2) \qquad \boxed{H_{2n+1}(x) = (-1)^{n+1} 2^{2n+1} n! x L_n^{(1/2)}(x^2)}$$

(Cf. G. Szegö [107a] p. 102; [107b]. Cf. also G. Palamà [84c], L. Toscano [113a] with bibliography; [113b]).

2. We now wish to find an integral expression for the polynomials $L_n^{(\alpha)}(x)$ in terms of polynomials $H_n(x)$, $[\alpha > -\frac{1}{2}]$.

Letting $\beta = -\frac{1}{2}$, $u = t^2$ in the integral relation (15) of Sec. 1.6, we have

$$L_n^{(\alpha)}(x) = 2\frac{\Gamma(n+\alpha+1)}{\Gamma(n+1/2)\Gamma(\alpha+1/2)} x^{-\alpha} \int_0^{\sqrt{x}} (x-t^2)^{\alpha-\frac{1}{2}} L_n^{(-\frac{1}{2})}(t^2)\,dt,$$

$$x > 0, \quad \sqrt{x} > 0, \quad \alpha > -1/2;$$

and by (1_1)

$$L_n^{(\alpha)}(x) = \frac{\Gamma(n+\alpha+1)}{\Gamma(n+1/2)\Gamma(\alpha+1/2)} \frac{(-1)^n x^{-\alpha}}{2^{2n} n!} \int_{-\sqrt{x}}^{\sqrt{x}} (x-t^2)^{\alpha-\frac{1}{2}} H_{2n}(t)\,dt,$$

also

$$2^{2n} n! \Gamma(n+\tfrac{1}{2}) = 2^{2n} n! (n-\tfrac{1}{2})(n-\tfrac{3}{2}) \dots \tfrac{1}{2}\Gamma(\tfrac{1}{2})$$

$$= 2^n n! (2n-1)(2n-3) \dots 1\sqrt{\pi} = (2n)!\sqrt{\pi}$$

whence

$$(2_1)\qquad \boxed{L_n^{(\alpha)}(x) = \frac{(-1)^n\Gamma(n+\alpha+1)x^{-\alpha}}{\sqrt{\pi}(2n)!\Gamma(\alpha+1/2)}\int_{-\sqrt{x}}^{\sqrt{x}}(x-t^2)^{\alpha-1/2}H_{2n}(t)dt},$$

$$\alpha > -1/2,$$

and finally changing t into $\sqrt{x}\cos\varphi$ we obtain the formula of J. W. Uspensky [[115], p. 604]

$$(2_2)\qquad \boxed{L_n^{(\alpha)}(x) = \frac{(-1)^n\Gamma(n+\alpha+1)}{\sqrt{\pi}(2n)!\Gamma(\alpha+1/2)}\int_0^{\pi} H_{2n}(\sqrt{x}\cos\varphi)\sin^{2\alpha}\varphi\, d\varphi},$$

$$\alpha > -1/2.$$

5. *Formulas for Asymptotic Approximation of the Polynomials* $H_n(x)$:

1. We now present the procedure of Stone [106] and Kowallik [60] for the determination of the formulas of asymptotic approximation of the polynomials $H_n(x)$. The procedure is similar to the one used by Bonnet [11] for the Legendre polynomials.

Consider the orthonormal sequence $\{f_n(x)\}$

$$(1)\qquad f_n(x) = \pi^{-1/4}(2^n n!)^{-1/2}e^{-(1/2)x^2}H_n(x)$$

in $(-\infty, +\infty)$[Sec. 2.6]. The functions f_n satisfy the differential equation [Sec. 3.2 a]

$$(2_1)\qquad \boxed{f_n'' + (-x^2 + 2n + 1)f_n = 0},$$

$$(2_2)\qquad \boxed{\int_{-\infty}^{+\infty} f_n^2(x)\,dx = 1}.$$

Now, if we let

$$s_n = \sin(\sqrt{2n+1}\,x),\quad c_n = \cos(\sqrt{2n+1}\,x)$$

we have

$$(3_1)\qquad s_n'' + (2n+1)s_n = 0,$$

$$(3_2)\qquad c_n'' + (2n+1)c_n = 0,$$

and multiplying (2_1) by s_n and (3_1) by f_n, also (2_1) by c_n and (3_2) by f_n and subtracting the results we get

$$(f_n' s_n - f_n s_n')' = x^2 f_n s_n, \quad (f_n' c_n - f_n c_n')' = x^2 f_n c_n$$

whence integrating between 0 and x

$$(4_1) \qquad [f_n'(\alpha) s_n(\alpha) - f_n(\alpha) s_n'(\alpha)]_{\alpha=0}^{\alpha=x} = \int_0^x \alpha^2 f_n(\alpha) s_n(\alpha)\, d\alpha$$

$$(4_2) \qquad [f_n'(\alpha) c_n(\alpha) - f_n(\alpha) c_n'(\alpha)]_{\alpha=0}^{\alpha=x} = \int_0^x \alpha^2 f_n(\alpha) c_n(\alpha)\, d\alpha,$$

and since

$$s_n(0) = 0, \quad s_n'(0) = \sqrt{2n+1}; \quad c_n(0) = 1, \quad c_n'(0) = 0,$$

(4_1) and (4_2) give

$$(5) \quad \begin{cases} f_n'(x)s_n(x) - s_n'(x) f_n(x) = -\sqrt{2n+1}\, f_n(0) + \int_0^x \alpha^2 f_n(\alpha) s_n(\alpha)\, d\alpha \\ f_n'(x)c_n(x) - c_n'(x) f_n(x) = f_n'(0) + \int_0^x \alpha^2 f_n(\alpha) c_n(\alpha)\, d\alpha; \end{cases}$$

and since

$$-s_n c_n' + s_n' c_n = \sqrt{2n+1},$$

we get finally from (5)

$$(6) \quad \boxed{\begin{aligned} f_n(x) = f_n(0) \cos [\sqrt{2n+1}\, x] + f_n'(0) \frac{\sin [\sqrt{2n+1}\, x]}{\sqrt{2n+1}} \\ - \frac{1}{\sqrt{2n+1}} \int_0^x \alpha^2 f_n(\alpha) \sin [\sqrt{2n+1}(\alpha - x)]\, d\alpha \end{aligned}}\,.$$

1a. If n is even, then we have by (1), and by (10) of Sec. 2.3,

$$f_{2n}(0) = \frac{H_{2n}(0)}{\sqrt[4]{\pi}\sqrt{2^{2n}(2n)!}}$$

$$= (-1)^n \frac{(2n)!}{n!} \frac{1}{\sqrt[4]{\pi}\sqrt{2^{2n}(2n)!}} = \frac{(-1)^n}{\sqrt[4]{\pi}} \sqrt{\frac{(2n-1)!!}{(2n)!!}},$$

$$f_{2n}'(0) = 0,$$

but by Wallis' formula (Appendix, Th. 25) we have

$$\sqrt{\frac{(2n-1)!!}{(2n)!!}} = \sqrt[4]{\frac{1}{\pi n}}\left(1-\frac{\varepsilon_1}{8n}\right), \qquad 0 < \varepsilon_1 < 1,$$

and therefore from (6)

$$f_{2n}(x) = \frac{(-1)^n}{\sqrt{\pi}} \sqrt[4]{\frac{1}{n}}\left(1-\frac{\varepsilon_1}{8n}\right) \cos\,[x\sqrt{4n+1}]$$
$$-\frac{1}{\sqrt{4n+1}}\int_0^x \alpha^2 f_{2n}(\alpha) \sin\,[\sqrt{4n+1}(\alpha - x)]\, d\alpha.$$

By the Schwarz inequality [Ch. I, Sec. 2.2] we have

$$\left|\int_0^x \alpha^2 f_{2n}(\alpha) \sin\,[\sqrt{4n+1}(\alpha-x)]\, d\alpha\right|$$
$$\leqq \left[\int_0^x \alpha^4 \sin^2 \sqrt{4n+1}(\alpha - x)\, d\alpha\right]^{1/2} \left[\int_0^x f_{2n}^2(\alpha)\, d\alpha\right]^{1/2}$$
$$\leqq \left[\int_0^x \alpha^4\, d\alpha\right]^{1/2} \left[\int_0^{+\infty} f_{2n}^2(\alpha)\, d\alpha\right]^{1/2} < |x|^{5/2},$$

whence letting

$$h(n, x) = \int_0^x \alpha^2 f_n(\alpha) \sin\,[\sqrt{2n+1}(\alpha - x)]\, d\alpha \tag{7}$$

we have

$$f_{2n}(x) = \frac{(-1)^n}{\sqrt{\pi}}\,\frac{1}{n^{1/4}}\left(1-\frac{\varepsilon_1}{8n}\right)\cos\,[x\sqrt{4n+1}] - \frac{h(2n, x)}{\sqrt{4n+1}}. \tag{8$_1$}$$

$$0 < \varepsilon_1 < 1, \quad |h(2n, x)| < |x|^{5/2}. \tag{8$_2$}$$

1b. Now if n is odd; we have

$$f_{2n+1}(0) = 0;$$

and by (1), and (10) of Sec. 2.3,

$$f'_{2n+1}(0) = \frac{H'_{2n+1}(0)}{\sqrt[4]{\pi}\,\sqrt{2^{2n+1}(2n+1)!}}$$

$$= 2\frac{(-1)^{n+1}}{\sqrt[4]{\pi}}\frac{(2n)!}{n!}\frac{2n+1}{\sqrt{2^{2n+1}(2n+1)!}} = (-1)^{n+1}\frac{\sqrt{2(2n+1)}}{\sqrt[4]{\pi}}\sqrt{\frac{(2n-1)!!}{(2n)!!}},$$

and since

$$1 > \sqrt{\frac{4n+2}{4n+3}}\sqrt[4]{\frac{2n}{2n+\theta}} > \sqrt{\frac{4n+2}{4n+3}}\left(1-\frac{1}{8n}\right) > 1-\frac{1}{4n},$$

we have

$$\frac{1}{\sqrt{4n+3}}f'_{2n+1}(0) = (-1)^{n+1}\frac{1}{\sqrt{\pi}n^{1/4}}\sqrt{\frac{4n+2}{4n+3}}\sqrt[4]{\frac{2n}{2n+\theta}}$$

$$= \frac{(-1)^{n+1}}{\sqrt{\pi}n^{1/4}}\left(1-\frac{\varepsilon_2}{8n}\right), \qquad 0 < \varepsilon_2 < 2,$$

whence by (6)

$$(9_1) \quad \boxed{f_{2n+1}(x) = \frac{(-1)^{n+1}}{\sqrt{\pi}}\frac{1}{n^{1/4}}\left(1-\frac{\varepsilon_2}{8n}\right)\sin\,[x\sqrt{4n+3}] - \frac{h(2n+1,x)}{\sqrt{4n+3}}}$$

$$(9_2) \qquad 0 < \varepsilon_2 < 2, \qquad |\,h(2n+1,x)\,| < |\,x\,|^{5/2}.$$

For the convenience of the reader we give the graph of the first seven functions $f_0(x), \ldots, f_6(x)$.

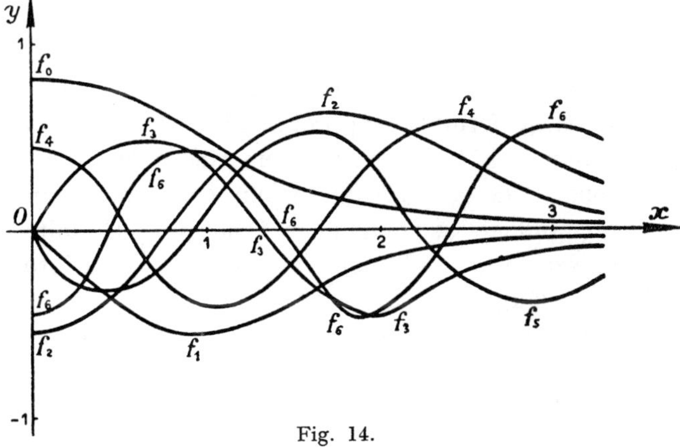

Fig. 14.

1c. From formulas (1), (8), (9) we get

$$(10)\quad \boxed{\begin{aligned} H_{2n}(x) &= e^{x^2/2}\sqrt{2^{2n}(2n)!}\,\pi^{-1/4}n^{-1/4} \\ &\times\left[(-1)^n\left(1-\frac{\varepsilon_1}{8n}\right)\cos\,[x\sqrt{4n+1}]-\frac{\pi^{1/2}n^{1/4}}{\sqrt{4n+1}}h(2n,x)\right] \\ H_{2n+1}(x) &= e^{x^2/2}\sqrt{2^{2n+1}(2n+1)!}\,\pi^{-1/4}n^{-1/4} \\ &\times\left[(-1)^{n+1}\left(1-\frac{\varepsilon_2}{8n}\right)\sin\,[x\sqrt{4n+3}]-\frac{\pi^{1/4}n^{1/4}}{\sqrt{4n+3}}h(2n+1,x)\right] \end{aligned}}$$

where

$$(10')\qquad 0<\varepsilon_1<1,\quad 0<\varepsilon_2<2;$$

and

$$|\,h(n,x)\,|<|\,x\,|^{5/2}$$

for every n.

By means of an integral of Cauchy, H. Cramer [25 b] derived from this expression for $H_n(x)$ the following inequality for $H_n(x)$

$$\boxed{|\,H_n(x)\,|<k\sqrt{n!}\,2^{n/2}e^{x^2/2}}$$

where, in accordance with a result of C. V. L. Charlier [20 a], the constant k is given by

$$k=1.086435.$$

2. In the study of expansions in series of Hermite polynomials of Sec. 10, a second formula of approximation of the polynomials $H_n(x)$ will be required. (For the subject matter of Secs. 5.2 and 5.3 cf. G. Sansone: "Due semplici limitazioni nel campo complesso delle funzioni associate ai polinomi di Tchebychef-Hermite e del termine complementare della loro rappresentazione asintotica," *Math. Z.* 52 (1950), pp. 593—598)

2a. Changing n to $2n$ in (6) and letting $N=4n+1$ we have

$$(11_1)\quad f_{2n}(x)=f_{2n}(0)\cos\,[N^{1/2}x]-\frac{1}{N^{1/2}}\int_0^x \alpha^2 f_{2n}(\alpha)\sin\,[N^{1/2}(\alpha-x)]\,d\alpha,$$

and substituting in the integral the expression for $f_{2n}(\alpha)$ which we get from (11_1) by changing x into α we get

$$(12_1)\quad f_{2n}(x) = f_{2n}(0)\cos[N^{1/2}x] - \frac{f_{2n}(0)}{N^{1/2}} I_1 \cos[N^{1/2}x] + \frac{f_{2n}(0)}{N^{1/2}} I_2 \sin[N^{1/2}x] + \frac{I_3}{N},$$

where

$$(13_1)\quad I_1 = \int_0^x \alpha^2 \sin[N^{1/2}\alpha]\cos[N^{1/2}\alpha]\,d\alpha,$$

$$(13_2)\quad I_2 = \int_0^x \alpha^2 \cos^2[N^{1/2}\alpha]\,d\alpha,$$

$$(13_3)\quad I_3 = \int_0^x \alpha^2 I_4(\alpha)\sin[N^{1/2}(\alpha - x)]\,d\alpha,$$

$$I_4(\alpha) = \int_0^\alpha \beta^2 f_{2n}(\beta)\sin[N^{1/2}(\beta - \alpha)]\,d\beta.$$

Applying the second theorem of the mean (Appendix. Th. 18) for $x > 0$ we have

$$I_1 = \frac{x^2}{2}\int_\xi^x \sin[2N^{1/2}\alpha]\,d\alpha = \frac{\theta_1 x^2}{2N^{1/2}}, \quad (0 \leqq \xi \leqq x),\ |\theta_1| < 1;$$

$$I_2 = \frac{x^3}{6} + \frac{1}{2}\int_0^x \alpha^2 \cos[2N^{1/2}\alpha]\,d\alpha = \frac{x^3}{6} + \frac{\theta_2 x^3}{2N^{1/2}}, \quad |\theta_2| < 1;$$

and since by (8_1) and (8_2) the inequality

$$|f_{2n}(\beta)| \leqq \frac{1}{\sqrt{\pi}\,n^{1/4}} + \frac{\beta^{5/2}}{N^{1/2}}$$

holds, we have also

$$|I_4(\alpha)| \leqq \alpha^3/(3\sqrt{\pi}\,n^{1/4}) + 2\alpha^{11/2}/(11N^{1/2}),$$

$$|I_3| \leqq x^6/(18\sqrt{\pi}\,n^{1/4}) + 4x^{17/2}/(187N^{1/2}),$$

and from (12_1) since $|f_{2n}(0)| < 1/(\sqrt{\pi}\,n^{1/4})$ we get

$$(14_1)\quad f_{2n}(x) = f_{2n}(0)\cos[N^{1/2}x] + \frac{f_{2n}(0)}{N^{1/2}}\frac{x^3}{6}\sin[N^{1/2}x] + \frac{t(2n, x)}{N},$$

where

$$(15_1) \qquad |\, t(2n, x) \,| \leq \frac{x^2}{\sqrt{\pi}\, n^{1/4}} \left[\frac{x^4}{18} + 1\right] + \frac{4}{187} \frac{x^{17/2}}{N^{1/2}},$$

and finally

$$(16_1) \qquad \boxed{\begin{aligned} H_{2n}(x) = \pi^{1/4}\sqrt{2^{2n}(2n)!}\, e^{x^2/2} \Big[& f_{2n}(0) \cos [N^{1/2} x] \\ & + \frac{f_{2n}(0)}{N^{1/2}} \frac{x^3}{6} \sin [N^{1/2} x] + \frac{t(2n, x)}{N}\Big] \end{aligned}}\,.$$

2b. Changing n to $2n+1$ in (6) and letting $N = 4n+3$, we have

$$(11_2) \qquad \begin{aligned} f_{2n+1}(x) = {} & \frac{f'_{2n+1}(0)}{N^{1/2}} \sin [N^{1/2} x] \\ & - \frac{1}{N^{1/2}} \int_0^x \alpha^2 f_{2n+1}(\alpha) \sin [N^{1/2}(\alpha - x)]\, d\alpha, \end{aligned}$$

and by the same procedure as in a), using the fact that

$$|\, f'_{2n+1}(0)/N^{1/2} \,| < 1/(\sqrt{\pi}\, n^{1/4})$$

we find

$$(14_2) \qquad \begin{aligned} f_{2n+1}(x) = {} & f'_{2n+1}(0) \frac{\sin [N^{1/2} x]}{N^{1/2}} \\ & - \frac{f'_{2n+1}(0)}{N} \frac{x^3}{6} \cos [N^{1/2} x] + \frac{t(2n+1, x)}{N}, \end{aligned}$$

where

$$(15_2) \qquad |\, t(2n+1, x) \,| \leq \frac{x^2}{\sqrt{\pi}\, n^{1/4}} \left[\frac{x^4}{18} + 1\right] + \frac{4}{187} \frac{x^{17/2}}{N^{1/2}},$$

and therefore

$$(16_2) \qquad \boxed{\begin{aligned} H_{2n+1}(x) = \pi^{1/4}\sqrt{2^{2n+1}(2n+1)!}\, e^{x^2/2} \Big[& f'_{2n+1}(0) \frac{\sin [N^{1/2} x]}{N^{1/2}} \\ & - \frac{f'_{2n+1}(0)}{N} \frac{x^3}{6} \cos [N^{1/2} x] + \frac{t(2n+1, x)}{N}\Big] \end{aligned}}\,.$$

2c. If we let

$$N = 2n + 1, \tag{17}$$

$$l_n = \Gamma(n/2 + 1)/\Gamma(n + 1), \qquad \text{if } n \text{ is even} \tag{18_1}$$

$$l_n = N^{1/2}\Gamma(n/2 + 3/2)/\Gamma(n + 2), \quad \text{if } n \text{ is odd} \tag{18_2}$$

then (16_1) and (16_2) can be incorporated into a single formula, namely

$$\boxed{\begin{aligned}(-1)^n l_n e^{-x^2/2} H_n(x) &= \cos\,[N^{1/2}x - n\pi/2] \\ &\quad + \frac{x^3}{6} N^{-1/2} \sin\,[N^{1/2}x - n\pi/2] + \frac{\bar{t}(n, x)}{N}\end{aligned}}, \tag{19_1}$$

where

$$\boxed{|\,\bar{t}(n, x)\,| \leqq \frac{5}{4} x^2 \left[1 + \frac{x^4}{18}\right] + \frac{36}{935} \frac{1}{n^{1/2}} x^{17/2}}. \tag{19_2}$$

(For (19_1) cf. G. Szegö [[107a] p. 194]. In Szegö's paper the symbol H_n corresponds to the symbol $(-1)^n H_n$ of the text).

2d. Differentiating with respect to x in (12_1) we get

$$\begin{aligned}f'_{2n}(x) = -N^{1/2} f_{2n}(0) \sin\,[N^{1/2}x] &+ f_{2n}(0) I_1 \sin\,[N^{1/2}x] \\ &+ f_{2n}(0) I_2 \cos\,[N^{1/2}x] + \frac{1}{N} \frac{dI_3}{dx},\end{aligned}$$

whence by the results derived in 2a

$$f'_{2n}(x) = -N^{1/2} f_{2n}(0) \sin\,[N^{1/2}x] + f_{2n}(0) \frac{x^3}{6} \cos\,[N^{1/2}x] + \frac{\sigma(2n, x)}{N^{1/2}}, \tag{20}$$

where

$$|\,\sigma(2n, x)\,| \leqq \frac{x^2}{\sqrt{\pi} n^{1/4}} \left[\frac{x^4}{18} + 1\right] + \frac{4}{187} \frac{x^{17/2}}{N^{1/2}}. \tag{21}$$

Differentiating (14_1) and comparing with (20) we get

$$N^{1/2} f_{2n}(0) \frac{x^2}{2} \sin\,[N^{1/2}x] + \frac{\partial t(2n, x)}{\partial x} = \sigma(2n, x) N^{1/2}$$

and therefore

$$(22)\quad \left|\frac{\partial t(2n, x)}{\partial x}\right| \leqq n^{1/4} A(x), \; A(x) = \sqrt{5}\left[\frac{1}{\sqrt{\pi}}\left\{\frac{3}{2}x^2 + \frac{x^6}{18}\right\} + \frac{4}{187}x^{17/2}\right].$$

Similarly we can show that

$$(23)\quad \left|\frac{\partial t(2n+1, x)}{\partial x}\right| \leq (n+1)^{1/4} A(x).$$

3. By (1), $f(x)$ is an entire transcendental function of x, as is also $t(n, x)$ by virtue of (14_1) and (14_2). For the asymptotic determination of the polynomials $L_n^{(\alpha)}(x)$ in Sec. 6 we shall need bounds for $f_n(x)$ and $t(n, x)$ also for complex values of x.

3a. If we let $x = u + iv$, and assume $v > 0$, and also let $N = 4n + 1$, then

$$\begin{aligned} f_{2n}(u + iv) &= f_{2n}(0) \cos [N^{1/2}(u + iv)] \\ &\quad - \frac{1}{N^{1/2}} \int_0^u \alpha^2 f_{2n}(\alpha) \sin [N^{1/2}(\alpha - u - iv] \, d\alpha \\ &\quad + \frac{1}{N^{1/2}} \int_0^v (u + it)^2 f_{2n}(u + it) \sinh [N^{1/2}(t - v)] \, dt, \end{aligned}$$

where the integrals are taken respectively along the segments $(0, u)$ and $(0, v)$ of the real axis.

If we also let

$$(24)\quad f_{2n}(u + iv) = \varphi(v),$$

$$(25)\quad \begin{aligned} &f_{2n}(0) \cos [N^{1/2}(u + iv)] \\ &\quad - \frac{1}{N^{1/2}} \int_0^u \alpha^2 f_{2n}(\alpha) \sin [N^{1/2}(\alpha - u - iv)] \, d\alpha = \omega(v), \end{aligned}$$

then

$$(26)\quad \varphi(v) = \omega(v) - \frac{1}{N^{1/2}} \int_0^v (u + it)^2 \sinh [N^{1/2}(v - t)] \varphi(t) dt,$$

which is a Volterra integral equation of the second kind in $\varphi(v)$ with the kernel

$$K(v, t) = -\frac{1}{N^{1/2}}(u + it)^2 \sinh [N^{1/2}(v - t)].$$

It follows that

$$|K(v, t)| \leqq \frac{1}{N^{1/2}}(u^2 + t^2)\, e^{\sqrt{N}(v-t)}, \qquad (0 \leqq t \leqq v)$$

and for the second iterated kernel $K^{(2)}(v, t)$ we have

$$|K^{(2)}(v, t)| \leqq \frac{1}{N}\int_t^v (u^2 + \xi^2)\, e^{\sqrt{N}(v-\xi)}(u^2 + t^2)\, e^{\sqrt{N}(\xi-t)}\, d\xi,$$

therefore

$$|K^{(2)}(v, t)| \leqq \frac{1}{N}(u^2 + t^2)(u^2 + v^2)\, e^{\sqrt{N}(v-t)}\frac{v - t}{1!},$$

and by induction, for the iterated kernel of rank l we have

$$|K^{(l)}(v, t)| \leqq \frac{1}{N^{l/2}}(u^2 + t^2)(u^2 + v^2)^{l-1}\, e^{\sqrt{N}(v-t)}\frac{(v - t)^{l-1}}{(l - 1)!},$$

and therefore for the resolvent kernel $\Gamma(v, t)$ of (26), we have the inequality

$$(27) \qquad |\Gamma(v, t)| \leqq \frac{1}{N^{1/2}}(u^2 + t^2)\, e^{\sqrt{N}(v-t) + N^{-1/2}(u^2+v^2)(v-t)}.$$

If we rewrite (25) in the form

$$\begin{aligned} \omega(v) = f_{2n}(0)\, \{&\cos [N^{1/2}u] \cosh [N^{1/2}v] - i \sin [N^{1/2}u] \sinh [N^{1/2}v]\} \\ &- \frac{\cosh [N^{1/2}v]}{N^{1/2}}\int_0^u \alpha^2 f_{2n}(\alpha) \sin [N^{1/2}(\alpha - u)]\, d\alpha \\ &+ i\frac{\sinh [N^{1/2}v]}{N^{1/2}}\int_0^u \alpha^2 f_{2n}(\alpha) \cos [N^{1/2}(\alpha - u)]\, d\alpha, \end{aligned}$$

and use (11_1) and the expression obtained from it by differentiating with respect to x, namely,

$$(11_3)\ f'_{2n}(x) = -N^{1/2}f_{2n}(0) \sin [N^{1/2}x] + \int_0^x \alpha^2 f_{2n}(\alpha) \cos [N^{1/2}(\alpha - x)]\, d\alpha$$

we get

$$\omega(v) = \cosh\,[N^{1/2} v]\, f_{2n}(u) + i\,\frac{\sinh\,[N^{1/2} v]}{N^{1/2}}\, f'_{2n}(u).$$

By the argument of Sec. 5.1, we get from (11_1) and (11_3)

$$(28)\quad |f_{2n}(u)| \leqq |f_{2n}(0)| + \frac{|u|^{5/2}}{N^{1/2}}, \qquad \left|\frac{f'_{2n}(u)}{N^{1/2}}\right| \leqq |f_{2n}(0)| + \frac{|u|^{5/2}}{N^{1/2}},$$

and since

$$|f_{2n}(0)| = \frac{1}{\sqrt[4]{\pi}}\sqrt{\frac{(2n-1)!!}{(2n)!!}} = \frac{\sqrt{2}}{\sqrt{\pi}}\,\frac{1}{\sqrt[4]{4n+2\theta}} < \frac{2^{3/4}}{\sqrt{\pi}}\,\frac{1}{N^{1/4}}$$

we have

$$|\omega(v)| \leqq \frac{e^{\sqrt{N}v}}{N^{1/4}} A(u^{1/2})$$

where

$$(29)\qquad \boxed{A(u^{1/2}) = 2\left[\frac{2^{3/4}}{\sqrt{\pi}} + \frac{|u|^{5/2}}{N^{1/4}}\right]},$$

but the $\varphi(v) = f_{2n}(u+iv)$ which satisfies (26) has the expression

$$f_{2n}(u+iv) = \omega(v) + \int_0^v \Gamma(v,t)\,\omega(t)\,dt,$$

whence by (27)

$$|f_{2n}(u+iv)| \leqq \frac{e^{\sqrt{N}v}}{N^{1/4}} A(u^{1/2}) \left[1 + \int_0^v \frac{u^2+v^2}{N^{1/2}}\, e^{\frac{(u^2+v^2)(v-t)}{N^{1/2}}}\, dt\right]$$

and finally

$$(30)\qquad \boxed{|f_{2n}(u+iv)| \leqq \frac{e^{\sqrt{N}|v|}}{N^{1/4}} A(u^{1/2})\, e^{\frac{(u^2+v^2)|v|}{N^{1/2}}}, \quad N = 4n+1},$$

where $A(u^{1/2})$ has the expression (29). If we use the inequality of H. Hille [47a] we could infer from (1) that

$$|f_n(u+iv)| \leqq \pi^{-1/4}(n!)^{1/2}[2^{1/2}R]^{-n}e^{-(u^2-v^2)/2+R(u^2+v^2)^{1/2}+n/2},$$

where

$$R=[(u^2+v^2+2n)^{1/2}-(u^2+v^2)^{1/2}]/2.$$

From (9_1) and noting that

$$\left|\frac{f'_{2n+1}(0)}{\sqrt{4n+3}}\right|=\frac{1}{\sqrt[4]{\pi}}\sqrt{\frac{4n+2}{4n+3}}\sqrt{\frac{(2n-1)!!}{(2n)!!}}<\frac{2^{3/4}}{\sqrt{\pi}}\frac{1}{(4n+3)^{1/4}}$$

we have in general

$$(30') \qquad \boxed{|f_n(u+iv)| \leqq \frac{e^{\sqrt{N}|v|}}{N^{1/4}}A(u^{1/2})\,e^{\frac{(u^2+v^2)|v|}{N^{1/2}}}, \quad N=2n+1},$$

where $A(u^{1/2})$ is given by (29).

3b. (15_1) gives a bound for the entire transcendental function $t(2n, x)$ which figures in (14_1), for real values of x, but another bound is necessary in the complex domain [cf. Sec. 4.6a].

To this end, we use the fact that in the Volterra integral equation of the second kind (11_1) the kernel $K(x, \alpha)$ has the expression $-N^{-1/2}\alpha^2 \sin[N^{1/2}(\alpha - x)]$, and by induction it can be proved that the iterated kernel of rank $l+1$, $K^{(l+1)}(x, \alpha)$, has the expression

$$(31) \qquad \begin{aligned} K^{(l+1)}(x,\alpha)=\frac{\alpha^2}{N^{(l+1)/2}}\,[(x-\alpha)Q^{(1)}_{3l-1}(x,\alpha)\cos[N^{1/2}(x-\alpha)] \\ +\,Q^{(2)}_{3l}(x,\alpha)\sin[N^{1/2}(x-\alpha)]] \end{aligned}$$

where $Q^{(1)}_{3l-1}(x,\alpha)$, $Q^{(2)}_{3l}(x,\alpha)$ denote polynomials in the variables x and α of degrees $3l-1$, $3l$, respectively, with coefficients uniformly bounded with respect to N.

From (11_1) follows

$$\begin{aligned} \frac{f_{2n}(x)}{f_{2n}(0)} &= \cos[N^{1/2}x]+\int_0^x K(x,\alpha)\cos[N^{1/2}\alpha]d\alpha \\ &\quad +\sum_{j=2}^{l}\int_0^x K^{(j)}(x,\alpha)\cos[N^{1/2}\alpha]d\alpha+\int_0^x K^{(l+1)}(x,\alpha)\frac{f_{2n}(\alpha)}{f_{2n}(0)}d\alpha, \end{aligned}$$

whence by (31) and the fact that

$$\int_0^x K(x, \alpha) \cos [N^{1/2}\alpha] d\alpha = \frac{x^3}{6N^{1/2}} \sin [N^{1/2}x] + \frac{1}{4N} \left[x^2 \cos [N^{1/2}x] - \frac{x \sin [N^{1/2} x]}{N^{1/2}} \right]$$

we have for the term $t(2n, x)$ which figures in (14_1) the expression

$$(32) \qquad \boxed{t(2n, x) = \frac{1}{N^{1/4}} \{P^{(1)}_{3l+3}(x) \cos [N^{1/2}x] + P^{(2)}_{3l+3}(x) \sin [N^{1/2}x]\} + \frac{\bar{t}(2n, x)}{N^{(l+1)/2}}},$$

where

$$\bar{t}(2n, x) = N^{(l+1)/2} \int_0^x K^{(l+1)}(x, \alpha) f_{2n}(\alpha) \, d\alpha, \qquad N = 4n + 1$$

and $P^{(1)}_{3l+3}(x)$, $P^{(2)}_{3l+3}(x)$ denote polynomials in x of degree $3l + 3$, with coefficients uniformly bounded with respect to n.

For $x = u + iv$, u and v real, we have

$$\bar{t}(2n, u + iv) = N^{(l+1)/2} \int_0^u K^{(l+1)}(u + iv, \alpha) f_{2n}(\alpha) d\alpha + iN^{(l+1)/2} \int_0^v K^{(l+1)}(u + iv, u + i\alpha) f_{2n}(u + i\alpha) \, d\alpha$$

where the two integrals are taken respectively along the segments $(0, u)$ and $(0, v)$ of the real axis.

Now by (31), and the first of (28) applied to the first integral and (30) applied to the second integral on the right side of (32_2), we get

$$(33) \qquad \boxed{| \bar{t}(2n, u+iv) | \leq \frac{e^{\sqrt{N}|v|}}{N^{1/4}} \left[\sum_{r=0}^{6l+11} c_r (u^2+v^2)^{r/4} \right] e^{\frac{(u^2+v^2|v|}{N^{1/2}}}}$$

where the c_r are positive constants, depending only on l.

3c. Substituting (14_1) in (11_1) we have for $t(2n, x)$ the integral equation

$$t(2n, x) = f_{2n}(0) \cos [N^{1/2}x] \left\{-\frac{x^6}{72} + \frac{5x^4}{48N} - \frac{15x^2}{48N^2} + \frac{x^2}{4}\right\}$$

$$+ f_{2n}(0) \frac{\sin [N^{1/2}x]}{N^{1/2}} \left\{\frac{x^5}{24} - \frac{5x^3}{24N} + \frac{15x}{48N^2} - \frac{x}{4}\right\}$$

$$- \frac{1}{N^{1/2}} \int_0^x \alpha^2 t(2n, x) \sin [N^{1/2}(\alpha - x)]\, d\alpha,$$

and substituting (32) for $t(2n, x)$ on the left side we get

$$(34) \quad \boxed{\bar{t}(2n, x) = \bar{t}_1(2n, x) \cos [N^{1/2}x] + \bar{t}_2(2n, x) \sin [N^{1/2}x]}$$

where $\bar{t}_1(2n, x)$, $\bar{t}_2(2n, x)$ are two entire transcendental functions in x, which both satisfy the inequality (33). Formulas (32), (33), (34) also hold if we change $2n$ to $2n + 1$ and take $N = 4n + 3$.

6. *Formulas for Asymptotic Approximation of the Polynomials* $L_n^{(\alpha)}(x)$

1a. We proceed to determine a bound for $L_n^{(\alpha)}(x)$ for $\alpha > -\frac{1}{2}$ and $x > 0$.

If in (2_1) of Sec. 4 we let $t = u\sqrt{x}$ we get

$$(1) \quad L_n^{(\alpha)}(x) = \frac{(-1)^n \Gamma(n + \alpha + 1)}{\sqrt{\pi}(2n)!\,\Gamma(\alpha + 1/2)} \int_{-1}^{1} (1 - u^2)^{\alpha - \frac{1}{2}} H_{2n}(u\sqrt{x})\, du, \quad (\alpha > -1/2),$$

whence by (10) of Sec. 5.1

$$H_{2n}(u\sqrt{x}) = e^{(1/2)xu^2} [2^{2n}(2n)!]^{1/2} \left[(-1)^n \frac{(2n-1)!!}{(2n)!!}\right]^{1/2} \cos [u\sqrt{(4n+1)x}] - \frac{\pi^{1/4}}{\sqrt{4n+1}} h(2n, u\sqrt{x})\Big]$$

where

$$|\, h(2n, u\sqrt{x}) \,| \leqq x^{5/4} \,|\, u \,|^{5/2}.$$

Applying Stirling's formula (cf. Appendix Th. 20) we have

$$\lim_{n\to\infty} \frac{1}{n^\alpha} \frac{\Gamma(n+\alpha+1)}{(2n)!} [2^{2n}(2n)!]^{1/2} \left[\frac{(2n-1)!!}{(2n)!!}\right]^{1/2} = 1,$$

$$\lim_{n\to\infty} \frac{1}{n^{\alpha-\frac{1}{4}}} \frac{\Gamma(n+\alpha+1)}{(2n)!} [2^{2n}(2n)!]^{1/2} \frac{\pi^{1/4}}{\sqrt{4n+1}} \frac{1}{\sqrt{\pi}} = \tfrac{1}{2},$$

and therefore there exist two constants $A(\alpha)$ and $B(\alpha)$, depending only on α such that for any n we have

$$\frac{\Gamma(n+\alpha+1)}{\pi^{1/2}(2n)!\Gamma(\alpha+1/2)} [2^{2n}(2n)!]^{1/2} \left[\frac{(2n-1)!!}{(2n)!!}\right]^{1/2} \leqq A(\alpha)n^\alpha,$$

$$\frac{\Gamma(n+\alpha+1)}{\pi^{1/2}(2n)!\Gamma(\alpha+1/2)} [2^{2n}(2n)!]^{1/2} \frac{\pi^{1/4}}{\sqrt{4n+1}} \leqq B(\alpha)n^{\alpha-\frac{1}{4}}.$$

We have also

$$\left|\int_{-1}^{1} e^{xu^2/2}(1-u^2)^{\alpha-\frac{1}{2}} \cos[u\sqrt{(4n+1)x}]\, du\right|$$

$$= 2\int_0^1 e^{xu^2/2}(1-u^2)^{\alpha-\frac{1}{2}} \cos[u\sqrt{(4n+1)x}]\, du$$

$$\leqq 2e^{x/2}\int_0^1 (1-u^2)^{\alpha-\frac{1}{2}}\, du = 2e^{x/2}C(\alpha)$$

where $C(\alpha)$ is a constant depending only on α. Moreover,

$$\left|\int_{-1}^{1} e^{xu^2/2}(1-u^2)^{\alpha-\frac{1}{2}} h(2n, u\sqrt{x})\, du\right|$$

$$\leqq 2x^{5/4}e^{x/2}\int_0^1 (1-u^2)^{\alpha-\frac{1}{2}} u^{5/2}\, du = 2D(\alpha)x^{5/4}e^{x/2}$$

where $D(\alpha)$ is a constant depending only on α, and therefore

(2) $$|L_n^{(\alpha)}(x)| \leqq 2e^{x/2}n^\alpha[A(\alpha)C(\alpha) + n^{-1/4}B(\alpha)D(\alpha)x^{5/4}]$$

and from this inequality it follows that for all x in the finite interval $[0, l]$, there exists a constant T such that

(3) $$|L_n^{(\alpha)}(x)| \leqq Tn^\alpha, \quad [0 \leqq x \leqq l, \quad n = 0, 1, \ldots, \alpha > -1/2].$$

1b. From (3) it follows that $\lim_{n\to\infty} L_n^{(\alpha)}(x) = 0$ for $-\frac{1}{2} < \alpha < 0$. We can also show that for a fixed $x > 0$ and for $\alpha > -\frac{1}{2}$,

we have

$$\lim_{n\to\infty} L_n^{(\alpha)}(x)/n^\alpha = 0, \qquad (x > 0,\ \alpha > -1/2). \tag{4}$$

In fact, by the second theorem of the mean (Appendix, Th. 18) if we assume that x is in the interval $[\varepsilon, l], 0 < \varepsilon < l$, $[0 \leqq \xi \leqq 1]$, we have

$$\left|\int_{-1}^{1} e^{xu^2/2}(1-u^2)^{\alpha-\frac{1}{2}} \cos\,[u\sqrt{(4n+1)x}]\,du\right|$$
$$= 2e^{x/2}\left|\int_{\xi}^{1} (1-u^2)^{\alpha-\frac{1}{2}} \cos\,[u\sqrt{(4n+1)x}]\,du\right| = 2e^{x/2}o(1)$$

as $n \to \infty$ uniformly for all x in $[\varepsilon, l]$, (Chap. II, Sec. 4.3, Theorem 12), whence

$$|\,L_n^{(\alpha)}(x)\,| \leqq 2e^{x/2} n^\alpha\,[A(\alpha)o(1) + n^{-1/4}\,B(\alpha)D(\alpha)x^{5/4}]. \tag{5}$$

Therefore, (4) holds uniformly for all x in $[\varepsilon, l]$.

2. In Sec. 6.4, a more precise asymptotic formula for $L_n^{(\alpha)}(x)$ will be derived. To this end, in this article and the next one, two formulas will be derived for the asymptotic evaluation of a certain integral, which will figure in the derivation of Sec. 4. (For the material of Sections 6.2 and 6.3 cf. G. Sansone; "La formula di approssimazione asintotica dei polinomi di Tchebychef-Laguerre col procedimento di J. V. Uspensky," *Math. Z.*, 53 (1950) pp. 97—105).

2a. Let $\alpha > -1/2$, $m > 0$, $\lambda > 0$, let $f(x)$ be a polynomial in x of degree not exceeding three (actually $f(x)$ could be assumed to be of any degree) and define $G(x)$ by

$$G(x) = e^{m^2x^2/2} f(mx)(1-x^2)^{\alpha-\frac{1}{2}}\,e^{i\lambda mx} \tag{6}$$

Then we seek an asymptotic expression for the integral

$$I = \int_{-1}^{1} G(x)dx. \tag{7}$$

For every $T > 1$ consider the polygonal contour $C(T)$ with consecutive vertices at points -1, $-1+iT$, $1+iT$, 1, which together with the segment $(-1, 1)$, determines a rectangle R. Recalling that $(1-x^2)^{\alpha-\frac{1}{2}}$ denotes the principal value (i.e., the value which $= 1$ for $x = 0$), then $G(x)$ is holomorphic except possibly at the points $x = \pm 1$. Since $\alpha - 1/2 > -1$ we may

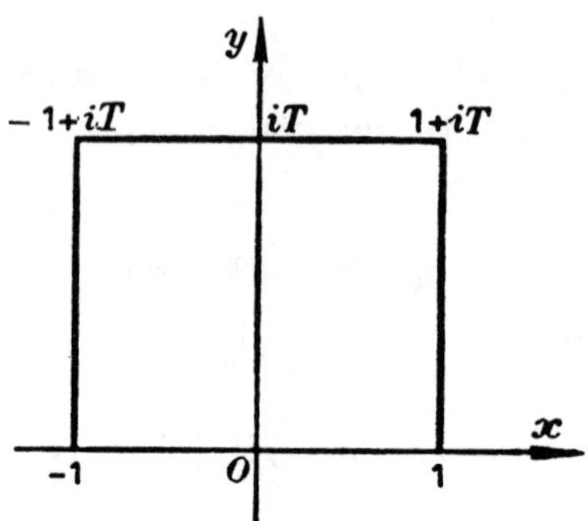

Fig. 15.

apply Cauchy's integral theorem to get

$$I = \int_{C(T)} G(x)\,dx,$$

or

$$I = \bar{I}_1 + \bar{I}_2 + \bar{I}_3 \tag{8}$$

where

$$\bar{I}_1 = \int_{-1}^{-1+iT} G(x)dx, \quad \bar{I}_2 = \int_{1+iT}^{1} G(x)dx, \quad \bar{I}_3 = \int_{-1+iT}^{1+iT} G(x)dx.$$

Clearly

$$\bar{I}_3 = e^{-\lambda mT} \int_{-1}^{1} e^{m[x^2-T^2+2ixT]/2} f[m(x+iT)]\,[1-(x+iT)^2]^{\alpha-\frac{1}{2}}\, e^{i\lambda mx} dx.$$

Then, making use of the fact that

$$e^{m^2(x^2-T^2)/2} \leqq 1, \quad | e^{im^2xT} | = 1, \quad | e^{i\lambda mx} | = 1,$$

that for $\alpha - \frac{1}{2} \geqq 0$

$$| [1-(x+iT)^2]^{\alpha-\frac{1}{2}} | = | 1-(x+iT)^2 |^{\alpha-\frac{1}{2}} \leqq [1+(1+T)^2]^{\alpha-\frac{1}{2}},$$

that for $-\frac{1}{2} < \alpha < \frac{1}{2}$

$$\begin{aligned} | [1-(x+iT)^2]^{\alpha-\frac{1}{2}} | &= | (1-x^2)+T^2-2ixT |^{\alpha-\frac{1}{2}} \\ &= [\{(1-x^2)+T^2\}^2+4x^2T^2]^{(\alpha-\frac{1}{2})/2} < T^{2(\alpha-\frac{1}{2})} < 1, \end{aligned}$$

and the fact that $\int_{-1}^{1} | f[m(x+iT)] | \, dx$ is majorized by a polynomial in T, we get

(9) $$\lim_{T\to\infty} \bar{I}_3 = 0.$$

If we let $x = -1 + it$, then $1 - x^2 = (1 - x)(1 + x) = 2it(1 - it/2)$, and as x describes the segment from -1 to $-1 + iT$, we have

$$\bar{I}_1 = 2^{\alpha-\frac{1}{2}} e^{m^2/2-i[\lambda m-(2\alpha+1)\pi/4]} \times \int_0^T e^{-im^2t-m^2t^2/2} e^{-\lambda mt} f(-m + imt) t^{\alpha-\frac{1}{2}} (1 - it/2)^{\alpha-\frac{1}{2}} \, dt,$$

whence by the transformation $\lambda mt = u$

$$\bar{I}_1 = \frac{2^{\alpha-\frac{1}{2}}}{(\lambda m)^{\alpha+\frac{1}{2}}} e^{m^2/2-i[\lambda m-(2\alpha+1)\pi/4]}$$

$$\times \int_0^{\lambda mT} e^{-imu/\lambda-u^2/(2\lambda^2)} e^{-u} f[-m + iu/\lambda] u^{\alpha-\frac{1}{2}} (1-iu/(2\lambda m))^{\alpha-\frac{1}{2}} x \, du$$

and since this last integral is convergent as $T \to \infty$ it follows from (8) and (9) that

(10) $$I = I_1 + I_2$$

where

(11_1) $$I_1 = \frac{2^{\alpha-\frac{1}{2}}}{(\lambda m)^{\alpha+\frac{1}{2}}} e^{m^2/2-i[\lambda m-(2\alpha+1)\pi/4]}$$

$$\times \int_0^\infty e^{-imu/\lambda-u^2/(2\lambda^2)} e^{-u} f[- m + iu / \lambda] u^{\alpha-\frac{1}{2}} (1 - iu / (2 \lambda m))^{\alpha-\frac{1}{2}} \, du,$$

(11_2) $$I_2 = \frac{2^{\alpha-\frac{1}{2}}}{(\lambda m)^{\alpha+\frac{1}{2}}} e^{m^2/2+i[\lambda m-(2\alpha+1)\pi/4]}$$

$$\times \int_0^\infty e^{imu/\lambda-u^2/(2\lambda^2)} e^{-u} f[m + iu/\lambda] u^{\alpha-\frac{1}{2}} (1 + iu/(2\lambda m))^{\alpha-\frac{1}{2}} \, du.$$

2b. Using Maclaurin's series with the remainder, we have

$$e^{imu/\lambda} = \cos\frac{mu}{\lambda} + i \sin\frac{mu}{\lambda} = 1 - \frac{m^2u^2}{2\lambda^2} \cos\frac{m\bar{u}}{\lambda} + i\frac{mu}{\lambda} - i\frac{m^3u^3}{6\lambda^3} \sin\frac{m\bar{\bar{u}}}{\lambda},$$

whence

$$(12)\qquad e^{imu/\lambda} = 1 + i\frac{mu}{\lambda} + \frac{u^2}{2\lambda^2} g_1(m, u, \lambda), \qquad | g_1(m, u, \lambda) | \leqq m^2 \left(1 + \frac{mu}{3\lambda}\right).$$

Clearly also

$$(13)\qquad e^{-u^2/2\lambda^2} = 1 - \frac{u^2}{2\lambda^2} g_2(u, \lambda), \qquad | g_2(u, \lambda) | < 1;$$

$$(14)\qquad f(m + iu/\lambda) = f(m) + i\frac{u}{\lambda} f'(m) + \frac{u^2}{2\lambda^2} g_3(u, \lambda, m), \qquad | g_3(u, \lambda, m) | \leqq | f''(m) | + \frac{u}{3\lambda} | f'''(m) | .$$

For $x > 0$, the principal value of $(1 + ix)^\mu$ is given by

$$(15)\quad (1+ix)^\mu = (1 + x^2)^{\mu/2} e^{i\mu \arctan x}, \quad (x > 0,\ 0 < \arctan x < \pi/2);$$

now $(1 + x^2)^{\mu/2} = 1 + (\mu/2)x^2(1 + \theta_1 x^2)^{\mu/2-1}$, $(0 < \theta_1 < 1)$, also $(1 + \theta_1 x^2)^{\mu/2-1} \leqq 1$ for $\mu \leqq 2$ and $(1 + \theta_1 x^2)^{\mu/2-1} < (1 + x^2)^{\mu/2-1}$ for $\mu > 2$. Consequently

$$(16)\quad (1+x^2)^{\mu/2} = 1 + \theta_2 \frac{|\mu|}{2} x^2 (1+x^2)^{|\mu/2-1|}, \text{ where } 0 < |\theta_2| < 1.$$

Moreover, if $0 < \arctan x < x$, we have by (12)

$$e^{i\mu \arctan x} = 1 + i\mu \arctan x + \frac{\mu^2}{2}\theta_3 x^2 \left[1 + \frac{|\mu|\, x}{3}\right], \quad 0 < |\theta_3| < 1,$$

and since

$$\arctan x = x - x^3 \frac{\bar{\theta}_3}{(1 + \bar{\theta}_3^2 x^2)^2} = x + \theta_4 x, \quad 0 < \bar{\theta}_3,\ |\theta_4| < 1,$$

$$(17)\quad e^{i\mu \arctan x} = 1 + i\mu x + \theta_5 \frac{\mu^2 x^2}{2} \left\{\frac{2x}{|\mu|} + 1 + |\mu| \frac{x}{3}\right\}, \qquad 0 < |\theta_5| < 1,$$

whence by (15), (16), (17)

$$(18)\quad (1+ix)^\mu = \left[1 + i\mu x + \theta_5 \frac{\mu^2 x^2}{2}\left\{\frac{2x}{|\mu|} + 1 + |\mu|\frac{x}{3}\right\}\right] \times \left[1 + \theta_2 \frac{|\mu|}{2} x^2 (1+x^2)^{|\mu/2-1|}\right],$$

and, in particular, denoting by ν_1 an integer not less than $|(\alpha - \frac{1}{2})/2 - 1|$, we have

$$(1 + iu/(2\lambda m))^{\alpha-\frac{1}{2}} = \left[1 + i(\alpha - \tfrac{1}{2})\frac{u}{2\lambda m} + \theta_5 |\alpha - \tfrac{1}{2}| \frac{u^2}{8\lambda^2 m^2}\left\{\frac{u}{\lambda m} + |\alpha - \tfrac{1}{2}| + (\alpha - \tfrac{1}{2})^2 \frac{u}{6\lambda m}\right\}\right] \times \left[1 + \tfrac{1}{2}|\alpha - \tfrac{1}{2}|\theta_1 \frac{u^2}{4\lambda^2 m^2}\left(1 + \frac{u^2}{4\lambda^2 m^2}\right)^{\nu_1}\right]$$

and therefore for $\lambda m \geqq 1$

$$(14')\quad (1 + iu/(2\lambda m))^{\alpha-\frac{1}{2}} = 1 + i(\alpha - \tfrac{1}{2})\frac{u}{2\lambda m} + \frac{u^2}{\lambda^3 m^2} g(u, \lambda, m)$$

$$|g(u, \lambda, m)| \leqq g(u)$$

where $g(u)$ is a polynomial in u of degree $2\nu_1 + 3$ with non-negative coefficients depending only on α.

If we let

$$(19)\quad \varphi(u) = e^{imu/\lambda - u^2/(2\lambda^2)} f(m + iu/\lambda)(1 + iu/(2\lambda m))^{\alpha-\frac{1}{2}}$$

then

$$(20)\quad \varphi(0) = f(m), \quad \varphi'(0) = \frac{i[mf'(m) + \{m^2 + (2\alpha - 1)/4\} f(m)]}{\lambda m}$$

and if $\lambda m \geqq 1$, $\lambda \geqq 1$, we have by (12), (13), (14), (14'),

$$(21)\quad \varphi(u) = \varphi(0) + u\varphi'(0) + \tau \frac{u^2 \omega(u)}{2\lambda^2}, \qquad |\tau| < 1,$$

$$(22)\quad \omega(u) = \frac{1}{m^2} \sum_{k=0}^{6} \varphi_k(u) m^k$$

where the $\varphi_k(u)$ denote polynomials in u of degree not exceeding $2\nu_1 + 11$ with nonnegative coefficients depending only on α.

By (19) and (21) the integral (11_2) becomes

$$I_2 = \frac{2^{\alpha-\frac{1}{2}}}{(\lambda m)^{\alpha+\frac{1}{2}}} e^{m^2/2+i[\lambda m-(2\alpha+1)\pi/4]}$$
$$\times \int_0^\infty e^{-u} u^{\alpha-\frac{1}{2}} \left[\varphi(0) + u\varphi'(0) + \tau \frac{u^2\omega(u)}{2\lambda^2}\right] du;$$

but

$$\int_0^\infty e^{-u} u^{\alpha-\frac{1}{2}} du = \Gamma(\alpha + 1/2), \quad \int_0^\infty e^{-u} u^{\alpha+\frac{1}{2}} du = (\alpha+1/2)\Gamma(\alpha+1/2),$$

$$\frac{1}{2}\int_0^\infty e^{-u} u^{\alpha+3/2} \omega(u)\, du = \frac{1}{2m^2} \sum_{k=0}^{6} m^k \int_0^\infty e^{-u} u^{\alpha+3/2} \varphi_k(u)\, du$$
$$= \frac{\Gamma(\alpha + 1/2)}{2m^2} \sum_{k=0}^{6} \bar{q}_k(\alpha)\, m^k,$$

where the $\bar{q}_k(\alpha)$ are nonnegative constants depending only on α. Consequently

$$I_2 = \frac{2^{\alpha-\frac{1}{2}}\Gamma(\alpha + 1/2)}{(\lambda m)^{\alpha+\frac{1}{2}}} e^{m^2/2+i[\lambda m-(2\alpha+1)\pi/4]}$$
$$\times \left[f(m) + \frac{i[mf'(m) + \{m^2 + (2\alpha - 1)/4\} f(m)](\alpha + 1/2)}{\lambda m}\right.$$
$$\left. + \frac{r_1}{2\lambda^2 m^2} \sum_{k=0}^{6} \bar{q}_k(\alpha)\, m^k\right], \qquad |r_1| < 1.$$

Similarly,

$$I_1 = \frac{2^{\alpha-\frac{1}{2}}\Gamma(\alpha + 1/2)}{(\lambda m)^{\alpha+\frac{1}{2}}} e^{m^2/2-i[\lambda m-(2\alpha+1)\pi/4]}$$
$$\times \left[f(-m) - \frac{i[-mf'(-m) + \{m^2 + (2\alpha-1)/4\} f(-m)]\,(\alpha+1/2)}{\lambda m}\right.$$
$$\left. + \frac{r_2}{2\lambda^2 m^2} \sum_{k=0}^{6} \bar{\bar{q}}_k(\alpha) m^k\right], \qquad |r_2| < 1.$$

Now, if we let $x = \cos\varphi$ in the integral (7) then by (10) we have for $\alpha > -\frac{1}{2}$, $m > 0$, $\lambda \geqq 1$, $\lambda m \geqq 1$, the formula of Uspensky:

$$\int_0^\pi e^{(m^2/2)\cos^2\varphi} f(m\cos\varphi)\, e^{i\lambda m\cos\varphi} \sin^{2\alpha}\varphi\, d\varphi$$

$$= \frac{2^{\alpha+\frac{1}{2}}\Gamma(\alpha+1/2)}{(\lambda m)^{\alpha+\frac{1}{2}}} e^{m^2/2}\left\{\frac{f(m)+f(-m)}{2}\cos[\lambda m-(2\alpha+1)\pi/4]\right.$$

$$(23)\qquad + i\frac{f(m)-f(-m)}{2}\sin[\lambda m-(2\alpha+1)\pi/4]$$

$$+\frac{F(-m)-F(m)}{2\lambda}\sin[\lambda m-(2\alpha+1)\pi/4]$$

$$\left.+ i\frac{F(m)+F(-m)}{2\lambda}\cos[\lambda m-(2\alpha+1)\pi/4]+\frac{r}{\lambda^2 m^2}\sum_{k=0}^{6} q_k(\alpha)m^k\right\},$$

(in the formula of Uspensky [[115], p. 607] the complementary term is given in a different form) where $|r| < 1$, and the $q_k(\alpha)$ are constants (positive or zero) depending only on α, and

$$(24)\qquad F(m) = \frac{\{mf'(m) + [m^2 + (2\alpha-1)/4]f(m)\}(\alpha+1/2)}{m}.$$

2c. If, instead of assuming $f(x)$ is a polynomial of the third degree as we did in 2a, it is assumed that $f(x)$ is of degree p in x, then the last term on the right side of (23) must be written in the form

$$\frac{r}{\lambda^2 m^2}\sum_{k=0}^{p+3} q_k(\alpha)\, m^k.$$

3. Another inequality for the integral

$$I = \int_{-1}^{1} e^{m^2x^2/2} f(mx)(1-x^2)^{\alpha-\frac{1}{2}} e^{i\lambda m x}\, dx$$

will now be established under the assumption that

$$(25)\qquad \alpha > -1/2,\quad m > 0,\quad \lambda \geqq 1,\quad 1/\lambda \leqq m \leqq \lambda^{\frac{1}{3}-\varepsilon},$$
$$0 < \varepsilon < 1/3,$$

and that $f(x)$ is transcendental and satisfies the inequality [u, v real]

$$(26) \qquad |f(u+iv)| \leqq \frac{e^{\lambda|v|}}{\lambda^{l+1}} P_{6l+11}[(u^2+v^2)^{1/4}]\, e^{(u^2+v^2)|v|/\lambda}$$

where $P_{6l+11}[(u^2+v^2)^{1/4}]$ is a polynomial in the variable $(u^2+v^2)^{1/4}$ with coefficients positive or zero, independent of λ and m, of degree $6l+11$, and the integer l satisfies the inequality

$$(27) \qquad l \geqq \frac{1}{3\tau}\left[\frac{13}{2}\left(\frac{1}{3}-\tau\right)+\nu\left(\frac{4}{3}-\tau\right)\right],$$

where $0 < \tau < \varepsilon$, and $\nu \geqq 2\,|\,\alpha - \frac{1}{2}|$. We shall prove that under these assumptions there exists a constant C independent of λ and m such that

$$(28) \qquad |I| \leqq C\,\frac{2^{\alpha+\frac{1}{2}}\Gamma(\alpha+1/2)\, e^{m^2/2}}{(\lambda m)^{\alpha+\frac{1}{2}}}.$$

Consider the polygonal contour $C(T)$, $T = \lambda^{\frac{1}{3}-\tau}/m$, with the consecutive vertices at the points -1, $-1+iT$, $1+iT$, 1. The polygonal line $C(T)$ together with the interval $[-1, 1]$ determine a closed contour [cf. fig. 15]. Applying Cauchy's theorem it follows that I can be split up into the sum of three terms

$$(29_1) \quad e^{-\lambda mT}\int_{-1}^{1} e^{m^2(x^2-T^2)/2}\, e^{im^2xT} f(mx+imT)[1-(x+iT)^2]^{\alpha-\frac{1}{2}}\, e^{i\lambda mx}dx,$$

$$(29_2) \quad \frac{2^{\alpha-\frac{1}{2}}}{(\lambda m)^{\alpha+\frac{1}{2}}}\, e^{m^2/2-i[\lambda m-(2\alpha+1)\pi/4]} \times \int_0^{\lambda^{4/3-\tau}} e^{-imu/\lambda-u^2/(2\lambda^2)}\, e^{-u} f[-m+iu/\lambda]u^{\alpha-\frac{1}{2}}(1-iu/2\lambda m)^{\alpha-\frac{1}{2}}du,$$

$$(29_3) \quad \frac{2^{\alpha-\frac{1}{2}}}{(\lambda m)^{\alpha+\frac{1}{2}}}\, e^{m^2/2+i[\lambda m-(2\alpha+1)\pi/4]} \times \int_0^{\lambda^{4/3-\tau}} e^{imu/\lambda-u^2/(2\lambda^2)}\, e^{-u} f[m+iu/\lambda]\, u^{\alpha-\frac{1}{2}}(1+iu/(2\lambda m))^{\alpha-\frac{1}{2}}du.$$

Since for $|x| \leqq 1$

$$m^2(x^2 - T^2) \leqq m^2 - m^2T^2 \leqq \lambda^{\frac{2}{3}-2\varepsilon} - \lambda^{\frac{2}{3}-2\tau} = -\lambda^{\frac{2}{3}-2\tau}(1 - \lambda^{-2(\varepsilon-\tau)}),$$

$$(m^2x^2 + m^2T^2)^{\frac{1}{4}} \leqq (m^2 + m^2T^2)^{\frac{1}{4}} \leqq (\lambda^{\frac{2}{3}} + \lambda^{\frac{2}{3}})^{\frac{1}{4}} = 2^{\frac{1}{4}}\lambda^{\frac{1}{6}},$$

$$(m^2x^2 + m^2T^2)\,mT/\lambda \leqq 2\lambda^{\frac{2}{3}+\frac{1}{3}-\tau-1} = 2/\lambda^{\tau},$$

$$T \leqq \lambda^{\frac{1}{3}-\tau}/(1/\lambda) < \lambda^{\frac{4}{3}}, \quad T \geqq \lambda^{\frac{1}{3}-\tau}/\lambda^{\frac{1}{3}-\varepsilon} > 1,$$

$$|1 - (x + iT)^2|^{\alpha-\frac{1}{2}} \leqq [1 + (1 + T)^2]^{|\alpha-\frac{1}{2}|} \leqq [1 + (1 + \lambda^{\frac{4}{3}})^2]^{|\alpha-\frac{1}{2}|},$$

we get, in view of (26), that the term (29_1) is majorized by

$$\frac{2}{\lambda^{l+1}}\, e^{-\lambda^{\frac{2}{3}-2\tau}(1-\lambda^{-2(\varepsilon-\tau)})}\, P_{6l+11}[2^{\frac{1}{4}}\lambda^{\frac{1}{6}}]\, e^{2/\lambda^{\tau}}[1 + (1+\lambda^{\frac{4}{3}})^2]^{|\alpha-\frac{1}{2}|}$$

which is an infinitesimal exponential as $\lambda \to \infty$ and the result of multiplying it by $(\lambda m)^{\alpha+\frac{1}{2}}/e^{m^2/2} < \lambda^{(4/3-\varepsilon)\,|\alpha+\frac{1}{2}|}$ is likewise an infinitesimal.

Since for $u \leqq \lambda^{4/3-\tau}$ we have

$$m^2 + \frac{u^2}{\lambda^2} \leqq \lambda^{\frac{2}{3}-2\varepsilon} + \lambda^{\frac{2}{3}-2\tau} < 2\lambda^{\frac{2}{3}-2\tau},$$

$$P_{6l+11}\left[\left(m^2 + \frac{u^2}{\lambda^2}\right)^{1/4}\right] \leqq C_1\lambda^{(\frac{1}{3}-\tau)(6l+11)/2},$$

$$\left(m^2 + \frac{u^2}{\lambda^2}\right)\frac{u}{\lambda^2} \leqq 2\lambda^{\frac{2}{3}-2\tau}\,\lambda^{-\frac{2}{3}-\tau} = 2\lambda^{-3\tau} \leqq 2,$$

it follows by (26) that the term (29_2) does not exceed the product of $2^{\alpha+\frac{1}{2}}\, e^{m^2/2}(\lambda m)^{-[\alpha+\frac{1}{2}]}$ by the sum

$$C_1\frac{e^2}{2}\,\frac{\lambda^{(\frac{1}{3}-\tau)(6l+11)/2}}{\lambda^{l+1}}\int_0^1 u^{\alpha-\frac{1}{2}}\left(1 + \frac{u^2}{4}\right)^{|\alpha-\frac{1}{2}|/2} du$$
$$+ C_2\lambda^{(\frac{1}{3}-\tau)(6l+11)/2+(\frac{4}{3}-\tau)(\nu+1)}/\lambda^{l+1},$$

where C_1 and C_2 are constants.

By (27) we have $(1/3 - \tau)(6l + 11)/2 + (4/3 - \tau)(\nu + 1) \leqq l + 1$, and, therefore, an inequality of the form (28) also holds for (29_2). A similar inequality holds for the term (29_3). Thus (28) is proved.

4a. Uspensky's formula (2_2) of Sec. 4.2 relative to the representation of $L_n^{(\alpha)}(x)$ in terms of the functions $H_{2n}(x)$, the formula of asymptotic approximation (16_1) of $H_{2n}(x)$ of Sec. 5.2a, and formulas (23), (24), (28) yield the desired asymptotic expression for $L_n^{(\alpha)}(x)$ for the case $\alpha > -\frac{1}{2}$.

In fact we have by (2_2) of Sec. 4.2 and (16_1) of Sec. 5.2a.

$$
\begin{aligned}
L_n^{(\alpha)}(x) = {} & \frac{\Gamma(n+\alpha+1)}{\sqrt{\pi}\, n!\, \Gamma(\alpha+1/2)} \int_0^\pi e^{(x/2)\cos^2\varphi} \Big\{ \cos[\sqrt{Nx}\cos\varphi] \\
& + \frac{x^{3/2}\cos^3\varphi}{6N^{1/2}} \sin[\sqrt{Nx}\cos\varphi] + \frac{1}{f_{2n}(0)} \frac{t(2n, \sqrt{x}\cos\varphi)}{N} \Big\} \sin^{2\alpha}\varphi\, d\varphi, \\
& (N = 4n+1).
\end{aligned} \tag{30}
$$

If we let $m = \sqrt{x}$, $\lambda = N^{1/2}$, and $f \equiv 1$ in (23) and then equate the real parts of the two sides of the equation, we get for $\lambda m > 1$ or $x > 1/N$.

$$
\begin{aligned}
& \int_0^\pi e^{(x/2)\cos^2\varphi} \cos[\sqrt{Nx}\cos\varphi] \sin^{2\alpha}\varphi\, d\varphi \\
& = \frac{2^{\alpha-\frac{1}{2}}\Gamma(\alpha+1/2)\, e^{x/2}}{[(4n+1)x]^{(\alpha+\frac{1}{2})/2}} [\cos\sqrt{Nx} - (2\alpha+1)\pi/4] \\
& - \frac{1}{\sqrt{Nx}} [x + (2\alpha-1)/4](\alpha+1/2) \sin[\sqrt{Nx} - (2\alpha+1)\pi/4] \\
& + \frac{R_1}{Nx} \sum_{k=0}^{6} x^{k/2} q_{1,k}(\alpha), \qquad |R_1| \leqq 1.
\end{aligned} \tag{31_1}
$$

Similarly, if we let $m = \sqrt{x}$, $\lambda = N^{1/2}$, $f(m) \equiv m^3$ in (23) and then equate the imaginary parts of the two sides of the equation, we get

$$
\begin{aligned}
& \frac{1}{6N^{1/2}} \int_0^\pi e^{(x/2)\cos^2\varphi} x^{3/2} \cos^3\varphi \sin[\sqrt{Nx}\cos\varphi] \sin^{2\alpha}\varphi\, d\varphi \\
& = \frac{x^{3/2}}{6N^{1/2}} \frac{2^{\alpha+\frac{1}{2}}\Gamma(\alpha+1/2)\, e^{x/2}}{[(4n+1)x]^{(\alpha+\frac{1}{2})/2}} \sin[\sqrt{Nx} - (2\alpha+1)\pi/4] \\
& + \frac{R_2}{Nx} \sum_{k=0}^{6} x^{k/2} q_{2,k}(\alpha), \qquad |R_2| \leqq 1.
\end{aligned} \tag{31_2}
$$

By (32), (34) of Sec. 5.3d and by (23), (28), noting that $|1/f_{2n}(0)| = \sqrt{\pi}\sqrt[4]{4n+1}\sqrt[4]{(2n+\theta)/(4n+1)} < \sqrt{\pi}N^{1/4}$, we have

$$\left|\frac{1}{f_{2n}(0)}\int_0^\pi e^{(x/2)\cos^2\varphi}\, t(2n, \sqrt{x}\cos\varphi)\sin^{2\alpha}\varphi\, d\varphi\right|$$

$$= \left|\int_0^\pi e^{(x/2)\cos^2\varphi}\left[\frac{1}{f_{2n}(0)N^{1/4}}\{\cos[\sqrt{Nx}\cos\varphi]\, P^{(1)}_{3l+3}[x^{1/2}\cos\varphi]\right.\right.$$

(31₃) $$+\sin[\sqrt{Nx}\cos\varphi]\, P^{(2)}_{3l+3}[x^{1/2}\cos\varphi]\}$$

$$\left.\left.+\frac{1}{f_{2n}(0)N^{1/4}}\frac{N^{1/4}\bar{t}(2n, x^{1/2}\cos\varphi)}{N^{(l+1)/2}}\right]\sin^{2\alpha}\varphi\, d\varphi\right|$$

$$\leqq C\frac{2^{\alpha+\frac{1}{2}}\Gamma(\alpha+1/2)\, e^{x/2}}{[(4n+1)x]^{(\alpha+\frac{1}{2})/2}}\sum_{k=0}^{3(l+2)} q_{3,k}(\alpha)\, x^{k/2}, \qquad (C = \text{constant});$$

and by (30), (31_1), (31_2), (31_3) we obtain, for

$$\alpha > -1/2, \qquad \frac{1}{4n+1} \leqq x \leqq (4n+1)^{\frac{2}{3}-2\varepsilon}, \qquad 0 < \varepsilon < 1/3,$$

and for every integer l such that

(27) $$l \geqq \frac{1}{3\tau}\left[\frac{13}{2}\left(\frac{1}{3}-\tau\right)+\nu\left(\frac{4}{3}-\tau\right)\right]$$

where

$$0 < \tau < \varepsilon, \quad \nu \geqq 2\,|\,\alpha - 1/2\,|,$$

the following formula of asymptotic approximation for the polynomials $L_n^{(\alpha)}(x)$

(32) $$\boxed{\begin{aligned} &L_n^{(\alpha)}(x) \\ &= \frac{\Gamma(n+\alpha+1)}{\sqrt{\pi}\, n!}\frac{2^{\alpha+\frac{1}{2}}\, e^{x/2}}{[(4n+1)x]^{(\alpha+\frac{1}{2})/2}}\left\{\cos\left[\sqrt{(4n+1)x}-\frac{2\alpha+1}{4}\pi\right]\right. \\ &+\frac{\frac{x^2}{6}-(\alpha+\frac{1}{2})x-\frac{1}{2}(\alpha^2-\frac{1}{4})}{\sqrt{(4n+1)x}}\sin\left[\sqrt{(4n+1)x}-\frac{2\alpha+1}{4}\pi\right] \\ &\left.+\frac{R(n, x, \alpha)}{(4n+1)x}\sum_{k=0}^{3l+8} q_k(\alpha)\, x^{k/2}\right\}\end{aligned}},$$

where the $q_k(\alpha)$ are constants (positive or zero) depending only on α, and for fixed α, $| R(n, x, \alpha) | \leqq 1$ for

$$1/(4n+1) \leqq x \leqq (4n+1)^{\frac{2}{3}-2\epsilon}, \qquad n = 1, 2, \ldots .$$

4b. By (32) we have in particular for all x in the closed interval $[a, b]$, $0 < a < b$, and for $\alpha > -\frac{1}{2}$ Uspensky's formula [[115], p. 609]

$$\boxed{\begin{aligned} & L_n^{(\alpha)}(x) \\ & = \frac{\Gamma(n+\alpha+1)}{\sqrt{\pi}\, n!} \frac{2^{\alpha+\frac{1}{2}} e^{x/2}}{[(4n+1)x]^{(\alpha+\frac{1}{2})/2}} \left\{ \cos\left[\sqrt{(4n+1)x} - \frac{2\alpha+1}{4}\pi \right] \right. \\ & \left. + \frac{\frac{x^2}{6} - (\alpha+\frac{1}{2})x - \frac{1}{2}(\alpha^2 - \frac{1}{4})}{\sqrt{(4n+1)x}} \sin\left[\sqrt{(4n+1)x} - \frac{2\alpha+1}{4}\pi \right] \right\} \\ & + \frac{R(n, x, \alpha)}{4n+1} \end{aligned}} \tag{33}$$

where for α fixed, $R(n, x, \alpha)$ is uniformly bounded with respect to n, for all x in $[a, b]$, $[0 < a \leqq x \leqq b]$.

5. We shall now extend formula (32) and therefore also Uspensky's formula (33) to the case of any real α. The proof is by induction.

If, for example, $\alpha > -3/2$; we have [Sec. 1.5, formula (14_1)]

$$\frac{d}{dx} L_n^{(\alpha)}(x) = -L_{n-1}^{(\alpha+1)}(x), \qquad \frac{d^2}{dx^2} L_n^{(\alpha)}(x) = L_{n-2}^{(\alpha+2)}(x)$$

and therefore [Sec. 1.5, formula (11)]

$$L_n^{(\alpha)}(x) = -\frac{x}{n} L_{n-2}^{(\alpha+2)} + \frac{\alpha - x + 1}{n} L_{n-1}^{(\alpha+1)}(x)$$

and since $\alpha + 1 > -\frac{1}{2}, \alpha + 2 > -\frac{1}{2}$, the right side of the equation can be evaluated by (32) to obtain

$$
\begin{aligned}
L_n^{(\alpha)}(x) = {} & \frac{\Gamma(n+\alpha+1)}{\sqrt{\pi}\, n!} \frac{2^{\alpha+\frac{1}{2}} e^{x/2}}{[(4n+1)x]^{(\alpha+\frac{1}{2})/2}} \\
& \times \left[\left(\frac{4n+1}{4n-7}\right)^{(\alpha+\frac{1}{2})/2} \frac{4(n-1)}{4n-7} \left\{ \cos\left[\sqrt{(4n-7)x} - \frac{2\alpha+1}{4}\pi\right] \right.\right. \\
& \left. + \frac{\frac{x^2}{6} - (\alpha+\frac{5}{2})x - \frac{1}{2}(\alpha+\frac{5}{2})(\alpha+\frac{3}{2})}{\sqrt{(4n-7)x}} \sin\left[\sqrt{(4n-7)x} - \frac{2\alpha+1}{4}\pi\right] \right\} \\
& + \left(\frac{4n+1}{4n-3}\right)^{(\alpha+\frac{1}{2})/2} \frac{2(\alpha - x + 1)}{\sqrt{(4n-3)x}} \sin\left[\sqrt{(4n-3)x} - \frac{2\alpha+1}{4}\pi\right] \\
& \left. + \frac{R_1(n,x,\alpha)}{(4n+1)x} \right].
\end{aligned} \tag{34}
$$

Clearly

$$
\left(\frac{4n+1}{4n-7}\right)^{(\alpha+\frac{1}{2})/2} = 1 + O\left(\frac{1}{4n+1}\right), \quad \frac{4(n-1)}{4n-7} = 1 + O\left(\frac{1}{4n+1}\right),
$$

$$
\frac{1}{\sqrt{4n-7}} = \frac{1}{\sqrt{4n+1}} + O\left(\frac{1}{(4n+1)^{3/2}}\right),
$$

$$
\begin{aligned}
\sin\left[\sqrt{(4n-3)x} - \frac{2\alpha+1}{4}\pi\right] = {} & \sin\left[\sqrt{(4n+1)x} - \frac{2\alpha+1}{4}\pi\right] \\
& + O\left(\frac{\sqrt{x}}{\sqrt{4n+1}}\right),
\end{aligned}
$$

$$
\begin{aligned}
\sin\left[\sqrt{(4n-7)x} - \frac{2\alpha+1}{4}\pi\right] = {} & \sin\left[\sqrt{(4n+1)x} - \frac{2\alpha+1}{4}\pi\right] \\
& + O\left(\frac{\sqrt{x}}{\sqrt{4n+1}}\right),
\end{aligned}
$$

$$
\begin{aligned}
\cos\left[\sqrt{(4n-7)x} - \frac{2\alpha+1}{4}\pi\right] = {} & \cos\left[\sqrt{(4n+1)x} - \frac{2\alpha+1}{4}\pi\right] \\
& + \frac{4\sqrt{x}}{\sqrt{4n+1}} \sin\left[\sqrt{(4n+1)x} - \frac{2\alpha+1}{4}\pi\right] + O\left(\frac{\sqrt{x}+x}{4n+1}\right),
\end{aligned}
$$

and if ν denotes a positive number such that $\nu \geqq 2[(\alpha+2)-1/2]$, and l is given by (27), we have

$$R_1(n, x, \alpha) = \sum_{k=0}^{3l+8} q_k(\alpha)\, x^{k/2},$$

and (34) changes to (32).

Accordingly, formula (32) is established for $\alpha \leqq -1/2$, $-(2s+1)/2 < \alpha \leqq -(2s-1)/2$, $(s = 1, 2, \ldots)$, $n+\alpha+1>0$ and $\nu \geqq 2[(\alpha+2s)-\frac{1}{2}]$.

Uspensky's formula (33) is therefore valid for all real values of α, $[n+\alpha+1>0]$.

6a. From Uspensky's formula (33) for a fixed α, Fejér's inequality [33b] is readily deduced. Namely, for α real, $0 < a \leqq x \leqq b$,

$$(35)\qquad \begin{aligned} L_n^{(\alpha)}(x) &= \frac{\Gamma(n+\alpha+1)}{\sqrt{\pi}\, n!} \frac{2^{(\alpha+\frac{1}{2})}\, e^{x/2}}{[(4n+1)x]^{(\alpha+\frac{1}{2})/2}} \\ &\times \left\{ \cos\left[\sqrt{(4n+1)x} - \frac{2\alpha+1}{4}\pi\right] + O\left(\frac{x^{3/2}}{\sqrt{4n+1}}\right)\right\}, \end{aligned}$$

holds uniformly for all x in a finite interval $[a, b]$, $[a < b$, $n = 1, 2, \ldots;\ n+\alpha+1>0]$.

6b. Evaluating the ratio $\Gamma(n+\alpha+1)/n!$ in (35) by Stirling's formula (Appendix, Th. 20), we deduce that the following formula holds uniformly for all x in every finite interval $[a, b]$ $0 < a < b$,

$$(36)\qquad L_n^{(\alpha)}(x) = O\left[\frac{e^{x/2}}{\sqrt{x}}\left(\frac{n}{x}\right)^{\alpha/2-\frac{1}{4}}\right].$$

From (32) we deduce that (36) holds for

$$\frac{1}{4n+1} \leqq x \leqq n^{2/(3l+6)}.$$

E. Kogbetliantz [[58b], p. 149] has observed that by the arguments of Perron [87] it follows that the same formula (36) holds for $x < n^{\frac{1}{3}-\varepsilon}$, $0 < \varepsilon < 1/3$; and has extended it to the case $x \leqq kn$ [pp. 149—158]. Subsequently, Ottaviani [[83], p. 19] stated that if $\alpha \geqq -1$, (36) holds for $1/n < x \leqq kn$, and if

$\alpha < -1$ it holds for $0 < a \leqq x \leqq kn$, (a constant) where the constant k is less than 4, and that between such limits there does not exist a better approximation. (For further study on the asymptotic representation of the polynomials $L_n^{(\alpha)}(x)$ cf. E. Moecklin [77] and F. Tricomi [[114c, d, e, f, g]]. Particularly important are the memoirs of Tricomi where the asymptotic behavior of the polynomials $L_n^{(\alpha)}(x)$ is obtained by the adroit use of the theory of Bessel's function, thus establishing again the results of Chapter VIII, pp. 221—229 of G. Szegö's treatise [107 a] *Orthogonal Polynomials*, for the case $1 < x \leqq (4-\eta)n$, $(4+\eta)n \leqq x \leqq nA$, $x = 4n + O(n^{1/3})$. The reader is referred to Szegö's work for the complete bibliography).

7. Completeness of the Polynomials $L_n^{(\alpha)}(x)$ and $H_n(x)$ with Respect to Square Integrable Functions [1]

1a. We begin by proving the completeness of the system $\{e^{-x/2}L_n(x)\}$. To this end we observe that if we let

$$I_n(x) = \int_0^x e^{-x/2}L_n(x)\,dx, \tag{1}$$

then Vitali's equation of completeness [Ch. I, Th. 26] becomes

$$\sum_{n=0}^{\infty} I_n^2(x) = x \quad \text{for} \quad x \geqq 0. \tag{2}$$

By Bessel's inequality [Ch. I, Th. 12] we have

$$\sum_{n=0}^{\infty} I_n^2(x) \leqq x, \tag{2'}$$

and we must now show that the equality sign holds. To this end the integrals $I_n(x)$ and the sum $S_n(x)$ of the first $n+1$ terms of the series (2) will be expressed as functions of the polynomials $L_n(x)$.

1b. It is easily verified, by integrating by parts and using (10), that the following relation holds between $I_n(x)$ and $L_n(x)$

$$I_n(x) + I_{n-1}(x) = -\,2e^{-x/2}[L_n(x) - L_{n-1}(x)]. \tag{3}$$

[1] Cf. G. Sansone [98d] and R. Caccioppoli [11].

Differentiating (3), we get

$$e^{-x/2}[L_n(x)+L_{n-1}(x)] = e^{-x/2}[L_n(x)-L_{n-1}(x)] - 2e^{-x/2}[L'_n(x)-L'_{n-1}(x)]$$

which by (10) of Sec. 1 reduces to the identity

$$e^{-x/2}[L_n(x)+L_{n-1}(x)] = e^{-x/2}[L_n(x)-L_{n-1}(x)] + 2e^{-x/2}L_{n-1}(x).$$

From (3) we get the recurrence formula

$$(-1)^n\tfrac{1}{2}I_n(x) = (-1)^{n-1}\tfrac{1}{2}I_{n-1}(x) - e^{-x/2}[(-1)^nL_n(x) + (-1)^{n-1}L_{n-1}(x)]$$

and therefore

$$(4)\quad (-1)^n\tfrac{1}{2}I_n(x) = 1 - e^{-x/2}[(-1)^nL_n(x) + 2\sum_{r=0}^{n-1}(-1)^rL_r(x)]$$

$$(5)\quad 2e^{-x/2}\sum_{r=0}^{n}(-1)^rL_r(x) = 1 - (-1)^n[\tfrac{1}{2}I_n(x) - e^{-x/2}L_n(x)].$$

1c. Clearly

$$S_n(x) = \sum_{r=0}^{n} I_r^2(x) = \sum_{r=0}^{n}\int_0^x\left[\frac{d}{dx}I_r^2(x)\right]dx = 4\int_0^x e^{-x/2}\sum_{r=0}^{n}(\tfrac{1}{2}I_r(x))L_r(x)dx$$

$$= 4\int_0^x e^{-x/2}\sum_{r=0}^{n}[(-1)^r - e^{-x/2}\{L_r(x) + 2\sum_{s=0}^{r-1}2(-1)^{r+s}L_s(x)\}]L_r(x)\,dx$$

$$= 4\int_0^x e^{-x/2}\left[\sum_{r=0}^{n}(-1)^rL_r(x) - e^{-x/2}\left\{\sum_{r=0}^{n}(-1)^rL_r(x)\right\}^2\right]dx$$

$$= \int_0^x 2e^{-x/2}\sum_{r=0}^{n}(-1)^rL_r(x)[2 - 2e^{-x/2}\sum_{s=0}^{n}(-1)^sL_s(x)]\,dx,$$

whence by (5) we have

$$S_n(x) = \int_0^x\{1 - [\tfrac{1}{2}I_n(x) - e^{-x/2}L_n(x)]^2\}\,dx$$

and finally the identity

$$(6)\qquad S_n(x) = \sum_{r=0}^{n} I_r^2(x) = x - \int_0^x[\tfrac{1}{2}I_n(t) - e^{-t/2}L_n(t)]^2\,dt.$$

1d. By (3) of Sec. 6.1 we have $|L_n(t)| < T$ for all t in $[0, x]$. By (4) of Sec. 6.1 if $t > 0$ then $\lim_{n\to\infty} L_n(t) = 0$ since $\alpha = 0$. The series $\sum_{n=0}^{\infty} I_n^2(x)$ converges in $[0, x]$ and therefore $\lim_{n\to\infty} I_n(t) = 0$. Finally by (6), passing to the limit under the integral sign, we have $\lim_{n\to\infty} S_n(x) = x$, which is precisely (2).

2. The completeness of the system

$$\{\Gamma^{1/2}(n+1)\Gamma^{-1/2}(n+\alpha+1)e^{-x/2}x^{\alpha/2}L_n^{(\alpha)}(x)\}, \qquad \alpha > -1,$$

with respect to square integrable functions reduces to the preceding case by the following considerations. Suppose that there exists a square integrable function $f(x)$ which satisfies the equation

$$\int_0^{+\infty} f(x)e^{-x/2}x^{\alpha/2}L_n^{(\alpha)}(x)dx = 0, \qquad (n = 0, 1, 2, \ldots; \alpha > -1).$$

This implies that

$$\int_0^{+\infty} f(x)e^{-x/2}x^{\alpha/2}x^n\,dx = 0, \qquad (n = 0, 1, 2, \ldots),$$

whence by substituting $x = 2t$ we get

$$\int_0^{+\infty} f(2t)e^{-t}t^{\alpha/2}t^n\,dt = 0, \qquad (n = 0, 1, 2, \ldots),$$

and letting

$$F(t) = f(2t)t^{1+\alpha/2}e^{-t/2}$$

we have

$$\int_0^{+\infty} F(t)e^{-t/2}t^{n-1}dt = 0, \qquad (n = 1, 2, \ldots),$$

whence

$$\int_0^{+\infty} F(t)[e^{-t/2}L_n(t)]\,dt = 0 \qquad (n = 0, 1, 2 \ldots).$$

Now the function $F(t)$ is square integrable in $[0, +\infty)$ [the factor $t^{1+\alpha/2}e^{-t/2}$ is bounded in $[0, +\infty)$], whence $F(t)$ and therefore $f(t)$ vanishes almost everywhere in $[0, +\infty)$.

3a. We now wish to prove the completeness of the system $\{e^{-x^2/2}H_n(x)/\sqrt{2^n n!\sqrt{\pi}}\}$ with respect to square integrable functions.

If we let

$$J_n(x) = \int_0^x e^{-x^2/2} H_n(x)\, dx, \tag{7}$$

then Vitali's equation of completeness becomes

$$\sum_{n=0}^{\infty} \frac{1}{2^n n!} J_n^2(x) = \sqrt{\pi}\, x, \tag{8}$$

for every nonnegative x. Now by Bessel's inequality the left side of (8) is convergent for every x. Consequently, if $S_n(x)$ denotes the sum of the first $2n + 1$ terms of (8), it will be sufficient to prove that

$$\lim_{n\to\infty} S_{2n}(x) = \sqrt{\pi} x. \tag{9}$$

3b. We begin by finding a convenient expression for $S_{2n}(x)$. The following two relations hold between the functions $H_n(x)$ and $J_n(x)$

$$(10_1) \quad -2e^{-x^2/2} H_{2n-1}(x) = -J_{2n}(x) + 2(2n-1)J_{2n-2}(x),$$

$$(10_2) \quad -2e^{-x^2/2} H_{2n}(x) = -J_{2n+1}(x) + 2(2n)J_{2n-1}(x) - 2(-1)^n \frac{(2n)!}{n!}$$

$$(n = 1, 2, \ldots).$$

For $x = 0$, they clearly hold by (10) of Sec. 2.3. For $x \neq 0$, the relations are obtained by taking the derivatives of $H_n(x)$ and $J_n(x)$ and using (4) and (5) of Sec. 2.2.

By (7) and (10_1) we have

$$J_{2n-1}^2(x) = \int_0^x \left[\frac{d}{dx} J_{2n-1}^2(x)\right] dx = 2\int_0^x e^{-x^2/2} J_{2n-1}(x) H_{2n-1}(x)\, dx$$
$$= \int_0^x J_{2n-1}(x) [J_{2n}(x) - 2(2n-1) J_{2n-2}(x)]\, dx,$$

and therefore

$$\frac{J_{2n-1}^2(x)}{2^{2n-1}(2n-1)!} = \int_0^x \left[\frac{1}{2^{2n-1}(2n-1)!} J_{2n}(x) J_{2n-1}(x) \right.$$
$$\left. - \frac{1}{2^{2n-2}(2n-2)!} J_{2n-1}(x) J_{2n-2}(x)\right] dx, \quad (n = 1, 2, \ldots).$$

Similarly, by (10_2), if we define $H_{-1} = 0$ and $J_{-1} = 0$, we get

$$\frac{J_{2n}^2(x)}{2^{2n}(2n)!} = \int_0^x \left[\frac{1}{2^{2n}(2n)!} J_{2n+1}(x) J_{2n}(x) - \frac{1}{2^{2n-1}(2n-1)!} J_{2n}(x) J_{2n-1}(x)\right] dx + (-1)^n \frac{1}{2^{2n-1}(n!)} \int_0^x J_{2n}(x)\, dx,$$

$$(n = 0, 1, 2, \ldots),$$

whence

$$S_{2n}(x) = \sum_{r=0}^{2n} \frac{1}{2^r r!} J_r^2(x) \tag{11}$$

$$= 2 \sum_{r=0}^{n} (-1)^r \frac{1}{2^{2r} r!} \int_0^x J_{2r}(u)\, du + \frac{1}{2^{2n}(2n)!} \int_0^x J_{2n+1}(u) J_{2n}(u)\, du.$$

By formula (5) of Sec. 2.2 we have

$$(-1)^r \frac{H_{2r}(t)}{2^{2r} r!} = \frac{1}{t}\left[(-1)^{r+1} \frac{H_{2r+1}(t)}{2^{2r+1} r!} - (-1)^r \frac{H_{2r-1}(t)}{2^{2r-1}(r-1)!}\right]$$

$$(r = 0, 1, 2, \ldots),$$

whence

$$2\int_0^x \sum_{r=0}^{n} \frac{(-1)^r}{2^{2r} r!} J_{2r}(u) du = 2\int_0^x du \int_0^u e^{-t^2/2} \sum_{r=0}^{n} (-1)^r \frac{H_{2r}(t)}{2^{2r} r!}\, dt$$

$$= \int_0^x du \int_0^u \frac{(-1)^{n+1}}{2^{2n}\, n!} e^{-t^2/2} \frac{H_{2n+1}(t)}{t}\, dt,$$

and (11) becomes finally

$$S_{2n}(x) = \int_0^x du \int_0^u \frac{(-1)^{n+1}}{2^{2n}\, n!} e^{-t^2/2} \frac{H_{2n+1}(t)}{t}\, dt + \frac{1}{2^{2n}(2n)!} \int_0^x J_{2n+1}(u) J_{2n}(u) du. \tag{12}$$

3c. By (10), (10′) of Sec. 5.1, and Wallis' formula we have

$$\int_0^u \frac{(-1)^{n+1}}{2^{2n}n!} e^{-t^2/2} \frac{H_{2n+1}(t)}{t} dt$$

$$= \frac{2}{\sqrt{\pi}}\left[1 + O\left(\frac{1}{n}\right)\right]\int_0^u \frac{\sin t\sqrt{4n+3}}{t} dt + O\left(\frac{1}{\sqrt[4]{n}}\right)$$

uniformly for all u in $[0, x]$ and therefore the first integral that figures in the right side of (12) is bounded with respect to all u in $[0, x]$ and uniformly bounded with respect to n. Moreover,

$$\lim_{n\to\infty}\int_0^u \frac{(-1)^{n+1}}{2^{2n}n!} e^{-t^2/2} \frac{H_{2n+1}(t)}{t} dt$$

$$= 2\pi^{-1/2} \lim_{n\to\infty}\int_0^u \frac{\sin t\sqrt{4n+3}}{t} dt = \sqrt{\pi},$$

whence

$$\text{(13)} \qquad \lim_{n\to\infty}\int_0^x du \int_0^u \frac{(-1)^{n+1}}{2^{2n}n!} e^{-t^2/2} \frac{H_{2n+1}(t)}{t} dt = \sqrt{\pi}\, x.$$

We next prove that the limit as $n \to \infty$ of the second integral on the right side of (12) is zero, namely

$$\text{(14)} \qquad \lim_{n\to\infty} \frac{1}{2^{2n}(2n)!}\int_0^x J_{2n+1}(u)\, J_{2n}(u)\, du = 0.$$

In fact we have by (10) of Sec. 5.1

$$J_{2n}(u) = \int_0^u e^{-t^2/2} H_{2n}(t)\, dt$$

$$= 2^n\sqrt{(2n)!}\int_0^u \left[(-1)^n(n\pi)^{-1/4}\left(1 - \frac{\varepsilon_1}{8n}\right)\cos\,[t\sqrt{4n+1}] - \frac{h(2n,t)}{\sqrt{4n+1}}\right] dt,$$

whence, integrating, we obtain uniformly for every u in $[0, x]$

$$J_{2n}(u) = O(2^n\sqrt{(2n)!}\; n^{-1/2})$$

and similarly by (10) of Sec. 5.1,

$$J_{2n+1}(u) = O(2^n\sqrt{(2n+1)!}\,n^{-1/2}),$$

whence

$$\frac{1}{2^{2n}(2n)!}\int_0^x J_{2n}(u)\,J_{2n+1}(u)\,du = O(n^{-1/2}),$$

and therefore (14).

Finally, by virtue of (12), (13), and (14), we have (8).

3d. The theorem just proved is equivalent to the following: if $f(x)$ is square integrable between $-\infty$ and $+\infty$, and all the moments of $f(x)e^{-x^2/2}$

$$\int_{-\infty}^{+\infty} f(x)e^{-x^2/2}x^n\,dx, \qquad (n = 0, 1, 2, \ldots), \tag{15}$$

are zero, then $f(x)$ vanishes almost everywhere in $(-\infty, +\infty)$.

Noting that if $k > 0$ and $x = t/\sqrt{2k}$ we have

$$\int_{-\infty}^{+\infty} x^n f(x)e^{-kx^2}\,dx = (2k)^{-(n+1)/2}\int_{-\infty}^{+\infty} f\left(\frac{t}{\sqrt{2k}}\right)e^{-t^2/2}t^n\,dt,$$

we can also assert that: if $k > 0$, and $f(x)$ is square integrable between $-\infty$ and $+\infty$ and all the moments of $f(x)e^{-kx^2}$ exist and are zero, namely

$$\int_{-\infty}^{+\infty} x^n f(x)e^{-kx^2}\,dx = 0, \qquad (n = 0, 1, 2, \ldots),$$

then $f(x)$ vanishes almost everywhere in $(-\infty, +\infty)$.

8. Bessel's Equality for Infinite Intervals

1. THEOREM. If $\{\varphi_n(t)\}$ is a sequence of orthonormal functions in the infinite interval I which satisfies Vitali's equation of completeness

$$|\alpha - a| = \sum_{n=1}^{\infty}\left[\int_a^{\alpha}\varphi_n(t)\,dt\right]^2, \tag{1}$$

where a is a fixed point of I and α is a variable point in I, then if $f(t)$ is a square integrable function in I, and a_n is the sequence of its Fourier constants, namely

$$a_n = \int_I f(t)\varphi_n(t)\, dt,$$

then we have Bessel's equality (G. Sansone [98e]):

$$\sum_{n=1}^{\infty} a_n^2 = \int_I f^2\, dt. \tag{2}$$

REMARK. This theorem could be omitted provided it is assumed that Theorem 22 of Ch. I is also valid for sets of infinite measure.

Proof. When the interval I is finite, (2) is a consequence of Theorem 27 of Ch. I. Here we assume that $I = [a, +\infty)$. Similar reasoning applies to $I = (-\infty, +\infty)$.

To prove (2), it suffices to prove that for $\sigma > 0$ there exists a linear combination with constant coefficients of $\varphi_1, \varphi_2, \ldots, \varphi_n$, with n sufficiently large, such that

$$F_n(t) = \gamma_1\varphi_1(t) + \gamma_2\varphi_2(t) + \ldots + \gamma_n\varphi_n(t) \tag{3}$$

$$\int_I |f - F_n|^2\, dt < \sigma. \tag{4}$$

Indeed, from Theorem 11 of Ch. I, follows

$$0 \leqq \int_I f^2\, dt - \sum_{k=1}^{n} a_k^2 = \int_I [f - \sum_{k=1}^{n} a_k\varphi_k]^2\, dt \leqq \int_I |f - F_n|^2\, dt < \sigma$$

and therefore (2).

To prove (4) we shall proceed step by step.

(a) Let $f(t) = c =$ constant, $a \leqq t \leqq \alpha$, and $f(t) = 0$ for $t \geqq \alpha$. Then

$$a_n = c\int_a^\alpha \varphi_n(t)\, dt, \qquad \int_I f^2\, dt = c^2(\alpha - a)$$

and if in (3) we let $\gamma_k = a_k$, namely

$$F_n(t) = a_1\varphi_1(t) + a_2\varphi_2(t) + \ldots + a_n\varphi_n(t)$$

we have

$$\int_I |f - F_n|^2\, dt = \int_I f^2\, dt - \sum_{k=1}^{n} a_k^2 = c^2\left[(\alpha - a) - \sum_{k=1}^{n}\left(\int_a^\alpha \varphi_k(t)dt\right)^2\right]$$

and then (4) follows from (1) which is satisfied by hypothesis.

(b) Let $f(t) = c$ for $\alpha > t < \beta$ and $f(t) = 0$ for $a \leqq t \leqq \alpha$ or $\beta \leqq t$. Let

$$f_1(t) = c \quad \text{for} \quad a \leqq t < \beta \quad \text{and} \quad f_1(t) = 0 \quad \text{for} \quad \beta \leqq t,$$
$$f_2(t) = c \quad \text{for} \quad a \leqq t \leqq \alpha \quad \text{and} \quad f_2(t) = 0 \quad \text{for} \quad \alpha < t,$$

and determine as in (a) two linear combinations F_{n_1}, F_{n_2} of $\{\varphi_n\}$ such that

$$\int_I | f_1 - F_{n_1} |^2 \, dt < \frac{\sigma}{4}, \qquad \int_I | f_2 - F_{n_2} |^2 \, dt < \frac{\sigma}{4}.$$

Then since $f(t) = f_1(t) - f_2(t)$, we have

$$\int_I [f - (F_{n_1} - F_{n_2})]^2 \, dt = \int_I [(f_1 - F_{n_1}) - (f_2 - F_{n_2})]^2 \, dt$$
$$< 2\int_I | f_1 - F_{n_1} |^2 \, dt + 2\int_I | f_2 - F_{n_2} |^2 \, dt < \sigma,$$

and therefore (4) is satisfied with $F(t) = F_{n_1}(t) - F_{n_2}(t)$.

(c) Let $a = \alpha_0 < \alpha_1 < \alpha_2 < \ldots < \alpha_s$ and let

$$f(t) = c_k \quad \text{for} \quad \alpha_{k-1} < t < \alpha_k, \quad (k = 1, 2, \ldots s),$$

and $f(t) = 0$ at $t = a, \alpha_1, \alpha_2, \ldots, \alpha_s$ and for $t > \alpha_s$.

Consider now the function $f_k(t) = c_k$ for $\alpha_{k-1} < t < \alpha_k$ and zero for $t \leqq \alpha_{k-1}$ and $t \geqq \alpha_k$ and determine as in (b) s linear combinations $F_{n_k}(t)$, $[k = 1, 2, \ldots, s]$, of the $\{\varphi_n\}$ such that

$$\int_I | f_k - F_{n_k} |^2 \, dt < \frac{\sigma}{s}; \qquad (k = 1, 2, \ldots, s).$$

Then, if we let $f = f_1 + f_2 + \ldots + f_s$, we have

$$\int_I \left[f - \sum_{k=1}^{s} F_{n_k} \right]^2 dt = \int_I \left[\sum_{k=1}^{s} (f_k - F_{n_k}) \right]^2 dt \leqq s \sum_{k=1}^{s} \int_I |f_k - F_{n_k}|^2 dt < \sigma.$$

(d) Let $f(t)$ be continuous in $[a, \alpha]$ and $f(t) = 0$ for $t > \alpha$. Divide the interval $[a, \alpha]$ by the points $a = \alpha_0 < \alpha_1 < \alpha_2 < \ldots < \alpha_s$ in such a way that in every subinterval $[\alpha_{k-1}, \alpha_k]$ the oscillation of $f(t)$ is less than $\sigma^{1/2}/(2\sqrt{|\alpha - a|})$ and let $c_k = f[(\alpha_{k-1} + \alpha_k)/2]$. Then define the function $F(t)$ such that $F(t) = 0$ for $t = \alpha_0, \alpha_1, \alpha_2, \ldots, \alpha_s$

and for $t > \alpha_s$, and $F(t) = c_k$ for t interior to (α_{k-1}, α_k). Then by (c) there exists a linear combination $F_n(t)$ of the $\{\varphi_n\}$ such that

$$\int_I | F - F_n |^2 \, dt < \frac{\sigma}{4};$$

whence

$$\int_I | f - F_n |^2 \, dt \leqq 2 \int_I | f - F |^2 \, dt + 2 \int_I | F - F_n |^2 \, dt$$

$$\leqq 2 \sum_{k=1}^{s} \int_{\alpha_{k-1}}^{\alpha_k} | f - F |^2 \, dt + \frac{\sigma}{2} < 2 \sum_{k=1}^{s} \frac{\sigma}{4 | \alpha - a |} \int_{\alpha_{k-1}}^{\alpha_k} dt + \frac{\sigma}{2} = \sigma.$$

(e) Finally let $f(t)$ be square integrable in $[a, +\infty)$ and take α sufficiently large to insure

$$\int_{\alpha}^{+\infty} f^2 dt < \frac{\sigma}{6}.$$

Since $f(t)$ is square integrable in the finite interval $[a, \alpha]$ we can determine a function $F(t)$ continuous in $[a, \alpha]$ such that

$$\int_a^{\alpha} | f(t) - F(t) |^2 \, dt < \frac{\sigma}{6}.$$

It is sufficient, for example, to take for $F(t)$ a partial sum of the Legendre series of $f(t)$ relative to $[a, \alpha]$ [Ch. III, Sec. 9.3, and Sec. 16].

Let $F(t)$ vanish for $t > \alpha$ and determine as in (d) a linear combination F_n of the $\{\varphi_n\}$ such that

$$\int_a^{+\infty} | F - F_n |^2 \, dt < \frac{\sigma}{6}.$$

Then

$$\int_a^{+\infty} (f - F_n)^2 \, dt \leqq 2 \int_a^{+\infty} | f - F |^2 \, dt + 2 \int_a^{+\infty} | F - F_n |^2 \, dt$$

$$\leqq 2 \int_a^{\alpha} | f - F |^2 \, dt + 2 \int_{\alpha}^{+\infty} f^2 \, dt + 2 \int_a^{+\infty} | F - F_n |^2 \, dt < \sigma,$$

which proves the theorem.

2. For the particular orthogonal systems of Laguerre and of Hermite we can now extend Theorem 22 of Ch. I.

THEOREM. Let $f(x)$ be square integrable in $[0 + \infty)$. Then if $\{a_n\}$ is the sequence of its Fourier coefficients

$$(5)\qquad a_n = \frac{\Gamma^{1/2}(n+1)}{\Gamma^{1/2}(n+\alpha+1)} \int_0^{+\infty} e^{-x/2} x^{\alpha/2} L_n^{(\alpha)}(x) f(x)\, dx, \quad (\alpha > -1),$$

with respect to the orthonormal system

$$\{e^{-x/2} x^{\alpha/2} L_n^{(\alpha)}(x)\, \Gamma^{1/2}(n+1)\, \Gamma^{-1/2}(n+\alpha+1)\},$$

Bessel's equality holds, namely

$$(6)\qquad \sum_{n=0}^{\infty} a_n^2 = \int_0^{+\infty} f^2(x)\, dx.$$

(a) If $\alpha = 0$, (6) follows from (2) of Sec. 7.1 and the theorem just proved.

(b) If $\alpha \neq 0$, $\alpha > -1$, we proceed as in the proof of the preceding theorem to show that for $\sigma > 0$ we can find a polynomial $P(x)$ such that

$$(7)\qquad \int_0^{+\infty} | f(x) - e^{-x/2} x^{\alpha/2} P(x) |^2\, dx < \sigma.$$

Let $\alpha > 0$ and let s be a nonnegative integer such that $\alpha/2 = s/4 + \varrho$ where $0 \leqq \varrho < 1/4$; then if $P(x)$, $P_0(x)$, $P_1(x)$, ..., $P_s(x)$ denote arbitrary polynomials, we have

$$\int_0^{+\infty} | f(x) - e^{-x/2} x^{\alpha/2} P(x) |^2\, dx = \int_0^{+\infty} \Big| [f(x) - e^{-x/2} P_0(x)]$$
$$+ \sum_{k=1}^{s} e^{-x/2} x^{(k-1)/4} [P_{k-1} - x^{1/4} P_k] + e^{-x/2} x^{s/4} [P_s - x^{\rho} P] \Big|^2 dx$$
$$(8)\qquad \leqq (s+2) \Big[\int_0^{+\infty} [f(x) - e^{-x/2} P_0(x)]^2\, dx$$
$$+ \sum_{k=1}^{s} \int_0^{+\infty} [e^{-x/2} x^{(k-1)/4} P_{k-1}(x) - e^{-x/2} x^{k/4} P_k(x)]^2\, dx$$
$$+ \int_0^{+\infty} [e^{-x/2} x^{s/4} P_s(x) - e^{-x/2} x^{s/4+\rho} P(x)]^2\, dx \Big].$$

By (a) we can find a polynomial $P_0(x)$ such that

$$\int_0^{+\infty} | f(x) - e^{-x/2} P_0(x) |^2 dx < \frac{\sigma}{s+2}.$$

For such a polynomial P_0 we have

$$\int_0^{+\infty} | e^{-x/2} P_0(x) - e^{-x/2} x^{1/4} P_1(x) |^2 dx$$

$$= 2 \int_0^{+\infty} | e^{-t} P_0(2t) - e^{-t} t^{1/4} 2^{1/4} P_1(2t) |^2 dt$$

$$= 2 \int_0^{+\infty} e^{-t} t^{1/2} | e^{-t/2} t^{-1/4} P_0(2t) - 2^{1/4} e^{-t/2} P_1(2t) |^2 dt.$$

There exists a constant M such that $| e^{-t} t^{1/2} | < M$ in $[0, +\infty)$, whence

$$\int_0^{+\infty} | e^{-x/2} P_0(x) - e^{-x/2} x^{1/4} P_1(x) |^2 dx$$

$$< 2M \int_0^{+\infty} | e^{-t/2} t^{-1/4} P_0(2t) - 2^{1/4} e^{-t/2} P_1(2t) |^2 dt,$$

and since the function $e^{-t/2} t^{-1/4} P_0(2t)$ is square integrable in $[0, +\infty)$ we can find by (a) a polynomial $P_1(2t)$ such that

$$\int_0^{+\infty} | e^{-t/2} t^{-1/4} P_0(2t) - 2^{1/4} e^{-t/2} P_1(2t) |^2 dt < \sigma/2M(s+2).$$

Therefore

$$\int_0^{+\infty} | e^{-x/2} P_0(x) - e^{-x/2} x^{1/4} P_1(x) |^2 dx < \frac{\sigma}{s+2}.$$

Having determined $P_1(x)$; $P_2(x)$, ..., $P_s(x)$, $P(x)$ can be successively determined so that each of the terms that are in brackets on the right side of (8) is less than $\sigma/(s+2)$, whence (7) follows.

(c) Finally let $-1 < \alpha < 0$; then by (b) a polynomial $P(x)$ can be found such that

$$\int_0^{+\infty} | f(x) - e^{-x/2} x^{1+\alpha} P(x) |^2 dx < \sigma,$$

and also

$$\int_0^{+\infty} | f(x) - e^{-x/2} x^{\alpha} [xP(x)] |^2 dx < \sigma.$$

THEOREM. Let $f(x)$ be square integrable in $(-\infty, +\infty)$. Then if $\{a_n\}$ is the sequence of its Fourier coefficients

$$a_n = \frac{1}{\sqrt[4]{\pi}\sqrt{2^n n!}} \int_{-\infty}^{+\infty} e^{-x^2/2} f(x) H_n(x)\, dx,$$

with respect to the orthonormal system $\{e^{-x^2/2} H_n(x)/\sqrt{\sqrt{\pi}\, 2^n n!}\}$, Bessel's equality holds

$$\sum_{k=0}^{\infty} a_k^2 = \int_{-\infty}^{+\infty} f^2(x)\, dx.$$

Proof. The theorem follows from (8) of Sec. 7 and the theorem of the preceding section.

9. Criteria for Uniform Convergence of the Series of Polynomials $L_n^{(\alpha)}(x)$ and $H_n(x)$

1. THEOREM of Nasarow [80] and Picone [88d]. Let

$$F(x) = F(0) + \int_0^x f(x)\, dx \tag{1}$$

where $f(x)$ is defined for $x \geqq 0$ and integrable in any finite interval, and let the following functions be square integrable in $[0, +\infty)$

$$e^{-x/2} x^{\alpha/2} F(x), \quad e^{-x/2} x^{(\alpha+1)/2} f(x), \qquad \alpha > -1.$$

In other words, the following two integrals are convergent ($A \geqq 0$, $G \geqq 0$)

$$A^2 = \int_0^{+\infty} e^{-x} x^{\alpha} F^2(x)\, dx, \qquad G^2 = \int_0^{+\infty} e^{-x} x^{\alpha+1} f^2(x)\, dx.$$

Let

$$l_k^{(\alpha)}(x) = \frac{\Gamma^{1/2}(k+1)}{\Gamma^{1/2}(k+\alpha+1)} L_k^{(\alpha)}(x), \quad k = 0, 1, 2, \ldots; \; p(x) = e^{-x} x^{\alpha},$$

and

$$a_k = \int_0^{+\infty} p(x) F(x)\, l_k^{(\alpha)}(x)\, dx, \quad k = 0, 1, 2, \ldots; \quad \alpha > -1.$$

Then the Fourier series of $p^{1/2}(x)F(x)$ with respect to the system $\{p^{1/2}(x)\,l_k^{(\alpha)}(x)\}$, namely

$$p^{1/2}(x)\,F(x) \sim \sum_{k=0}^{\infty} p^{1/2}(x)\,a_k\,l_k^{(\alpha)}(x),$$

converges uniformly and absolutely to $p^{1/2}(x)F(x)$ in any interval $[c, +\infty)$ where $c > 0$, and therefore in any interval $[c, d]$ interior to $[0, \infty)$ we have uniformly

$$F(x) = \sum_{k=0}^{\infty} a_k\,l_k^{(\alpha)}(x). \tag{2}$$

Proof. From (1) we have, for any points a and b, $a > 0$, $b > 0$

$$F(b) = F(a) + \int_a^b f(x)\,dx,$$

whence

$$|\,F(b)\,| \leqq |\,F(a)\,| + \left|\int_a^b f(x)\,dx\right| = |\,F(a)\,|$$
$$+ \left|\int_a^b e^{x/2}x^{-(\alpha+1)/2}\,e^{-x/2}x^{(\alpha+1)/2}f(x)\,dx\right|$$

and therefore

$$|\,F(b)\,| \leqq |\,F(a)\,| + G\left|\int_a^b e^x\,x^{-(\alpha+1)}\,dx\right|^{1/2}.$$

Consequently, for $x \geqq 1$

$$|\,F(x)\,| \leqq |\,F(1)\,| + G\left[\int_1^x e^x\,dx\right]^{1/2} \leqq |\,F(1)\,| + G e^{x/2},$$

whence

$$|\,F(x)\,| \leqq e^{x/2}[\,|\,F(1)\,| + G].$$

For $-1 < \alpha < 0$, and $0 < x < 1$

$$|F(x)| \leqq |F(0)| + G\left[\int_0^x \frac{e^x}{x^{1+\alpha}}\,dx\right]^{1/2} \leqq |F(0)| + Ge^{x/2}\left[\int_0^x \frac{1}{x^{1+\alpha}}\,dx\right]^{1/2},$$

$$|\,F(x)\,| \leqq |\,F(0)\,| + G e^{x/2}\,|\,\alpha\,|^{-1/2} \leqq e^{x/2}[|F(0)\,| + G\,|\,\alpha\,|^{-1/2}],$$

and therefore for $-1 < \alpha < 0$ and for arbitrary x, there exists a positive contant k such that

$$(3_1) \qquad |F(x)| < ke^{x/2}.$$

For $\alpha = 0$, and $0 < x < 1$

$$|F(x)| \leqq |F(1)| + G\left|\int_x^1 \frac{e^x}{x}\,dx\right|^{1/2} \leqq |F(1)| + Ge^{1/2}\,|\lg x|^{1/2}$$

whence for $\alpha = 0$ there exists a positive constant k such that

$$(3_2) \qquad \begin{cases} |F(x)| < k[1 + |\lg x|^{1/2}] & \text{for } 0 < x < 1, \\ |F(x)| < ke^{x/2} & \text{for } x \geqq 1. \end{cases}$$

Finally for $\alpha > 0$, and $0 < x < 1$

$$|F(x)| \leqq |F(1)| + Ge^{1/2}\left[\int_x^1 x^{-(\alpha+1)}\,dx\right]^{1/2},$$

$$|F(x)| \leqq |F(1)| + Ge^{1/2}\,|\alpha|^{-1/2}x^{-\alpha/2},$$

and therefore there exists a positive constant k such that

$$(3_3) \qquad \begin{cases} |F(x)| < kx^{-\alpha/2} & \text{for } x < 1, \\ |F(x)| < ke^{x/2} & \text{for } x \geqq 1. \end{cases}$$

Let [Sec. 1.5, (11)]

$$l_k^{(\alpha+1)}(x) = \frac{\Gamma^{1/2}(k+1)}{\Gamma^{1/2}(k+\alpha+2)}\,L_k^{(\alpha+1)}(x) = -\frac{\Gamma^{1/2}(k+1)}{\Gamma^{1/2}(k+\alpha+2)}\,\frac{d}{dx}L_{k+1}^{(\alpha)}$$

and

$$a'_k = \int_0^{+\infty} xp\,l_k^{(\alpha+1)} f\,dx,$$

then

$$f(x) \sim \sum_{k=0}^{\infty} a'_k l_k^{(\alpha+1)}(x).$$

Now, integrating by parts and using (3_1), (3_2), (3_3) and (14_1) of Secs. 1.1 and 1.5, we have

$$a_k' = -\frac{\Gamma^{1/2}(k+1)}{\Gamma^{1/2}(k+\alpha+2)}\int_0^{+\infty}\left[xp\frac{d}{dx}L_{k+1}^{(\alpha)}(x)\right]f(x)\,dx$$

$$= \frac{\Gamma^{1/2}(k+1)}{\Gamma^{1/2}(k+\alpha+2)}\int_0^{+\infty}F(x)\left[p(\alpha-x)\frac{d}{dx}L_{k+1}^{(\alpha)}(x)\right.$$

$$\left.+\,p\frac{d}{dx}L_{k+1}^{(\alpha)}(x)+px\frac{d^2}{dx^2}L_{k+1}^{(\alpha)}(x)\right]dx$$

$$= -\frac{(k+1)\Gamma^{1/2}(k+1)}{\Gamma^{1/2}(k+\alpha+2)}\int_0^{+\infty}pF(x)L_{k+1}^{(\alpha)}\,dx,\quad [xp'=p(\alpha-x)],$$

whence

$$(4)\qquad \begin{aligned} a_k' l_k^{(\alpha+1)}(x) &= \frac{\Gamma(k+2)}{L(\alpha+k+2)}\left[\int_0^{+\infty}pF(x)L_{k+1}^{(\alpha)}(x)\,dx\right]\frac{d}{dx}L_{k+1}^{(\alpha)}(x)\\ &= a_{k+1}\frac{d}{dx}l_{k+1}^{(\alpha)}(x),\end{aligned}$$

and therefore

$$(4')\qquad F(x)\sim\sum_{k=1}^{\infty}a_k l_k^{(\alpha)}(x),\qquad f(x)\sim\sum_{k=0}^{\infty}a_{k+1}\frac{d}{dx}l_{k+1}^{(\alpha)}(x).$$

Now by (4′) and Theorem 32 of Ch. I it follows for $x \geqq c > 0$ that

$$(5)\qquad \begin{aligned} &-\int_x^{+\infty}F(\xi)e^{-\xi}\,d\xi\\ &\qquad = -a_0\int_x^{+\infty}e^{-\xi}l_0^{(\alpha)}(\xi)\,d\xi-\sum_{k=0}^{\infty}a_{k+1}\int_x^{+\infty}e^{-\xi}l_{k+1}^{(\alpha)}(\xi)\,d\xi.\end{aligned}$$

To establish (5), we refer to Ch. I, Th. 32, Remark 1, and take $f_1 = F$; $f_2 = e^{-\xi}$, $\theta(\xi) = e^{-\xi}\xi^{\alpha}$, $\theta_1 = e^{-\xi}$, $\theta_2 = \xi^{\alpha}$, $\xi^{\alpha} > \varrho$; then we have for $X \geqq \xi \geqq x \geqq c > 0$

$$-\int_x^{X}F(\xi)e^{-\xi}d\xi = -a_0\int_x^{X}e^{-\xi}l_0^{(\alpha)}(\xi)\,d\xi-\sum_{k=0}^{\infty}a_{k+1}\int_x^{X}e^{-\xi}l_{k+1}^{(\alpha)}(\xi)\,d\xi.$$

Clearly,

$$\left|\sum_{k=N+1}^{\infty}a_{k+1}\int_x^{X}e^{-\xi}l_{k+1}^{(\alpha)}(\xi)\,d\xi\right| \leqq \left[\sum_{k=N+1}^{\infty}a_{k+1}^2\right]^{1/2}\left[\sum_{k=N+1}^{\infty}\left[\int_x^{X}e^{-\xi}l_{k+1}^{(\alpha)}(\xi)d\xi\right]^2\right]^{1/2}.$$

Now if we take the function $\varphi(\xi)$ equal to $e^{-\xi/2}/\xi^{\alpha/2}$ for $x \leqq \xi \leqq X$, and zero in $(0, X)$ and $(X, +\infty)$ then

$$\sum_{k=N+1}^{\infty}\left[\int_x^X e^{-\xi} l_{k+1}^{(\alpha)}(\xi)d\xi\right]^2 \leqq \sum_{k=0}^{\infty}\left[\int_x^X e^{-\xi} l_{k+1}^{(\alpha)}(\xi)\,d\xi\right]^2 = \int_x^X \frac{e^{-\xi}}{\xi^\alpha}\,d\xi < \int_x^\infty \frac{e^{-\xi}}{\xi^\alpha}d\xi,$$

and the convergence of the series $\sum_{k=0}^\infty a_k^2$ implies the uniform convergence of the series on the right side of (5).

From (5) we get

$$(5_1)\qquad -\int_x^{+\infty} F(\xi)\,e^{-\xi}\,d\xi = -\,a_0\,e^{-x}\,l_0^{(\alpha)} - \sum_{k=0}^{\infty} a_{k+1}\int_x^{+\infty} e^{-\xi}\,l_{k+1}^{(\alpha)}(\xi)\,d\xi.$$

Again referring to Ch. I, Th. 32, Remark 1, and taking $f_1 = f$; $f_2 = e^{-\xi}$, $\theta(\xi) = e^{-\xi}\,\xi^{\alpha+1}$, $\theta_1 = e^{-\xi}$, $\theta_2 = \xi^{\alpha+1}$ we get

$$\int_x^{+\infty} f(\xi)\,e^{-\xi}\,d\xi = \sum_{k=0}^{\infty} a_k'\int_x^{+\infty} e^{-\xi}\,l_k^{(\alpha+1)}(\xi)\,d\xi,$$

whence by (4) we have

$$(5_2)\qquad \int_x^{+\infty} f(\xi)\,e^{-\xi}\,d\xi = \sum_{k=0}^{\infty} a_{k+1}\int_x^{+\infty} e^{-\xi}\,\frac{d}{d\xi}\,l_{k+1}^{(\alpha)}(\xi)\,d\xi.$$

The series on the right side of (5_1) and (5_2) are absolutely and uniformly convergent in $[c, +\infty)$. Now, clearly,

$$\begin{aligned} -&\int_x^{+\infty} F(\xi)\,e^{-\xi}\,d\xi \\ &= [e^{-\xi}F(\xi)]_x^{+\infty} - \int_x^{+\infty} e^{-\xi}f(\xi)\,d\xi = -\,e^{-x}F(x) - \int_x^{+\infty} e^{-\xi}f(\xi)\,d\xi, \end{aligned}$$

$$\begin{aligned} -\int_x^{+\infty} e^{-\xi}\,l_{k+1}^{(\alpha)}(\xi)\,d\xi + \int_x^{+\infty} e^{-\xi}\,\frac{d}{d\xi}\,l_{k+1}^{(\alpha)}(\xi)\,d\xi& \\ = [e^{-\xi}\,l_{k+1}^{(\alpha)}(\xi)]_x^{+\infty} = -\,e^{-x}\,l_{k+1}^{(\alpha)}(x)&, \end{aligned}$$

whence, adding (5_1) and (5_2), we have

$$-\,e^{-x}F(x) = -\,e^{-x}\sum_{k=0}^{\infty} a_k\,l_k^{(\alpha)}(x)$$

and therefore (2) holds uniformly in any interval $[c, d]$ interior to $(0, \infty)$.

2. A criterion of convergence at the point 0 is given by the following

THEOREM. If

$$F(x) = F(0) + \int_0^x f(x)\,dx,$$

where $f(x)$ is integrable in any finite interval, if $F(x) = O(e^{kx/2})$, $k < 1$, and, finally, if the integral $\int_0^{+\infty} p(x) f^2(x) dx = A^2$ is convergent, where $p(x) = e^{-x}$, $[\alpha = 0]$, then

$$F(0) = \sum_{k=0}^{\infty} a_k l_k(0), \quad a_k = \int_0^{+\infty} p(x) F(x)\, l_k(x)\, dx; \quad l_k(x) = L_k^{(0)}(x).$$

Proof. We let

$$F(x) = \sum_{k=0}^{n} a_k l_k(x) + \varrho_{n+1}(x) \tag{6}$$

and then derive an inequality for $\varrho_{n+1}^2(0)$.

From (6) follows

$$f(x) = \sum_{k=0}^{n} a_k l_k'(x) + \varrho_{n+1}'(x) \tag{7}$$

almost everywhere in $[0, \infty)$.

Since $F(x) = O(e^{kx/2})$, $k < 1$, the integral $\int_0^{+\infty} pF^2(x) dx$ is convergent. Now, if we let

$$S_n = \int_0^{+\infty} p(x)\, F^2(x)\, dx - \sum_{k=0}^{n} a_k^2 \tag{8}$$

then by the first theorem of Sec. 8.2 we have

$$\lim_{n\to\infty} S_n = 0. \tag{8'}$$

Moreover, by (6)

$$S_n = \int_0^{+\infty} p(x) \Big[F(x) - \sum_{k=0}^{n} a_k l_k(x)\Big]^2 dx = \int_0^{+\infty} p(x) \varrho_{n+1}^2(x) dx. \tag{9}$$

We note that if $P(x)$ is a polynomial of degree not greater than n we get

$$\int_0^{+\infty} p(x)\, P(x) \varrho_{n+1}(x)\, dx = 0. \tag{10}$$

In fact, $P(x)$ is a linear combination of $l_0(x)$, $l_1(x)$, . . ., $l_n(x)$ and

$$\int_0^{+\infty} p(x)\, l_k(x)\, \varrho_{n+1}(x)\, dx = 0 \quad \text{for} \quad k = 0,\ 1,\ 2, \ldots, n.$$

Now

$$\int_0^{+\infty} p(x)\, \varrho_{n+1}^2(x)\, dx = \int_0^{+\infty} e^{-x} \varrho_{n+1}^2(x)\, dx$$

$$= -\left[e^{-x} \varrho_{n+1}^2(x)\right]_0^{+\infty} + 2\int_0^{+\infty} e^{-x} \varrho_{n+1}(x)\, \varrho_{n+1}'(x)\, dx,$$

whence

$$\varrho_{n+1}^2(0) = -2\int_0^{+\infty} e^{-x} \varrho_{n+1}(x)\, \varrho_{n+1}'(x)\, dx + \int_0^{+\infty} e^{-x} \varrho_{n+1}^2(x)\, dx.$$

Multiplying (7) by $e^{-x}\varrho_{n+1}(x)dx$, integrating between 0 and $+\infty$ and using (10) we have

$$\int_0^{+\infty} e^{-x} \varrho_{n+1}(x) f(x)\, dx = \int_0^{+\infty} e^{-x} \varrho_{n+1}'(x)\, \varrho_{n+1}(x)\, dx$$

whence

$$\varrho_{n+1}^2(0) = -2\int_0^{+\infty} e^{-x} \varrho_{n+1}(x) f(x)\, dx + \int_0^{+\infty} e^{-x} \varrho_{n+1}^2(x)\, dx,$$

$$\varrho_{n+1}^2(0) \leqq 2\left[\int_0^{+\infty} e^{-x} f^2(x)\, dx\right]^{1/2} \left[\int_0^{+\infty} e^{-x} \varrho_{n+1}^2(x)\, dx\right]^{1/2}$$

$$+ \int_0^{+\infty} e^{-x} \varrho_{n+1}^2(x)\, dx \leqq 2\,|\,A\,|\, S_n^{1/2} + S_n$$

and, finally, by (8′) $\lim_{n\to\infty} \varrho_{n+1}(0) = 0$.

3. Before studying the expansion of a function in series of polynomials $H_n(x)$ we give the following criterion of Stone [106] for the uniform convergence of the series of polynomials $H_n(x)$ relative to a function $F(x)$ under quite general conditions. (For a more general criterion cf. A. Ghizzetti: "Sugli Sviluppi in serie di Funzioni di Hermite," *Ann. Sc. Norm. Sup. of Pisa*, (3), 5 (1951), pp. 29—37).

THEOREM. Let $f(x)$ be defined for all finite values of x and integrable in any finite interval. Also let

$$F(x) = F(0) + \int_0^x f(x)\, dx, \qquad -\infty < x < \infty \tag{11}$$

and

$$G(x) = -2xF(x) + f(x). \tag{13}$$

Then, if

$$F(x) = O(e^{kx^2}), \quad k < 1 \tag{12}$$

and if the integral

$$\int_{-\infty}^{+\infty} e^{-x^2} G^2(x)\, dx \tag{14}$$

is convergent, it follows that the series

$$\sum_{n=0}^{\infty} c_n e^{-x^2} H_n(x), \tag{15_1}$$

$$c_n = \frac{1}{2^n n! \sqrt{\pi}} \int_{-\infty}^{+\infty} e^{-x^2} F(x) H_n(x)\, dx \tag{15_2}$$

converges uniformly to $e^{-x^2}F(x)$ in $(-\infty, +\infty)$ and the series

$$c_0 H_0(x) + c_1 H_1(x) + \ldots + c_n H_n(x) + \ldots \tag{16}$$

converges uniformly to $F(x)$ in any finite interval.

(The series (15_1) and (16) are called of type A and type H respectively, cf. C. V. L. Charlier [[20], p. 57]. We note in passing that if we let

$$\varphi_n(x) = e^{-x^2} H_n(x), \qquad f_n(x) = \frac{e^{-x^2/2} H_n(x)}{\sqrt{2^n n! \sqrt{\pi}}},$$

then, if $f(x)$ is an arbitrary function, the series

$$\sum_{n=0}^{\infty} \alpha_n H_n(x), \quad \sum_{n=0}^{\infty} \beta_n f_n(x), \quad e^{\sum_{n=0}^{\infty} \gamma_n H_n(x)}, \quad \sum_{n=0}^{\infty} \delta_n \varphi_n(x)$$

where

$$\alpha_n = \int_{-\infty}^{+\infty} \frac{e^{-x^2} H_n(x) f(x)}{2^n n! \sqrt{\pi}}\, dx, \qquad \beta_n = \int_{-\infty}^{+\infty} f_n(x) f(x)\, dx,$$

$$\gamma_n = \int_{-\infty}^{+\infty} \frac{e^{-x^2} H_n(x) \log f(x)}{2^n n! \sqrt{\pi}}\, dx, \quad \delta_n = \int_{-\infty}^{+\infty} \frac{H_n(x) f(x)}{2^n n! \sqrt{\pi}}\, dx$$

are called respectively of type H, h, C, A, according to Charlier).

Proof. Integrating by parts and using (4) of Sec. 2 and hypothesis (12), we have

$$\begin{aligned} c_n &= \frac{1}{2^n n!\sqrt{\pi}} \int_{-\infty}^{+\infty} e^{-x^2} F(x) H_n(x)\, dx \\ &= -\frac{1}{2^{n+1}(n+1)!\sqrt{\pi}} \int_{-\infty}^{+\infty} e^{-x^2} F(x) H'_{n+1}(x)\, dx \\ &= \frac{1}{2^{n+1}(n+1)!\sqrt{\pi}} \int_{-\infty}^{+\infty} H_{n+1}(x) \left[\frac{d}{dx}\left(e^{-x^2}F(x)\right)\right] dx \\ &= \frac{1}{2^{n+1}(n+1)!\sqrt{\pi}} \int_{-\infty}^{+\infty} e^{-x^2} G(x) H_{n+1}(x)\, dx = C_{n+1}. \end{aligned}$$

By (10) of Sec. 5 we have for any x,

$$| H_n(x) | < K e^{x^2/2} 2^{n/2} (n!)^{1/2} n^{-1/4} (1 + | x |^{5/2}), \quad K \text{ constant,}$$

whence

$$\begin{aligned} | c_n e^{-x^2} H_n(x) | &< K e^{-x^2/2}\, 2^{n/2} (n!)^{1/2} n^{-1/4} (1 + | x |^{5/2}) \,| C_{n+1} | \\ &< K' e^{-l(x^2/2)} 2^{(n+1)/2} [(n+1)!]^{1/2} (n+1)^{-3/4} \,| C_{n+1} | \\ &< K' e^{-l(x^2/2)} \{2^{n+1}(n+1)! C_{n+1}^2 + (n+1)^{-3/2}\} = e^{-l(x^2/2)} T_n \end{aligned}$$

where $0 < l < 1$, K' is a positive constant independent of n, and

$$T_n = 2^{n+1}(n+1)! C_{n+1}^2 + (n+1)^{-3/2}.$$

The assumed convergence of $\int_{-\infty}^{+\infty} e^{-x^2} G^2(x) dx$ together with Bessel's equality imply the convergence of the series of the sum of the squares of the Fourier coefficients of $e^{-x^2/2} G(x)$ with respect to the orthonormal system $\{e^{-x^2/2} H_n(x)/\sqrt{2^n n! \sqrt{\pi}}\}$, namely of the series

$$\sum_{n=0}^{\infty} \frac{1}{2^n n!\sqrt{\pi}} \left[\int_{-\infty}^{+\infty} e^{-x^2} H_n(x)\, G(x)\, dx\right]^2 = \sum_{n=0}^{\infty} 2^n n! \sqrt{\pi} C_n^2 .$$

Consequently, the numerical series $\sum_{n=0}^{\infty} T_n$ is convergent. (Obviously the series $\sum_{n=0}^{\infty} (n+1)^{-3/2}$ is convergent) and therefore the series $\sum_{n=0}^{\infty} c_n e^{-x^2} H_n(x)$ is uniformly convergent in $(-\infty, +\infty)$ to a continuous function $F^*(x)$, namely

$$F^*(x) = \sum_{n=0}^{\infty} c_n e^{-x^2} H_n(x) \tag{17_1}$$

and

$$F^*(x) = O(e^{-l(x^2/2)}). \tag{17_2}$$

By virtue of the convergence of the series $\sum_{n=0}^{\infty} T_n$, we have for $N \geqq m$

$$\left| \frac{1}{2^m m! \sqrt{\pi}} \int_{-\infty}^{+\infty} F^*(x) H_m(x)\, dx - c_m \right|$$

$$= \left| \frac{1}{2^m m! \sqrt{\pi}} \int_{-\infty}^{+\infty} [F^*(x) - \sum_{n=0}^{N} c_n H_n(x) e^{-x^2}] H_m(x)\, dx \right|$$

$$\leqq \left| \int_{-\infty}^{+\infty} H_m(x) \sum_{N+1}^{\infty} c_n e^{-x^2} H_n(x)\, dx \right| \leqq \int_{-\infty}^{+\infty} e^{-l(x^2/2)} |H_m(x)| \sum_{N+1}^{\infty} T_n dx$$

$$= o(1) \left(\int_{-\infty}^{+\infty} e^{-l(x^2/2)} \mid H_m(x) \mid dx \right) = o(1),$$

therefore

$$c_m = \frac{1}{2^m m! \sqrt{\pi}} \int_{-\infty}^{+\infty} F^*(x) H_m(x)\, dx,$$

whence by (15_2) we have

$$\int_{-\infty}^{+\infty} [F^*(x) e^{(l/p)x^2} - e^{-(1-l/p)x^2} F(x)] e^{-(l/p)x^2} H_n(x)\, dx = 0,$$

where p denotes a positive integer > 2, such that

$$k + \frac{l}{p} - 1 < 0.$$

Since the difference $F^*(x) e^{(l/p)x^2} - e^{-(1-l/p)x^2} F(x)$ is square integrable in $(-\infty, +\infty)$ it follows from Sec. 7.3d that the function $e^{(l/p)x^2}(F^*(x) - e^{-x^2} F(x))$ vanishes almost everywhere, and by the continuity of F and F^*, $F^*(x) = e^{-x^2} F(x)$, and therefore

$$e^{-x^2} F(x) = \sum_{n=0}^{\infty} c_n e^{-x^2} H_n(x),$$

uniformly in $(-\infty, +\infty)$, and, finally,

$$F(x) = \sum_{n=0}^{\infty} c_n H_n(x)$$

uniformly in every finite interval.

(For an extension of this theorem to the case where $F(x)$ has derivatives up to a certain order, cf. C. V. L. Charlier, op. cit. [[20], pp. 71, 73].)

REMARK. By (17_2) we have $F(x) = e^{x^2}F^*(x) = O(e^{(1-l/2)x^2})$ which is a consequence of (12), for $k \leqq 1/2$.

10. Pointwise Convergence of the Series of Type h and Uspensky's Criterion for Convergence [115]

1. Suppose $f(x)$ is integrable in any finite interval and that $\int_{-\infty}^{+\infty} f(x)\,dx$ is convergent. We shall study the pointwise convergence of the Fourier series of type h of $f(x)$, namely

$$f(x) \sim \sum_{n=0}^{\infty} c_n f_n(x), \tag{1}$$

where

$$f_n(x) = \pi^{-1/4}(2^n n!)^{-1/2} e^{-x^2/2} H_n(x), \tag{2}$$

$$c_n = \int_{-\infty}^{\infty} f(x) f_n(x)\,dx. \tag{3}$$

(The convergence of the integral $\int_{-\infty}^{+\infty} f(x)dx$ implies that the integrals $\int_{-\infty}^{+\infty} ff_n dx$ are convergent; in fact for a sufficiently large, $e^{-x^2/2}H_n(x)$ is monotone in $[a, +\infty)$, and tends to zero as $n \to \infty$, whence for $a < k$ we have by the second theorem of the mean

$$\int_a^k ff_n\,dx = f_n(a)\int_a^{k_1} f\,dx + f_n(k)\int_{k_1}^k f\,dx, \qquad a < k_1 < k$$

and from this the convergence of $\int_{-\infty}^{+\infty} ff_n\,dx$ follows at once.)

Christoffel's formula of Sec. 2.4, (11) can be written

$$\sum_{k=0}^{n} f_k(x) f_k(y) = \sqrt{\frac{n+1}{2}}\,\frac{f_{n+1}(x)f_n(y) - f_{n+1}(y)f_n(x)}{y - x}$$

so that for the sum $S_n(x)$ of the first $n + 1$ terms of the series we have

$$S_n(x) = \sum_{k=0}^{n} c_k f_k(x) = \sum_{k=0}^{n} f_k(x) \int_{-\infty}^{+\infty} f(\alpha) f_k(\alpha)\, d\alpha$$

$$= \int_{-\infty}^{+\infty} f(\alpha) \Big[\sum_{k=0}^{n} f_k(x) f_k(\alpha) \Big] d\alpha$$

whence

$$S_n(x) = \sqrt{\frac{n+1}{2}} \int_{-\infty}^{+\infty} f(\alpha) \frac{f_{n+1}(x) f_n(\alpha) - f_{n+1}(\alpha) f_n(x)}{\alpha - x}\, d\alpha,$$

or, letting

$$k_n(x, \alpha) = [f_{n+1}(x) f_n(\alpha) - f_{n+1}(\alpha) f_n(x)]/(\alpha - x) \tag{4}$$

we have

$$S_n(x) = \sqrt{\frac{n+1}{2}} \int_{-\infty}^{+\infty} k_n(x, \alpha) f(\alpha)\, d\alpha \tag{5}$$

2. To study the behavior of the integral (5) as $n \to \infty$ we first establish an important functional relationship for $k_n(x, \alpha)$.

Clearly, if $f(x) = e^{-x^2/2} P(x)$ where $P(x)$ is a polynomial of the nth degree, then $S_n(x) = f(x)$ so that if we take for $f(x)$

$$f(x) = \frac{f_{n+1}(\xi) f_n(x) - f_{n+1}(x) f_n(\xi)}{x - \xi} = k_n(\xi, x)$$

where ξ is a parameter, (5) becomes

$$\frac{f_{n+1}(\xi) f_n(x) - f_{n+1}(x) f_n(\xi)}{x - \xi} = \sqrt{\frac{n+1}{2}} \int_{-\infty}^{+\infty} k_n(x, \alpha)\, k_n(\xi, \alpha)\, d\alpha$$

whence, passing to the limit, as $\xi \to x$, we have

$$f_{n+1}(x) f_n'(x) - f_{n+1}'(x) f_n(x) = \sqrt{\frac{n+1}{2}} \int_{-\infty}^{+\infty} [k_n(x, \alpha)]^2\, d\alpha \tag{6}$$

3. We can now determine the asymptotic expression for $k_n(x, \alpha)$. From (14_1), (14_2) of Sec. 5.2 we have

$$\sqrt{\frac{2n+1}{2}}\,(x-\alpha)\,k_{2n}(x,\alpha)$$

$$\begin{aligned}
&= -C^{(n)}\left[\sin[\sqrt{4n+3}\,x] - \frac{x^3}{6}\frac{\cos[\sqrt{4n+3}\,x]}{\sqrt{4n+3}} + \frac{t(2n+1,x)}{f'_{2n+1}(0)\sqrt{4n+3}}\right]\\
&\quad\times\left[\cos[\sqrt{4n+1}\,\alpha] + \frac{x^3}{6}\frac{\sin[\sqrt{4n+1}\,x]}{\sqrt{4n+1}} + \frac{t(2n,\alpha)}{f_{2n}(0)(4n+1)}\right]\\
&\quad+ C^{(n)}\left[\sin[\sqrt{4n+3}\,\alpha] - \frac{\alpha^3}{6}\frac{\cos[\sqrt{4n+3}\,\alpha]}{\sqrt{4n+3}} + \frac{t(2n+1,\alpha)}{f'_{2n+1}(0)\sqrt{4n+3}}\right]\\
&\quad\times\left[\cos[\sqrt{4n+1}\,x] + \frac{x^3}{6}\frac{\sin[\sqrt{4n+1}\,x]}{\sqrt{4n+1}} + \frac{t(2n,x)}{f_{2n}(0)(4n+1)}\right],
\end{aligned} \tag{7}$$

where from the expressions for $f_{2n}(0)$, $f'_{2n+1}(0)$ of Sec. 5.1, we get

$$C^{(n)} = \sqrt{\frac{2n+1}{2}}\,\frac{f_{2n}(0)f'_{2n+1}(0)}{\sqrt{4n+3}} = -\frac{\sqrt{2n+1}}{\sqrt{2\pi}}\sqrt{\frac{4n+2}{4n+3}}\,\frac{(2n-1)!!}{(2n)!!}$$

$$= -\frac{1}{\pi}\sqrt{\frac{4n+2}{4n+3}}\sqrt{\frac{2n+1}{2n+\theta}}, \qquad 0<\theta<1,$$

and since

$$1+\frac{1}{4n} > \sqrt{1+\frac{1}{2n}} > \sqrt{\frac{4n+2}{4n+3}}\sqrt{\frac{2n+1}{2n+\theta}} > \sqrt{\frac{4n+2}{4n+3}}$$

$$= \sqrt{1-\frac{1}{4n+3}} > 1 - \frac{1}{2(4n+2)} \geqq 1 - \frac{1}{12n},$$

we have also

$$-C^{(n)}\pi = 1 + \frac{\varepsilon}{12n}, \qquad |\varepsilon| < 3. \tag{8}$$

Now from the expressions for $f_{2n}(0)$, $f'_{2n+1}(0)$ of Sec. 5.1 we get

$$\frac{1}{4n+1}\left|\frac{1}{f_{2n}(0)}\right| = \frac{\sqrt[4]{\pi}}{4n+1}\sqrt{\frac{(2n)!!}{(2n-1)!!}} < \frac{\pi^{1/2}}{4n}\sqrt[4]{\frac{2n+\theta}{2}} < \frac{\pi^{1/2}}{4}\sqrt[4]{\frac{3}{2}}\frac{1}{n^{3/4}},$$

$$\frac{1}{\sqrt{4n+3}}\left|\frac{1}{f'_{2n+1}(0)}\right| = \frac{\sqrt[4]{\pi}}{\sqrt{4n+3}}\sqrt[4]{\frac{(2n)!!}{(2n-1)!!}}\frac{1}{\sqrt{4n+2}}$$
$$< \frac{\pi^{1/2}}{4n}\sqrt[4]{\frac{2n+\theta}{2}} < \frac{4}{\pi^{1/2}}\sqrt[4]{\frac{3}{2}}\frac{1}{n^{3/4}},$$

and by the inequalities (15_1), (15_2) of Sec. 5.2 for $t(n, x)$ and the inequalities (22), (23) of the same section for $\partial t(2n, x)/\partial x$, we get from (7) and (8)

$$(9)\qquad \sqrt{\frac{2n+1}{2}}(x-\alpha)\,k_{2n}(x,\alpha) = M_1^{(n)} + M_2^{(n)} + M_3^{(n)} + M_4^{(n)} + \frac{(x-\alpha)T_5^{(n)}(x,\alpha)}{\sqrt{n}},$$

where

$$(10_1)\qquad M_1^{(n)} = \cos[\sqrt{4n+1}\,\alpha]\sin[\sqrt{4n+3}\,x] - \cos[\sqrt{4n+1}\,x]\sin[\sqrt{4n+3}\,\alpha],$$

$$(10_2)\qquad \sqrt{4n+1}\,M_2^{(n)} = -\frac{x^3}{6}\sin[\sqrt{4n+1}\,x]\sin[\sqrt{4n+3}\,\alpha] + \frac{\alpha^3}{6}\sin[\sqrt{4n+1}\,\alpha]\sin[\sqrt{4n+3}\,x],$$

$$(10_3)\qquad \sqrt{4n+3}\,M_3^{(n)} = \frac{\alpha^3}{6}\cos[\sqrt{4n+1}\,x]\cos[\sqrt{4n+3}\,\alpha] - \frac{x^3}{6}\cos[\sqrt{4n+1}\,\alpha]\cos[\sqrt{4n+3}\,x],$$

$$(10_4)\qquad \sqrt{(4n+1)(4n+3)}\, M_4^{(n)} = \frac{\alpha^3 x^3}{6}\left[-\cos\left[\sqrt{4n+3}\,x\right] - \sin\left[\sqrt{4n+1}\,\alpha\right] + \cos\left[\sqrt{4n+3}\,\alpha\right]\sin\left[\sqrt{4n+1}\,x\right]\right],$$

and

$$(11)\qquad |\,T_5^{(n)}(x, \alpha)\,| \leqq L_5 ,$$

where L_5 is a constant independent of n for all x and α in a finite interval.

Letting

$$N = \frac{\sqrt{4n+3} + \sqrt{4n+1}}{2} = \frac{1}{\sqrt{4n+3} - \sqrt{4n+1}}$$

we have

$$\begin{aligned} M_1^{(n)} &= \tfrac{1}{2}\left[\sin\left(\sqrt{4n+3}\,x + \sqrt{4n+1}\,\alpha\right)\right. \\ &\quad + \sin\left(\sqrt{4n+3}\,x - \sqrt{4n+1}\,\alpha\right) \\ &\quad \left. - \sin\left(\sqrt{4n+3}\,\alpha + \sqrt{4n+1}\,x\right) - \sin\left(\sqrt{4n+3}\,\alpha - \sqrt{4n+1}\,x\right)\right] \\ &= \cos(x+\alpha)\frac{\sqrt{4n+3}+\sqrt{4n+1}}{2}\sin(x-\alpha)\frac{\sqrt{4n+3}-\sqrt{4n+1}}{2} \\ &\quad + \cos(x+\alpha)\frac{\sqrt{4n+3}-\sqrt{4n+1}}{2}\sin(x-\alpha)\frac{\sqrt{4n+3}+\sqrt{4n+1}}{2} \\ &= \cos\left[N(x+\alpha)\right]\sin\frac{x-\alpha}{2N} + \cos\frac{x+\alpha}{2N}\sin\left[N(x-\alpha)\right] \end{aligned}$$

whence

$$M_1^{(n)} = \cos\left[N(x+\alpha)\right]\sin\frac{x-\alpha}{2N} - 2\sin^2\frac{x+\alpha}{4N}\sin\left[N(x-\alpha)\right] + \sin\left[N(x-\alpha)\right],$$

$$(12_1)\qquad M_1^{(n)} = \sin\left[N(x-\alpha)\right] + \frac{(x-\alpha)\,T_1^{(n)}(x, \alpha)}{N},$$

where, since

$$T_1^{(n)}(x, \alpha) = \tfrac{1}{2}\cos[N(x+\alpha)]\frac{\sin[(x-\alpha)/2N]}{(x-\alpha)/2N}$$
$$-\frac{1}{8}\frac{\sin^2[(x+\alpha)/4N]}{[(x+\alpha)/4N]^2}\frac{\sin[N(x-\alpha)]}{N(x-\alpha)}(x^2-\alpha^2)^2$$

for x and α varying in finite intervals, we have

$$(13_1) \qquad | T_1^{(n)}(x, \alpha) | < L_1,$$

with L_1 a constant independent of n.

By (10_2) we have

$$\sqrt{4n+1}\, M_2^{(n)} = \frac{\alpha^3 - x^3}{6}\sin[\sqrt{4n+1}\, x]\sin[\sqrt{4n+3}\,\alpha]$$
$$+\frac{\alpha^3}{6}\{\sin[\sqrt{4n+1}\,\alpha]\sin[\sqrt{4n+3}\, x]$$
$$-\sin[\sqrt{4n+1}\, x]\sin[\sqrt{4n+3}\,\alpha]\}$$
$$= \frac{\alpha^3 - x^3}{6}\sin[\sqrt{4n+1}\, x]\sin[\sqrt{4n+3}\,\alpha]$$
$$+\frac{\alpha^3}{12}[-\cos(\sqrt{4n+1}\,\alpha+\sqrt{4n+3}\, x)+\cos(\sqrt{4n+3}\, x-\sqrt{4n+1}\,\alpha)$$
$$+\cos(\sqrt{4n+1}\, x+\sqrt{4n+3}\,\alpha)-\cos(\sqrt{4n+3}\,\alpha-\sqrt{4n+1}\, x)$$
$$= \frac{\alpha^3 - x^3}{6}\sin[\sqrt{4n+1}\, x]\sin[\sqrt{4n+3}\,\alpha]$$
$$+\frac{\alpha^3}{6}\left[\sin\frac{x-\alpha}{2N}\sin[N(x+\alpha)] - \sin\frac{x+\alpha}{2N}\sin[N(x-\alpha)]\right],$$

$$\left|\frac{\sqrt{4n+1}}{x-\alpha}M_2^{(n)}\right| \leqq | x^2 + \alpha x + \alpha^2 |$$
$$+\frac{|\alpha|^3}{6}\left|\frac{1}{2}\frac{\sin[(x-\alpha)/2N]}{(x-\alpha)/2N}\frac{\sin[N(x+\alpha)]}{N(x+\alpha)}(x+\alpha)\right.$$
$$\left.-\frac{\sin[(x+\alpha)/2N]}{(x+\alpha)/2N}\frac{x+\alpha}{2}\frac{\sin[N(x-\alpha)]}{N(x-\alpha)}\right|$$

whence

$$M_2^{(n)} = \frac{(x-\alpha)\, T_2^{(n)}(x, \alpha)}{N} \tag{12$_2$}$$

where

$$|\, T_2^{(n)}(x, \alpha)\, | \leqq L_2, \tag{13$_2$}$$

with L_2 a constant independent of n for x and α varying in finite intervals.

Similarly,

$$M_3^{(n)} = \frac{(x-\alpha)\, T_3^{(n)}(x, \alpha)}{N}, \tag{12$_3$}$$

where

$$|\, T_3^{(n)}(x, \alpha)\, | \leqq L_3, \tag{13$_3$}$$

where again the constant L_3 is independent of n for x and α varying in finite intervals. Now applying the theorem of the mean value to the factor in brackets on the right side of (10_4) we get

$$M_4^{(n)} = (x-\alpha)\, T_4^{(n)}(x, \alpha) \tag{12$_4$}$$

where

$$|\, T_4^{(n)}(x, \alpha)\, | \leqq L_4, \tag{13$_4$}$$

with L_4 independent of n for x and α varying in finite intervals. Finally from (9), (11), (12_1), (13_1), . . . , (12_4), (13_4), and by analogous considerations for $k_{2n+1}(x, \alpha)$, we deduce

$$\boxed{\sqrt{\frac{n+1}{2}}\, k_n(x, \alpha) = \frac{1}{\pi}\left[\frac{\sin N(x-\alpha)}{x-\alpha} + \frac{T^{(n)}(x, \alpha)}{N}\right]} \tag{14}$$

where

$$N = \frac{\sqrt{2n+3} + \sqrt{2n+1}}{2}, \tag{15}$$

and

$$|\, T^{(n)}(x, \alpha)\, | \leqq L, \tag{16}$$

where L is a constant independent of N for x and α varying in finite intervals.

4. By (4)

$$k_n(x, \alpha) = \frac{-f_{n+1}(x) f_n(\alpha) + f_{n+1}(\alpha) f_n(x)}{x - \alpha},$$

whence passing to the limit in (14) as $\alpha \to x$ we get

$$\sqrt{\frac{n+1}{2}}\,[f_{n+1}(x) f_n'(x) - f_{n+1}'(x) f_n(x)] = \frac{N}{\pi} + \frac{T^{(n)}(x, x)}{N}$$

and by (6)

$$\sqrt{\frac{n+1}{2}} \int_{-\infty}^{+\infty} [k_n(x, \alpha)]^2 \, d\alpha = \sqrt{\frac{2}{n+1}} \left[\frac{N}{\pi} + \frac{T^{(n)}(x, x)}{N}\right];$$

but

$$N\sqrt{\frac{2}{n+1}} = 2 + O\left(\frac{1}{N}\right), \quad \frac{1}{N} = O\left(\frac{1}{\sqrt{n}}\right)$$

therefore

$$\text{(17)} \qquad \sqrt{\frac{n+1}{2}} \int_{-\infty}^{+\infty} [k_n(x, \alpha)]^2 \, d\alpha = \frac{2}{\pi} + \frac{E(x)}{\sqrt{n}}$$

where the absolute value of $E(x)$ is bounded for all x in a finite interval.

5. If α varies in a finite interval $[-g, g]$ and x varies in an interval $[a, b]$ interior to $[-g, g]$, then

$$\text{(18)} \qquad \begin{aligned} &\sqrt{\frac{n+1}{2}} \int_{-g}^{g} [k_n(x, \alpha)]^2 \, d\alpha \\ &= \frac{2}{\pi^2} \frac{1}{\sqrt{2n+2}} \int_{-g}^{g} \left[\frac{\sin N(x - a)}{x - \alpha} + \frac{T^{(n)}(x, \alpha)}{N}\right]^3 d\alpha. \end{aligned}$$

Clearly

$$\frac{1}{\sqrt{2n+2}} \int_{-g}^{g} \frac{\sin^2 N(x - \alpha)}{(x - \alpha)^2} \, d\alpha = \frac{N}{\sqrt{2n+2}} \int_{-g}^{g} \frac{\sin^2 N(x - \alpha)}{N^2 (x - \alpha)^2} N \, d\alpha,$$

but

$$\frac{N}{\sqrt{2n+2}} = 1 + O\left(\frac{1}{n}\right),$$

$$\int_{-g}^{g} \frac{\sin^2 N(x-\alpha)}{N^2(x-\alpha)^2} N d\alpha = \int_{-N(g+x)}^{N(g-x)} \frac{\sin^2 v}{v^2} dv = \int_{-\infty}^{+\infty} \frac{\sin^2 v}{v^2} dv$$

$$-\int_{-\infty}^{-N(g+x)} \frac{\sin^2 v}{v^2} dv - \int_{N(g-x)}^{+\infty} \frac{\sin^2 v}{v^2} dv = \pi + O\left(\frac{1}{N}\right)$$

uniformly for all x in $[a, b]$.

(Here we used the fact that [Cf. Ch. II, Sec. 10.3, (3)]

$$\int_{-\infty}^{+\infty} \frac{\sin^2 v}{v^2} dv = 2\int_{0}^{+\infty} \frac{\sin^2 v}{v^2} dv = \pi;$$

and

$$\left|\int_{N(g-x)}^{+\infty} \frac{\sin^2 v}{v^2} dv\right| < \int_{N(g-x)}^{+\infty} \frac{1}{v^2} dv = \left[-\frac{1}{v}\right]_{N(g-x)}^{+\infty} = \frac{1}{N(g-x)}\Bigg).$$

Moreover,

$$\left|\frac{1}{\sqrt{2n+2}}\int_{-g}^{g} \frac{\sin N(x-\alpha)}{N(x-\alpha)} T^{(n)}(x, \alpha)\, d\alpha\right|$$

$$< \frac{1}{\sqrt{2n+2}}\int_{-g}^{g} |\, T^{(n)}(x, \alpha)\,|\, d\alpha = O\left(\frac{1}{N}\right),$$

whence by (18)

$$\sqrt{\frac{n+1}{2}}\int_{-g}^{g} [k_n(x, \alpha)^2 d\alpha = \frac{2}{\pi} + O\left(\frac{1}{N}\right),$$

and therefore by (17)

$$\sqrt{\frac{n+1}{2}}\int_{+g}^{+\infty} [k_n(x, \alpha)]^2 d\alpha = O\left(\frac{1}{N}\right),$$

$$\sqrt{\frac{n+1}{2}}\int_{-\infty}^{-g} [k_n(x, \alpha)^2 d\alpha = O\left(\frac{1}{N}\right).$$

It follows that for a fixed interval $[a, b]$ interior to $[-g, g]$ there exists a positive constant Γ such that for any n and for all x in $[a, b]$ we have

$$\boxed{\int_{-\infty}^{-g} [k_n(x, \alpha)]^2 \, d\alpha < \frac{2\Gamma^2}{n+1}, \quad \int_{g}^{+\infty} [k_n(x, \alpha)]^2 \, d\alpha < \frac{2\Gamma^2}{n+1}} . \tag{19}$$

These inequalities obviously hold for any $g_1 > g$.

6. We are now in a position to prove Uspensky's criterion of pointwise convergence.

Assume that x varies in a finite interval $[a, b]$ and let $[-g, g]$ be an interval that contains $[a, b]$ in its interior.

Under the hypothesis that the integral

$$\int_{-\infty}^{+\infty} f^2(x) \, dx$$

is convergent and for $\sigma > 0$, determine $g_1 > g$ so that

$$\int_{-\infty}^{-g_1} f^2(x) \, dx < \sigma^2, \qquad \int_{g_1}^{+\infty} f^2(x) \, dx < \sigma^2.$$

Now by (5), the formula for the sum of the first $n + 1$ terms of series (1), we have

$$\begin{aligned} S_n(x) &= \sqrt{\frac{n+1}{2}} \int_{-g_1}^{g_1} k_n(x, \alpha) f(\alpha) \, d\alpha \\ &+ \sqrt{\frac{n+1}{2}} \int_{-\infty}^{-g_1} k_n(x, \alpha) f(\alpha) d\alpha + \sqrt{\frac{n+1}{2}} \int_{g_1}^{+\infty} k_n(x, \alpha) f(\alpha) \, d\alpha, \end{aligned} \tag{20}$$

and applying the Schwarz inequality to the second and third integrals we get

$$\begin{aligned} &\left| \sqrt{\frac{n+1}{2}} \int_{-\infty}^{-g_1} k_n(x, \alpha) f(\alpha) \, d\alpha \right| \\ &\leq \sqrt{\frac{n+1}{2}} \left[\int_{-\infty}^{-g_1} k_n^2(x, \alpha) \, d\alpha \right]^{1/2} \left[\int_{-\infty}^{-g_1} f^2(\alpha) \, d\alpha \right]^{1/2} < \Gamma\sigma, \end{aligned} \tag{21$_1$}$$

$$
(21_2)\qquad \begin{aligned}
&\left|\sqrt{\frac{n+1}{2}}\int_{g_1}^{+\infty} k_n(x, \alpha)\, f(\alpha)\, d\alpha\right| \\
&\leqq \sqrt{\frac{n+1}{2}}\left[\int_{g_1}^{+\infty} k_n^2(x, \alpha)\, d\alpha\right]^{1/2}\left[\int_{g_1}^{+\infty} f^2(\alpha)\, d\alpha\right]^{1/2} < \Gamma\sigma
\end{aligned}
$$

for any n and for all x in $[a, b]$.

On the other hand by (14) we have

$$
(22)\qquad \begin{aligned}
&\sqrt{\frac{n+1}{2}}\int_{-g_1}^{g_1} k_n(x, \alpha) f(\alpha)\, d\alpha \\
&= \frac{1}{\pi}\int_{-g_1}^{g_1} f(\alpha)\,\frac{\sin N(x-\alpha)}{x-\alpha}\, d\alpha + \frac{1}{\pi N}\int_{-g_1}^{g_1} T^{(n)}(x, \alpha) f(\alpha)\, d\alpha
\end{aligned}
$$

and by (16) there exists an integer n_0 such that for $n > n_0$ we get (as $n \to \infty$ also $N \to \infty$)

$$
(21_3)\qquad \frac{1}{\pi N}\left|\int_{-g_1}^{g_1} T^{(n)}(x, \alpha) f(\alpha)\, d\alpha\right| < \sigma,
$$

and from (20), (21_1), (21_2), (21_3), (22) we get for $n > n_0$ and for all x in $[a, b]$

$$
\left| S_n(x) - \frac{1}{\pi}\int_{-g_1}^{g_1} f(\alpha)\,\frac{\sin N(x-\alpha)}{x-\alpha}\, d\alpha\right| < \sigma(1 + 2\Gamma).
$$

Then by the results of Ch. II, Sec. 4, we have the following theorem due to Uspensky [115]:

If $f(x)$ is a function defined in $(-\infty, +\infty)$, integrable in any finite interval, and if the two integrals

$$
(23)\qquad \int_{-\infty}^{+\infty} f(x)\, dx, \quad \int_{-\infty}^{+\infty} f^2(x)\, dx
$$

are convergent, then at a finite point x_0 the series

$$
(24_1)\qquad \sum_{n=0}^{\infty} c_n f_n(x),
$$

$$
(24_2)\qquad c_n = \int_{-\infty}^{+\infty} f(x) f_n(x)\, dx,
$$

$$f_n(x) = \frac{e^{-x^2/2}H_n(x)}{\sqrt{2^n n!\sqrt[4]{\pi}}}, \tag{24_3}$$

behaves like the Fourier trigonometric series of a function which coincides with $f(x)$ in an arbitrarily small neighborhood $(x_0 - h, x_0 + h)$ of x_0.

In particular if $f(x)$ is of bounded variation in a neighborhood $(x_0 - h, x_0 + h)$ of x_0 we have

$$\sum_{n=0}^{\infty} c_n f_n(x_0) = \tfrac{1}{2}[f(x_0 +) + f(x_0 -)],$$

and again if $f(x)$ is continuous and of bounded variation in $[a, b]$, then series (24_1) converges uniformly to $f(x)$ in any interval interior to $[a, b]$. (For the application of (C, δ) summation to the series (24_1) the reader is referred to the memoir of Kogbetliantz [58b], which also contains an ample bibliography on expansions in series of polynomials $H_n(x)$).

For the application of the so-called transformation of Gauss to the series of Hermite polynomials cf. F. Tricomi [114b], G. Palamà [84a].

11. Series of Laguerre Polynomials

1. Let $\alpha > -1$, and consider as in Sec. 7.1 the polynomials

$$l_k^{(\alpha)}(x) = \frac{\Gamma^{1/2}(k+1)}{\Gamma^{1/2}(k+\alpha+1)} L_k^{(\alpha)}(x), \quad (k = 0, 1, \ldots), \tag{1}$$

then (16_1) and (16_2) of Sec. 1.7 become respectively

$$\int_0^{+\infty} e^{-x} x^{\alpha}\, l_n^2(x)\, dx = 1, \tag{2_1}$$

$$\int_0^{+\infty} e^{-x} x^{\alpha}\, l_n(x)\, l_m(x)\, dx = 0, \qquad (n \neq m). \tag{2_2}$$

(For typographic convenience, here and in the sequel, the polynomial $l_n^{(\alpha)}(x)$ will be denoted by $l_n(x)$ when there is no possibility of confusion). Also let $f(x)$ be a given function defined in $[0, +\infty)$, and assume its Fourier coefficients

$$a_k = \int_0^{+\infty} e^{-x} x^\alpha f(x) l_k(x)\, dx, \qquad (k = 0, 1, \ldots) \tag{3}$$

are convergent.

We now study the behavior of the series

$$\sum_{k=0}^{\infty} a_k l_k(x). \tag{4}$$

For the sum $S_n(x)$ of the first $n + 1$ terms we have

$$S_n(x) = \int_0^{+\infty} e^{-y} y^\alpha f(y) k_n(x, y) dy, \tag{5}$$

where

$$k_n(x, y) = \sum_{k=0}^{n} \frac{k!}{\Gamma(k + \alpha + 1)} L_k^{(\alpha)}(x)\, L_k^{(\alpha)}(y)$$
$$= \sum_{k=0}^{n} L_k^{(\alpha)}(x)\, L_k^{(\alpha)}(y)/\Gamma(\alpha + 1) \binom{k + \alpha}{k},$$

or by (8) of Sec. 1

$$\boxed{k_n(x, y) = \sqrt{(n+1)(n+\alpha+1)}\, \frac{l_{n+1}(x) l_n(y) - l_n(x)\, l_{n+1}(y)}{y - x}} \tag{6}$$

2a. For the polynomial $f(x) = [l_{n+1}(\xi)\, l_n(x) - l_{n+1}(x)\, l_n(\xi)]/(x - \xi) = k_n(\xi, x)$, of the nth degree in x, series (4) reduces to the sum $S_n(x)$ of its first $n + 1$ terms. Since the difference $f(x) - S_n(x)$ is orthogonal with respect to the weight function $e^{-x}x^\alpha$ to all the polynomials $l_k(x)$, it vanishes identically [Sec. 7.2]. Therefore

$$\frac{l_{n+1}(\xi)\, l_n(x) - l_{n+1}(x)\, l_n(\xi)}{x - \xi} = \sqrt{(n + 1)(n + \alpha + 1)} \int_0^{+\infty} e^{-y} y^\alpha\, k_n(\xi, y) k_n(x, y)\, dy, \tag{7}$$

and passing to the limit as $\xi \to x$ we obtain

$$l_{n+1}(x) l_n'(x) - l_{n+1}'(x) l_n(x) = \sqrt{(n + 1)(n + \alpha + 1)} \int_0^{+\infty} e^{-y} y^\alpha\, [k_n(x, y)]^2 dy. \tag{8}$$

2b. By formula (33) of Sec. 6.4, and the fact that $d\,[l_n^{(\alpha)}(x)]/dx = -\sqrt{n}\,l_{n-1}^{(\alpha+1)}(x)$ we get the following asymptotic representation

$$(9)\qquad l_{n+1}(x)l'_n(x) - l'_{n+1}(x)\,l_n(x) = \frac{x^{-\alpha-\frac{1}{2}}\,e^x}{\pi\sqrt{n}} + \frac{E_n}{n}$$

where E_n is uniformly bounded with respect to n, as x varies in a finite interval $[a, b]$, $0 < a < b$, (Cf. J. V. Uspensky, [[115], p. 611]), and also the existence of a constant Γ such that

$$(10)\quad \int_0^a y^\alpha\,e^{-y}\,[k_n(x, y)]^2\,dy < \frac{\Gamma}{n^2}, \quad \int_b^{+\infty} y^\alpha\,e^{-y}\,[k_n(x, y)]^2\,dy < \frac{\Gamma}{n^2},$$

for x variable in $[a, b]$, (Cf. J. V. Uspensky, [[115], p. 614]).

3a. On the basis of the preceding results, J. V. Uspensky [[115], p. 618], proceeding as in Sec. 10, has proved the following theorem on pointwise convergence.

Let $\alpha > -1$, and let $f(x)$ be defined in $[0, +\infty)$, integrable in every finite interval of the real positive axis and satisfy the following conditions:

(i). Let the integral $\int_\gamma^{+\infty} y^\alpha\,e^{-y}\,[f(y)]^2\,dy, \gamma > 0$ be convergent;

(ii). for $-1 < \alpha \leqq -\frac{1}{2}$ let the integral $\int_0^\beta y^\alpha\,e^{-y}\,[f(y)]^2\,dy$, be convergent, and for $\alpha > -\frac{1}{2}$, let the integral $\int_0^\beta y^{\alpha/2-1/4}\,|\,f(y)\,|\,dy$, $\beta > 0$, be convergent;

(iii). Let $f(x)$ be of bounded variation in $[x - \delta, x + \delta]$, $x > 0$, $0 < \delta < x$;

then the series (4) is convergent at the point x to $[f(x+) + f(x-)]/2$.

3b. The following theorem for pointwise equiconvergence of W. Rotach [95′] and G. Szegö [[107c]; [107a], pp. 239—240, 259—263] will be stated without proof:

Let $\alpha > -1$, let $f(x)$ be measurable in the sense of Lebesgue in $[0, +\infty)$ and let the integrals

$$\int_0^1 x^\alpha\,|\,f(x)\,|\,dx, \quad \int_0^1 x^{\alpha/2-1/4}\,|\,f(x)\,|\,dx, \quad \int_1^{+\infty} e^{-x/2}\,x^{\alpha/2-13/12}\,|\,f(x)\,|\,dx,$$

$$\int_n^{\infty} e^{-x/2}\,x^{\alpha/2-13/12}\,|\,f(x)\,|\,dx = o(n^{-1/2})$$

be convergent. Then at every point $x > 0$ we have

$$\lim_{n\to\infty}\left[S_n(x) - \frac{1}{\pi}\int_{x^{1/2}-\delta}^{x^{1/2}+\delta} f(\tau^2)\,\frac{\sin\{2n^{1/2}(x^{1/2}-\tau)\}}{x^{1/2}-\tau}\,d\tau\right] = 0,$$

$(0 < \delta < x^{1/2})$, and therefore (cf. Ch. II, Sec. 4) the behavior of series (4) at point x is the same as the behavior of the Fourier trigonometric series of $f(\tau^2)$ at the point $\tau = \sqrt{x}$.

Appendix

Definition 1. If a and b are two points of a line r, the set of points x such that $a \leqq x \leqq b$ will be called a closed interval of r and, will be denoted by $[a, b]$. The set of points x such that $a < x < b$ will be called an open interval of r and will be denoted by (a, b). $(-\infty, \infty)$ will denote the set of all the points of r. The number $b - a$ will be called the length of the interval.

Definition 2. If S is a set of points of r and Σ is a set of a finite or denumerably infinite number of intervals such that every point of S belongs to at least one set of Σ, then Σ will be said to cover S.

Definition 3. If Σ is a set of intervals, the sum of the lengths of the intervals will be called the length of the set Σ. The case where the sum may be infinite is not excluded.

Definition 4. If S is a set of points of r, the greatest lower bound (we admit $+\infty$ as such a value) of the lengths of all sets which cover S will be called the exterior (Lebesgue) measure of S, and will be denoted by $E(S)$.

Definition 5. If S is a set of points of r and $\Sigma - S$ denotes the complement of the set S with respect to a covering Σ of S, then if the greatest lower bound of $E(\Sigma - S)$ for all Σ covering S is zero, the set S will be called measurable and the exterior measure $E(S)$ will be called the measure of S and will be denoted by $\mu(S)$.

Definition 6. If g is a measurable set of points of r and $f(t)$ is a real function defined in g, then if f satisfies any of the following four equivalent conditions, f will be called a measurable function.

1. For every real finite number α, the set of points S of g for which $f > \alpha$ is measurable.

2. For every real finite number α, the set of points S of g for which $f \geqq \alpha$ is measurable.

3. For every real finite number α, the set of points S of g for which $f < \alpha$ is measurable.

4. For every real finite number α, the set of points S of g for which $f \leqq \alpha$ is measurable.

THEOREM 1. (Theorem of Egoroff-Severini). If g is a measurable set of finite measure, and $\{f_n(t)\}$ is a sequence of measurable functions finite almost everywhere in g, which converge as $n \to \infty$ almost everywhere in g to a finite function $f(t)$, then for every $\varepsilon > 0$, there exists a closed subset g^* of g such that the measure $\mu(g - g^*) < \varepsilon$, and the convergence of $f_n(t)$ to $f(t)$ is uniform in g^*.

DEFINITION 7. If g is a measurable set of points of a line, if $f(t)$ is a function defined almost everywhere in g having a finite or denumerably infinite number of distinct finite values λ_1, $\lambda_2, \ldots \lambda_n, \ldots$, if every set γ_n in which $f(t) = \lambda_n$ is measurable, then $f(t)$ will be called a step function in g.

DEFINITION 8. If $f(t)$ is a step function in g and if $f(t) = \lambda_n$ in γ_n, then $f(t)$, and therefore $|f(t)|$, will be called integrable in g, if and only if the sum

$$\sum \lambda_n \mu(\gamma_n) \tag{1}$$

(where it is understood that $\lambda_n \mu(\gamma_n) = 0$ if $\lambda_n = 0$, even if $\mu(\gamma_n)$ is infinite) has a finite number of terms or is an absolutely convergent series.

DEFINITION 9. If a step function is integrable in g, where $f(t) = \lambda_n$ in γ_n, then the sum (1) (which is independent of the order of the terms) will be called the (Lebesgue) integral of $f(t)$ in g and will be denoted by

$$\int_g f(t)\, dt.$$

DEFINITION 10. If $f(t)$ is a measurable function in g, let I' be the set of the integrals of all step functions $\varphi(t)$ satisfying almost everywhere the condition $\varphi(t) \leqq f(t)$ and let I'' be the set of the integrals of all step functions $\psi(t)$ satisfying the condition $f(t) \leqq \psi(t)$. If the greatest lower bound of I'' is equal to the least upper bound of I', then the common value of these two extrema will be called the (Lebesgue) integral of $f(t)$ in g and will be denoted by

$$\int_g f(t)\, dt$$

and the function will be called (Lebesgue) integrable.

THEOREM 2. If $\{g_n\}$ is a sequence of mutually distinct measurable sets of points whose sum is g, if $f(t)$ is integrable in each of the g_n and if the series

$$\sum_{n=1}^{\infty} \int_{g_n} | f(t) |\, dt,$$

is convergent, then $f(t)$ is integrable in g and

$$\int_g f(t)\, dt = \sum_{n=1}^{\infty} \int_{g_n} f(t)\, dt.$$

THEOREM 2a. If $\{g_n\}$ is a sequence of mutually distinct measurable sets of points whose sum is g, if $f(t)$ is integrable in each of the g_n, and the series

$$\Sigma_n \int_{g_n} | f(t) |\, dt$$

is convergent, then $f(t)$ is integrable in g, and

$$\left| \int_g f(t)\, dt \right| \leqq \Sigma_n \int_{g_n} | f(t) |\, dt.$$

THEOREM 3. If $f(t)$ is integrable in g, then

$$\left| \int_g f(t)\, dt \right| \leqq \int_g | f(t) |\, dt.$$

THEOREM 3a. If $\gamma_1, \gamma_2, \ldots, \gamma_n, \ldots$ is a sequence of measurable sets such that $\gamma_{n+1} \subseteqq \gamma_n$, if g is their product and if $f(t)$ is integrable in γ_1, then

$$\int_g f(t)\, dt = \lim_{n\to\infty} \int_{\gamma_n} f(t)\, dt.$$

THEOREM 4. Let $f(t)$ be integrable in g and let t be a real number. If g is the subset of the points of g preceding t and the integral

$$\int_{g_t} f(t)\,dt$$

is zero for every t, then $f(t) = 0$ almost everywhere in g.

THEOREM 5. If $\int_{g_t} f(t)\,dt = 0$ for all the rational values of t, then $f(t) = 0$ almost everywhere in g.

DEFINITION 11. A function $f(x)$ defined in the closed interval $[a, b]$, is said to be absolutely continuous in $[a, b]$ if corresponding to any $\varepsilon > 0$ there exists a $\delta > 0$ such that for any finite or denumerable set of nonoverlapping intervals $[x_n, x'_n]$ belonging to $[a, b]$ for which $\Sigma_n \mid x'_n - x_n \mid < \delta$ then $\Sigma_n \mid f(x'_n) - f(x_n) \mid < \varepsilon$.

THEOREM 6. If $f(x)$ is integrable in $[a, b]$ and $F(t) = \int_a^t f(x)\,dx$, $[F(a) = 0]$, then $F(t)$ is absolutely continuous in $[a, b]$.

More generally, if $f(t)$ is integrable in g and $F(t) = \int_{g_t} f(t)dt$, then for any $\varepsilon > 0$ there exists a $\delta > 0$ such that for any finite or denumerable set of nonoverlapping intervals $[x_n, x'_n]$ for which $\Sigma_n \mid x'_n - x_n \mid < \delta$ then $\Sigma_n \mid F(x'_n) - F(x_n) \mid < \varepsilon$.

DEFINITION 12. A sequence of functions $\{f_n(t)\}$, integrable in any subset γ of finite measure belonging to g, is called internally convergent in g to a function also integrable in g if for any γ

$$\lim_{n\to\infty} \int_\gamma \mid f_n(t) - f(t) \mid dt = 0.$$

THEOREM 7. If a sequence $\{f_n(t)\}$ of functions are internally convergent in g to two functions $f(t)$ and $\varphi(t)$, then $f(t) = \varphi(t)$, almost everywhere in g. Moreover if $\{f_n(t)\}$ converges internally in g to $f(t)$, it also converges internally in g to every function which equals $f(t)$ almost everywhere in g.

DEFINITION 13. Let $\{f_n(t)\}$ be a sequence of functions integrable in g and let $F_n(t) = \int_{g_t} f_n(t)dt$. If for any $\varepsilon > 0$ there exists a $\delta > 0$ such that for any finite or denumerable set of nonoverlapping intervals $[x'_i, x_i]$ for which $\Sigma_i \mid x'_i - x_i \mid < \delta$, it follows that $\Sigma_i \mid F_n(x_i') - F_n(x_i) \mid < \varepsilon$, then the integrals are called equi-absolutely continuous.

THEOREM 8. If $\{f_n(t)\}$ is a sequence of functions integrable in g, if $\{f_n(t)\}$ converges to $f(t)$ almost everywhere in g, and if the integrals $\int_g f_n(t)\,dt$ are equi-absolutely continuous, then $f_n(t)$ converges internally to $f(t)$ in g.

THEOREM 8a. If $\{f_n(t)\}$ is a sequence of functions measurable in g, and if there exists a function f integrable in g, such that

$$| f_n(t) | \leqq f(t) \qquad (n = 1, 2, \ldots)$$

almost everywhere in g, then the integrals $\int_g f_n(t)\,dt$ are equi-absolutely continuous in g.

THEOREM 9. A sequence $\{f_n(t)\}$ of functions integrable in every measurable subset of g of finite measure is internally convergent in g if and only if for every measurable subset γ of g of finite measure the following condition is satisfied

$$\lim_{\substack{n\to\infty \\ m\to\infty}} \int_\gamma | f_n(t) - f_m(t) | \, dt = 0.$$

THEOREM 10. If the sequence $\{f_n(t)\}$ is integrable in every measurable subset γ of g of finite measure and satisfies

$$\lim_{\substack{n\to\infty \\ m\to\infty}} \int_\gamma | f_n(t) - f_m(t) | \, dt = 0$$

then there exists a subsequence of $\{f_n(t)\}$ which converges to a function f almost everywhere in g and $\{f_n(t)\}$ converges internally in g to f. Conversely, if $\{f_n(t)\}$ converges internally in g to f, there exists a subsequence of $\{f_n(t)\}$ which converges to f almost everywhere in g.

THEOREM 11. Let $f(t)$ be integrable in a set g of points of a line r. Let g_t denote the set of points of g which precede t for any real number t. Finally, let $F(t)$ denote $\int_{g_t} f(t)\,dt$. Then $F(t)$ has a derivative almost everywhere on r which equals $f(t)$ almost everywhere in g and which vanishes almost everywhere in the complement g' of g with respect to r.

DEFINITION 14. Let $f(x)$ be defined in $[a, b]$ and let S be any finite or denumerable set of nonoverlapping intervals $[x_n, x'_n]$ belonging to $[a, b]$. If there exists a number $M > 0$ such that

$$\Sigma_n \mid f(x'_n) - f(x_n) \mid < M$$

for every such set S of intervals $[x'_n, x_n]$, then $f(x)$ is said to be of bounded variation in $[a, b]$.

THEOREM 12. If $f(x)$ is of bounded variation in $[a, b]$, there exist two nondecreasing (nonnegative) functions $f_1(x)$ and $f_2(x)$ such that

$$f(x) \equiv f_1(x) - f_2(x)$$

for all x in $[a, b]$. (See Jeffery, *Theory of Functions of a Real Variable*, p. 121).

THEOREM 13. A function of bounded variation in $[a, b]$ is continuous in $[a, b]$ except for at most a denumerable set of points. At every point x of discontinuity $f(x+)$ and $f(x-)$ exist, that is $f(x)$ is either continuous in $[a, b]$ or it has a jump discontinuity.

DEFINITION 15. $f(x)$ is called regular at a point x if $f(x)$ is continuous at x or if $f(x)$ has a discontinuity such that

$$f(x-) + f(x+) = 2f(x).$$

DEFINITION 16. A function $f(x)$ of bounded variation in $[a, b]$ and regular at every point of $[a, b]$ is called regular in $[a, b]$.

THEOREM 14. If a function $f(x)$ is absolutely continuous in $[a, b]$ then it is of bounded variation in $[a, b]$.

THEOREM 15. If $f(x)$ is of bounded variation in $[a, b]$ then $f'(x)$ exists and is finite almost everywhere in $[a, b]$.

THEOREM 16. If $f(x)$ is integrable in a finite interval $[a, b]$ and α is a constant, the function $\mid f(x) - \alpha \mid$ is the derivative of its indefinite integral except at most for a set of measure zero, not depending on α. That is if $F(x) = c + \int_a^x \mid f(t) - \alpha \mid dt$, ($c$ = constant), then $F'(x) = \mid f(x) - \alpha \mid$, except for a set of points x of measure zero, not depending on α.

THEOREM 17. If $u(t)$ and $v(t)$ are absolutely continuous in $[a, b]$ then

$$\int_a^b u(t)\,\frac{dv(t)}{dt}\,dt = u(b)\,v(b) - u(a)\,v(a) - \int_a^b v(t)\,\frac{du(t)}{dt}\,dt,$$

(integration by parts).

THEOREM 18. (Second theorem of the mean). If $f(t)$ is integrable in $[a, b]$ and $\varphi(t)$ is nonnegative, nondecreasing and bounded in $[a, b]$, with $0 < \varphi(t) \leqq L$, then there is a point ξ in $[a, b]$ such that

$$\int_a^b f(t)\,\varphi(t)\,dt = L\int_\xi^b f(t)\,dt.$$

THEOREM 19. Let $\Gamma(\alpha)$ denote the Gamma function defined for $\alpha > 0$ by the following

$$\Gamma(\alpha) = \int_0^{+\infty} x^{\alpha-1}\,e^{-x}\,dx.$$

It follows at once that $\Gamma(1) = 1$ and also

$$\Gamma(\alpha+1) = \int_0^{+\infty} x^{\alpha}\,e^{-x}\,dx = [-x^{\alpha}e^{-x}]_0^{+\infty} + \alpha\int_0^{+\infty} x^{\alpha-1}\,e^{-x}\,dx = \alpha\Gamma(\alpha)$$

whence the recurrence formula $\Gamma(\alpha + 1) = \alpha\Gamma(\alpha)$, or more generally,

$$\begin{aligned}\Gamma(\alpha+n+1) &= (\alpha+n)\,\Gamma(\alpha+n) = (\alpha+n)\,(\alpha+n-1)\,\Gamma(\alpha+n-1)\\ &= (\alpha+n)\ldots(\alpha+n-m+1)\,\Gamma(\alpha+n-m+1).\end{aligned}$$

In particular, for nonnegative integer values of n, $\Gamma(n + 1) = n!$

THEOREM 20. Stirling's formula is the following

$$\text{(a)}\qquad \Gamma(n+1) = \sqrt{2\pi n}\left(\frac{n}{e}\right)^n e^{\frac{\theta}{12n}}, \qquad 0 < \theta < 1.$$

(Cf. Wilson, *Advanced Calculus* p. 458. Uspensky, *Introduction to Mathematical Probability*, p. 352).

For convenience of the reader, we shall show directly that for $n > 0$ we have

$$\text{(a}'\text{)}\qquad \Gamma(n+1) = \sqrt{2\pi n}\,(n/e)^n[1 + \varepsilon/\sqrt{2\pi n}] \text{ where } 0 < \varepsilon < 1.$$

We have $(x^n e^{-x})' = e^{-x} x^{n-1}(n - x)$, whence, as x varies from 0 to n, $x^n e^{-x}$ increases from 0 to $n^n e^{-n}$, and as x varies from n to ∞, $x^n e^{-x}$ decreases from $n^n e^{-n}$ to 0. As t increases from $-\infty$ to 0, $n^n e^{-n} e^{-t^2}$ increases from 0 to $n^n e^{-n}$, and decreases from $n^n e^{-n}$ to 0 as t increases from 0 to $+\infty$. Consequently, we can make a change of variable by letting

$$\text{(b)} \qquad x^n e^{-x} = n^n e^{-n} e^{-t^2}$$

which establishes a correspondence between x and t so that as t varies from $-\infty$ to $+\infty$, x increases from 0 to $+\infty$. From (b) follows

$$\text{(c)} \qquad t^2 = x - n + n \log n - n \log x = x - n - n \log (x/n),$$

whence, if $0 \leqq x < n$, $t = -\sqrt{n \log (n/x) - (n - x)}$, if $x = n$, $t = 0$, and if $x > n$, $t = \sqrt{x - n - n \log (x/n)}$. If we let $x = n + z$, t has the same sign as z and (c) gives

$$t^2 = z - n \log (1 + z/n).$$

Developing $\log (1 + z/n)$ into a Maclaurin series with remainder, we have

$$t^2 = z - n\left[\frac{z}{n} - \frac{z^2}{2n^2}\frac{1}{[1 + \theta z/n]^2}\right] = \frac{nz^2}{2(n + \theta z)^2}, \qquad 0 < \theta < 1,$$

whence

$$t = z\sqrt{n}/\sqrt{2}(n + \theta z)$$

and therefore

$$\text{(d)} \qquad \frac{n}{z} + \theta = \frac{1}{t}\sqrt{\frac{n}{2}}.$$

Taking the derivative in (c) with respect to t, we have

$$2t = \frac{dx}{dt}\left[1 - \frac{n}{x}\right], \quad \frac{dx}{dt} = \frac{2tx}{x - n} = \frac{2t(n + z)}{z} = 2\left[\sqrt{\frac{n}{2}} + (1-\theta)t\right];$$

whence

$$\text{(e)}\quad \Gamma(n+1) = \int_0^{+\infty} x^n e^{-x}\,dx = 2\int_{-\infty}^{+\infty} n^n e^{-n} e^{-t^2}\left[\sqrt{\frac{n}{2}} + (1-\theta)t\right]dt$$

$$= 2\sqrt{\frac{n}{2}}\, n^n e^{-n}\int_{-\infty}^{+\infty} e^{-t^2}dt + 2n^n e^{-n}\int_{-\infty}^{+\infty} e^{-t^2}(1-\theta)t\,dt$$

$$= \sqrt{2n\pi}\, n^n e^{-n} + 2n^n e^{-n}\int_{-\infty}^{+\infty} e^{-t^2}(1-\theta)\,t\,dt.$$

The last integral cannot be evaluated numerically since θ is not known as a function of t. However, since

$$\int_{-\infty}^{+\infty} e^{-t^2}(1-\theta)t\,dt = \int_{-\infty}^{0} e^{-t^2}(1-\theta)t\,dt + \int_{0}^{+\infty} e^{-t^2}(1-\theta)t\,dt$$

and the first integral on the right side has a negative integrand and is less in absolute value than $\int_0^{+\infty} e^{-t^2}t\,dt = 1/2$, and the second integral is positive and less than $\frac{1}{2}$, we have

$$\left|2\int_{-\infty}^{+\infty} e^{-t^2}(1-\theta)t\,dt\right| < 1$$

and, therefore (a′) follows at once from (e).

THEOREM 21. If $\sum a_n$ converges, then

$$\lim_{x\to 1-} \sum a_n x^n = \sum a_n$$

(See Copson; *Functions of a Complex Variable* pp. 100—101).

DEFINITION 17. Let a finite interval $[a, b]$ be divided into a number of nonoverlapping subintervals $[x_0, x_1]$, $[x_1, x_2]$, . . . $[x_{r-1}, x_r]$, . . ., $[x_{n-1}, x_n]$ where $x_0 = a$, $x_n = b$. Consider all sums $\sum_{r=1}^{n} | f(x_r) - f(x_{r-1}) |$ for all possible subdivisions. If these sums are bounded we say that $f(x)$ is of bounded variation in $[a, b]$ and the least upper bound of these sums is called the total variation of $f(x)$ in $[a, b]$ (cf Def. 14).

For a given subdivision of $[a, b]$, consider the sum $\sum_1[f(x_r) - f(x_{r-1})]$ where those values of r are taken for which $f(x_r) - f(x_{r-1})$ is positive and the sum $\sum_2 |f(x_r) - f(x_{r-1})|$ where those values of r are taken for which $f(x_r) - f(x_{r-1})$ is negative.

If $f(x)$ is of bounded variation, clearly the sums $\sum_1$ and $\sum_2$ are bounded, and the least upper bounds of $\sum_1$ and $\sum_2$ for all possible subdivisions are the total positive variation and the total negative variation respectively of $f(x)$ in $[a, b]$. (See Hobson, *Theory of Real Variables*. Vol. I 3rd. ed., 1950, pp. 307—309).

THEOREM 22. For the proof of formula (13) of Ch. III, Sec. 19.2, cf. e.g. M. Picone [88b],: *Appunti di Analisi Superiore*, (Naples, 1940), p. 89].

The proof of the formula can also be obtained as follows.

Let T be a closed connected domain of three dimensional euclidian space with one or more smooth boundaries. Let FT denote the boundary of T, n the inner normal to FT, $d\sigma$ the surface element of FT. If X, Y, Z are continuous in T, and $\partial X/\partial x$, $\partial Y/\partial y$, $\partial Z/\partial z$ are continuous in the interior of T, then we have Ostrogradsky's formula:

$$\iiint_T \left[\frac{\partial X}{\partial x} + \frac{\partial Y}{\partial y} + \frac{\partial Z}{\partial z}\right] dx\, dy\, dz = -\int_{FT} [X \cos nx + Y \cos ny + Z \cos nz]\, d\sigma.$$

Let $X = \varphi\partial\psi/\partial x$, $Y = \varphi\partial\psi/\partial y$, $Z = \varphi\partial\psi/\partial z$, and

$$\Delta_2\psi = \frac{\partial^2\psi}{\partial x^2} + \frac{\partial^2\psi}{\partial y^2} + \frac{\partial^2\psi}{\partial z^2}, \quad \nabla(\varphi, \psi) = \frac{\partial\varphi}{\partial x}\frac{\partial\psi}{\partial x} + \frac{\partial\varphi}{\partial y}\frac{\partial\psi}{\partial y} + \frac{\partial\varphi}{\partial z}\frac{\partial\psi}{\partial z},$$

if φ, $\partial\psi/\partial x$, $\partial\psi/\partial y$, $\partial\psi/\partial z$, are continuous in T, and $\partial\varphi/\partial x$, $\partial\varphi/\partial y$, $\partial\varphi/\partial z$, $\partial^2\psi/\partial x^2$, $\partial^2\psi/\partial y^2$, $\partial^2\psi/\partial z^2$ are continuous in the interior of T, we have

$$\iiint_T \varphi\Delta_2\psi\, dx\, dy\, dz + \iiint_T \nabla(\varphi, \psi) dx\, dy\, dz = -\int_{FT} \varphi \frac{d\psi}{dn}\, d\sigma.$$

Interchanging φ with ψ and subtracting

$$\iiint_T [\psi\Delta_2\varphi - \varphi\Delta_2\psi] dx\, dy\, dz + \int_{FT} \left[\psi\frac{d\varphi}{dn} - \varphi\frac{d\psi}{dn}\right] d\sigma = 0,$$

so that if φ and ψ and their partial derivatives of the first order are continuous in T and harmonic in the interior of T, we have

$$(*) \qquad \int_{FT} \left[\psi \frac{d\varphi}{dn} - \varphi \frac{d\psi}{dn}\right] d\sigma = 0,$$

and, in particular, for $\psi \equiv 1$

$$(**) \qquad \int_{FT} \frac{d\varphi}{dn}\, d\sigma = 0.$$

Now let φ be a function such that φ and its first derivatives are continuous in T and such that φ is harmonic in the interior of T. Also, if (x_0, y_0, z_0) is a point in the interior of T, and r is defined by

$$r = [(x - x_0)^2 + (y - y_0)^2 + (z - z_0)^2]^{1/2},$$

then

$$\frac{\partial \dfrac{1}{r}}{\partial x} = -(x - x_0)r^{-3}, \qquad \frac{\partial^2 \dfrac{1}{r}}{\partial x^2} = -\frac{1}{r^3} + 3\,\frac{(x - x_0)^2}{r^5},$$

therefore in every domain which excludes the point (x_0, y_0, z_0); $\Delta_2\, 1/r = 0$.

Consider a sphere I with center at P_0 and radius ϱ in the interior of T; in the closed domain $T - I$ the functions $\varphi(x, y, z)$, $\psi = 1/r$ satisfy the conditions required for the validity of (*) and therefore

$$(***) \qquad \int_{FT} \left[\frac{d\varphi}{dn}\frac{1}{r} - \varphi \frac{d\dfrac{1}{r}}{dn}\right] d\sigma + \int_{FI} \left[\frac{d\varphi}{dn}\frac{1}{r} - \varphi \frac{d\dfrac{1}{r}}{dn}\right] d\sigma = 0.$$

On FI we have $1/r = 1/\varrho$ and by (**)

$$\int_{FI} \frac{d\varphi}{dn}\frac{1}{r}\, d\sigma = \frac{1}{\varrho}\int_{FI} \frac{d\varphi}{dn}\, d\sigma = 0.$$

Moreover $d(1/r)/dn = -1/\varrho^2$

$$-\int_{FI} \varphi \frac{d\dfrac{1}{r}}{dn}\, d\sigma = \frac{1}{\varrho^2}\int_{FI} \varphi\, d\sigma,$$

and by the theorem of the mean $\lim_{\rho\to 0} 1/\rho^2 \int \varphi d\sigma = 4\pi\varphi(x_0, y_0, z_0)$, and consequently by (***) we get at once Stokes' formula:

$$\varphi(x_0, y_0, z_0) = \frac{1}{4\pi}\int_{FT}\left[\varphi\frac{d\frac{1}{r}}{dn} - \frac{1}{r}\frac{d\varphi}{dn}\right]d\sigma.$$

THEOREM 23. The Beta function $B(a, b)$ is related to the Gamma function by the following formula

$$B(a, b) = \int_0^1 x^{a-1}(1-x)^{b-1}\,dx = \frac{\Gamma(a)\Gamma(b)}{\Gamma(a+b)}, \quad (a > 0, b > 0).$$

Proof.

$$\begin{aligned}\Gamma(a)\,\Gamma(b) &= \int_0^{+\infty} e^{-x}x^{a-1}\,dx \int_0^{+\infty} e^{-y}\,y^{b-1}\,dy\\ &= 4\int_0^{+\infty} e^{-t^2}\,t^{2a-1}\,dt \int_0^{+\infty} e^{-u^2}\,u^{2b-1}\,du\\ &= 4\int_0^{+\infty}\int_0^{+\infty} e^{-(t^2+u^2)}\,t^{2a-1}\,u^{2b-1}\,dt\,du.\end{aligned}$$

Changing coordinates from Cartesian to polar coordinates by $t = \varrho\cos\theta$, $u = \varrho\sin\theta$, we get

$$\Gamma(a)\,\Gamma(b) = 4\int_0^{+\infty} d\varrho \int_0^{\pi/2} e^{-\rho^2}\varrho^{2a+2b-1}\cos^{2a-1}\theta\,\sin^{2b-1}\theta\,d\theta:$$

but since

$$\Gamma(a+b) = \int_0^{+\infty} e^{-t}\,t^{a+b-1}\,dt = 2\int_0^{+\infty} e^{-\rho^2}\,\varrho^{2a+2b-1}\,d\varrho,$$

it follows that

$$\Gamma(a)\,\Gamma(b)/\Gamma(a+b) = 2\int_0^{\pi/2}\cos^{2a-1}\theta\,\sin^{2b-1}\theta\,d\theta.$$

Changing variables by $\cos^2\theta = x$, the right side becomes

$$\int_0^1 x^{a-1}(1-x)^{b-1}\,dx$$

whence the formula follows at once.

THEOREM 24. The determinant of the matrix product

$$\begin{Vmatrix} x_1, x_2, \ldots, x_m \\ y_1, y_2, \ldots, y_m \end{Vmatrix}^2 = \begin{Vmatrix} x_1, x_2, \ldots, x_m \\ y_1, y_2, \ldots, y_m \end{Vmatrix} \times \begin{Vmatrix} x_1\,, y_1, \\ \cdots\cdots \\ x_m\,, y_m \end{Vmatrix}$$

satisfies the identity

$$(*)\qquad \begin{vmatrix} \sum_{i=1}^{m} x_i^2, & \sum_{i=1}^{m} x_i y_i \\ \sum_{i=1}^{m} x_i y_i, & \sum_{i=1}^{m} y_i^2 \end{vmatrix} = \sum_{i>k}^{m} \begin{vmatrix} x_i, & x_k \\ y_i, & y_k \end{vmatrix}^2$$

whence we get the inequality

$$(**)\qquad \Big(\sum_{i=1}^{m} x_i y_i\Big)^2 \leqq \sum_{i=1}^{m} x_i^2 \sum_{i=1}^{m} y_i^2.$$

The identity (*) for $m = 3$ is due to Lagrange [62], and the inequality (**) for any m is due to Cauchy [17].

THEOREM 25. The formula of J. Wallis [*Opera Mathematica* Oxford (1695), p. 469].

If n is a positive integer, $0 \leqq x \leqq \pi/2$,

$$\sin^{2n+1} x \leqq \sin^{2n} x \leqq \sin^{2n-1} x,$$

whence integrating between 0 and $\pi/2$

$$\frac{(2n)!!}{(2n+1)!!} < \frac{(2n-1)!!}{(2n)!!}\frac{\pi}{2} < \frac{(2n-2)!!}{(2n-1)!!},$$

and multiplying by $(2n)!!/(2n-1)!!$

$$[(2n)!!/(2n-1)!!]^2/(2n+1) < \pi/2 < [(2n)!!/(2n-1)!!]^2/(2n),$$

which gives immediately the formula of Wallis

$$\frac{\pi}{2} = \left[\frac{(2n)!!}{(2n-1)!!}\right]^2 \frac{1}{2n+\theta}, \qquad 0 < \theta < 1.$$

Moreover, we have

$$\sqrt{\frac{(2n-1)!!}{(2n)!!}} = \sqrt[4]{\frac{2}{\pi(2n+\theta)}} = \sqrt[4]{\frac{1}{\pi n}}\sqrt[4]{\frac{2n}{2n+\theta}} = \sqrt[4]{\frac{1}{\pi n}}\left(1 - \frac{\varepsilon_1}{8n}\right),$$

$$0 < \varepsilon_1 < 1;$$

since

$$1 > \sqrt[4]{\frac{2n}{2n+\theta}} > \sqrt[4]{\frac{2n}{2n+1}} > 1 - \frac{1}{8n}.$$

BIBLIOGRAPHY

[1] ABEL, N. H., Sur une espèce particulière de fonctions entières nées du développement de la fonction $(1-v)^{-1}e^{-\frac{xv}{1-v}}$ suivant les puissances de v, *Oeuvres completes*, II Christiania, 1881, p. 284.

[2] ANDREOLI, G., and NALLI, P., Sui processi integrali di Stieltjes, *Ann. Mat. pura e appl.* (4), VII (1929), pp. 47–59.

[3] BACHMAN, P., *Zahlentheorie*, II, Leipzig, 1894, p. 401.

[4] BERNOULLI, D., Petr. N. Comm., 1772.

[5] BERNOULLI, GIAC, Solutio Problematum Fraternorum . . ., *Acta Eruditorum*, 1697, p. 214.

[6a] BERNSTEIN, S., *Leçons sur les propriétés extrémales et la meilleure approximation*, Paris, 1926.

[6b] ———, Sur l'ordre de la meilleure approximation des fonctions, *Mem. publiés Acad. Belg.* (2), 4 (1912), pp. 3–104.

[6c] ———, Démonstration du théorème de Weierstrass fondée sur le calcul des probabilités, *Comm. Soc. Math. Kharkow*, 13 (1912), pp. 1–2.

[6d] ———, Remarques sur l'inégalité de Wladimir Markoff, *Comm. Soc. Math. Kharkow*, 14 (1913–15), pp. 81–83.

[6e] ———, Quelques remarques sur l'interpolation; *Comm. Soc. Math. Kharkow*, 15 (1915–17), pp. 49–61.

[6f] ———, Sur la limitation des valeurs d'un polynome $P_n(x)$ de degré n sur tout un segment par des valeurs en $n+1$ points du segment, *Bull. Acad. Sci. U.R.S.S.* (1931), pp. 1025–1040.

[7] BIANCHI, L., *Lezioni sulla Teoria delle Funzioni di variabile complessa*, Pisa, 1901, pp. 135, 177.

[8a] BÔCHER, M., Introduction to the theory of Fourier Series, *Ann. Math.*, 7 (1906), p. 123.

[8b] ———, On Gibbs's phenomenon, *J. reine u. angew. Math.*, 144 (1914), pp. 41–47.

[9] BOCHNER, S., *Vorlesungen über Fouriersche Integrale*, Leipzig, 1932, pp. 66–67, 68–74.

[10] BONNESEN, T., *Les problèmes des isopérimètres et des isépiphanes*, Paris, 1929.

[11] BONNET, O., Sur le développement des Fonctions en Séries ordonnées suivant les Fonctions X_n, Y_n, *J. Math.* (2) XVII, (1852), pp. 265–300.

[12a] BOREL, E., *Leçons sur les fonctions de variables réelles*, Paris, 1905, pp. 74–82, p. 85.

[12b] ———, *Leçons sur les séries divergentes*, Paris, 1928, Ch. VI, p. 216.

[13] BRUNS, H., Zur Theorie der Kugelfunktionen, *J. reine u. angew. Math.*, 90 (1881), pp. 322–328.

[14] CACCIOPPOLI, R., Sull'approssimazione per polinomi delle funzioni definite in campi illimitati, *Giorn. Ist. It. Attuari*, III (1932), pp. 364–375.

[15a] Cantelli, F. P., Una nuova dimostrazione del secondo teorema limite del calcolo delle probabilità, *Rend. Circ. Mat. di Palermo,* LII (1928), pp. 151–174.

[15b] ———, Un nuovo teorema a proposito del secondo teorema limite del calcolo delle probabilità, *Rend. Circ. Mat. di Palermo,* LII (1928), pp. 416–424.

[15c] ———, Considerations sur la convergence dans le Calcul des Probabilités, *Ann. Sc. Inst. Poincaré,* 5 (1935), p. 7.

[16] Castelnuovo, G., *Calcolo delle Probabilità,* 2nd ed., Bologna, 1928, II, p. 149, pp. 185–187.

[17] Cauchy, A., *Cours d'Analyse,* 1821; *Oeuvres* (2), III, p· 372.

[18] Cesaro, E., Sur la multiplication des séries, *Bull. Sc. Math.* (2), 14 (1890), pp. 114–120.

[19] Chapmann, S., On non integral orders of summability of series and integrals, *Proc. London Math. Soc.,* (2), (1911), pp. 369–409.

[20a] Charlier, C. V. L., *Les applications de la théorie des probabilités aux sciences mathématiques et aux sciences physiques; Application de la théorie des probabilitiés à l'astronomie,* Paris, 1930, p. 52.

[20b] ———, ibid, p. 57.

[21] Chisini, O., Sulla teoria elementare degli isoperimetri, in F. Enriques, *Questioni riguardanti le matematiche elementari, raccolte e coordinate,* III, 3rd ed., Bologna, 1917, pp. 201–310.

[22] Christoffel, E. B., Über die Gaussische Quadratur und eine Gemeinerung derselben, *J. reine u. angew. Math.,* LV (1858), p. 73.

[23a] Cinquini, S., Sopra alcuni polinomi approssimativi, *Boll. Un. Mat. It.,* (2), 17 (1938), pp. 84–90.

[23b] ———, Sopra alcuni risultati relativi al problema dell'approssimazione delle funzioni, *Rend. Ist. Lombardo Sci. e Lettere,* 75 (1941), pp. 23–36.

[24] Cipolla, M., *Analisi Algebrica e introduzione al Calcolo Infinitesimale,* 2nd ed., Palermo, 1921 p. 145, p. 183.

[25a] Cramer, H., Études sur la sommation des séries de Fourier, *Arkiv Math., Astron. och Fysik,* (Stockholm), 13 (1919), no. 20, pp. 1–21.

[25b] ———, Valeurs Maxima des Polynômes d'Hermite, in C. V. L. Charlier [20a], pp. 50–53.

[26] Darboux, G., Mémoire sur l'approximation des fonctions de très-grands nombres, et sur une classe étendue de développement en séries, *J. de Math.* (3), IV (1878), p. 39.

[27a] de la Vallée,Poussin, Ch. J., *Cours d'Analyse Infinitésimale,* Vol. II, 2nd ed., 1912, pp. 126–137.

[27b] ———, *Leçons sur l'approximation des fonctions d'une variable réelle,* Paris, 1919, pp. 7–9, 39–42, 74–109, 94.

[27c] ———, Sur l'approximation des fonctions d'une variable réelle et leurs dérivées par des polynômes et des suites limitées de Fourier, *Bull. Sc. Acad. Belg.,* (1908) pp. 193–254.

[27d] ———, Sur la convergence des formules d'interpolation entre ordonnées équidistantes, *Bull. Sc. Acad. Belg.* (1908), pp. 373–376.

[27e] ———, Un nouveau cas de convergence de séries de Fourier, *Rend. Circ. Mat. Palermo,* XXXI (1911), pp. 296–299.

[27f] ———, Sur l'Integrale de Lebesgue, *Trans. Am. Math. Soc.*, XVI (1915), p. 446.

[28a] DINI, U., Sulla rappresentazione analitica delle funzioni di una variabile reale date arbitrariamente in certi intervalli, *Ann. Univ. Toscane*, 17 (1880), p. 199.

[28b] ———, *Serie di Fourier e altre rappresentazioni analitiche delle funzioni di una variabile reale*, Pisa, 1880.

[28c] ———, *Lezioni sulla teoria delle funzioni sferiche e delle funzioni di Bessel*, Pisa, 1912, pp. 113–128.

[28d] ———, *Lezioni di Analisi Infinitesimale*, II, Pt. I, Pisa, 1909, pp. 409–411.

[28e] ———, Sulle serie di funzioni sferiche, *Ann. Mat. pura e appl.* (2), VI (1874), pp. 112–140.

[28f] ———, Sulla unicità degli sviluppi delle funzioni di una variabile in serie di funzioni X_n, *Ann. Mat. pura e appl.*, (2), VI (1874), pp. 216–225.

[29a] LEJEUNE-DIRICHLET, P. G., Sur les séries dont le terme général dépend de deux angles, et qui servent à exprimer des fonctions arbitraires entre des limites données, *J. reine u. angew. Math.*, 17 (1837), pp. 35–56.

[29b] ———, Über eine neue Methode zur Bestimmung vielfacher Integrale, *Kgl. preuss. Akad. Wiss.*, (1839), pp. 18–25.

[30] EMDE, F., and JAHNKE E., *Funktionen-Tafeln*, Berlin (1933).

[31a] EULER, L., *Opera Omnia*, (1), XIV, p. 584.

[31b] ———, ibid., XIX, p. 227.

[31c] ———, *Methodus inveniendi lineas curvas maximi minimive proprietate gaudentes*, Losanne-Geneve, 1744, Ch. V, Art. 41.

[32] FABER, G., Über die interpolatorische Darstellung stetiger Funktionen, *Jahrb. deutsch. Math. Ver.*, 23 (1914), pp. 192–210.

[32′] FAVARD, J., Sur les multiplicateurs d'interpolation, *J. Math. pures et appl.*, (9), 23 (1944), pp. 219–247.

[33a] FEJÉR, L., Untersuchungen über Fouriersche Reihen, *Math. Ann.*, 58 (1904), pp. 51–69.

[33b] ———, Über die Bestimmung asymptotischer Werte, (Asymptotikus értéker maghtábozásáról), Math. és természettudományi értesitö, 27 (1909), pp. 1–33.

[33c] ———, Über die Laplacesche Reihe, *Math. Ann.*, 67 (1908), pp. 100–103.

[33d] ———, Über die Summabilität der Laplaceschen Reihe durch arithmetische Mittel, *Math. Z.*, 24 (1925), pp. 267–284.

[33e] ———, Abschätzungen für die Legendreschen und verwandte Polynome, *Math. Z.*, 24 (1925), pp. 285–298.

[33f] ———, Über Weierstraßsche Approximation, besonders durch Hermitesche Interpolation, *Math. Ann.*, 102 (1930), pp. 707–725.

[34] FELDHEIM, E., Théorie de la convergence des procédés d'interpolation et de quadrature mécanique, *Mem. Sci. Math.*, 95 (1939), pp. 1–95.

[35] FERRERS, N. M., *Spherical Harmonics*, (1877), p. 76.

[36] FISCHER, E., Sur la convergence en moyenne, *Compt. Rend. Acad. Sci. Paris*, 144 (1907), pp. 1022–1024.

[37a] FOURIER, J. B., Théorie Analytique de la Chaleur (1822), *Oeuvres*, I, Paris, 1888, p. 495, no. 415.

[37b] ———, ibid., p. 475, no. 404.

[37c] ———, ibid., p. 144.

[38] FRÉCHET, M., *Recherches théoriques modernes sur la théorie des probabilités*, Paris, 1937, p. 275.

[39] FROBENIUS, G., Über die Leibnitzsche Reihe, *J. reine u. angew. Math.*, 89 (1880), pp. 262–264.

[40] FUBINI, G., Sugli integrali multipli, *Rend. R. Acc. Naz. Lincei*, (6), 16_1 (1907), pp. 608–614.

[40′] GHIZZETTI, A., Sugli sviluppi in serie di funzioni di Hermite, *Ann. Sc. Norm. Sup. Pisa* (3), 5 (1951), pp. 29–37.

[41] GIBBS, J. W., Fourier Series, *Nature*, 59 (1899), p. 200, p. 606.

[42] GRAM, J. P., Über die Entwickelung reeller Functionen in Reihen mittelst der Methode der kleinsten Quadrate, *J. reine u. angew. Math.*, 94 (1883), pp. 21–73.

[43] GRONWALL, T. H., Zur Gibbschen Erscheinung, *Ann. Math.*, 31 (1930), pp. 233–240.

[44a] HARDY, G. H., Theorems relating to the summability and convergence of slowly oscillating series, *Proc. London Math. Soc.* (2), VIII (1910. pp. 301–320).

[44b] ———, On the summability of Fourier series, *Proc. London Math. Soc.* (2), 12 (1913), pp. 365–372.

[44c] ———, and LITTLEWOOD, J. E., Notes on the theory of series (XVII), Some new convergence criteria for Fourier series, *Proc. London Math. Soc.*, 7 (1932), pp. 252–256.

[45] HEINE, H. E., *Handbuch der Kugelfunktionen*, Berlin, 1878, pp. 171–187.

[46a] HERMITE, CH., Sur un nouveau développement en série de fonctions, *Oeuvres complètes*, II, Paris, 1908, pp. 293–308; (Compt. Rend. Acad. Sci., (Paris), LVIII (1864), pp. 93–100; 266–273); III, Paris, 1912, p. 432.

[46b] ———, Sur les polynômes de Legendre, *J. reine u. angew. Math.*, 107 (1891), pp. 80–83.

[46c] ———, Lettr. no. 385 de Stieltjes à Hermite (1893), *Corr. d'Hermite et de Stieltjes*, II, Paris, 1905, pp. 337–339.

[47a] HILLE, E., A class of reciprocal functions, *Ann. Math.*, 27 (1925), p. 431.

[47b] ———, SHOHAT, J. A. and WALSH, J. L., *A bibliography on orthogonal polynomials*, New York, 1940.

[48a] HOBSON, E. W., *The Theory of Functions of a Real Variable and the Theory of Fourier series* I, 3rd ed., Cambridge, 1927, p. 366, p. 538–561; II, 2nd ed., 1926, pp. 65–98; p. 422, p. 480, p. 761.

[48b] ———, *The Theory of Spherical and Ellipsoidal Harmonics*, Cambridge, 1931, pp. 318–329; pp. 347– 359.

[48c] ———, On a general convergence theorem, and the theory of the representation of a function by series of normal functions, *Proc. London Math. Soc.*, VI (1908), p. 388.

[48d] ———, On the representation of a function by a series of Legendre's functions, *Proc. London Math. Soc.*, VII (1909), pp. 24–39.

[49] HÖLDER, O., Über einen Mittelwertsatz, *Gottingen Nachr.*, 1889, p. 44.

[50a] HURWITZ, H., Sur quelques applications géométriques des séries de Fourier, *Ann. Ec. Norm. Sup.*, (3) 19 (1902), pp. 357–408.

[50b] ———, Über die Fourierschen Konstanten integrierbarer Funktionen, *Math. Ann.*, 57 (1903), pp. 425–446; *Math. Werke*, I, p. 556, II, pp. 509–554.

[51a] JACKSON, D., The Theory of Approximation. *Am. Math. Coll.*, XI, New York, 1930, p. 73, p. 76.

[51b] ———, *Über die Genauigkeit der Annäherung stetiger Funktionen durch ganze rationale Funktionen*, Inaug.-diss. Göttingen, 1911.

[51c] ———, A formula of trigonometric interpolation, *Rend. Circ. Mat. Palermo*, 37 (1914), pp. 371–375.

[52] JACOB, M., Sugli integrali di Stieltjes e sulla loro applicazione nella matematica attuariale, *Giorn. Ist. It. Attuari*, III (1932), pp. 160–181.

[53a] JACOBI, C. G. J., *Ges. Werke*, III, p. 11.

[53b] ———, Ges. Werke, VI, pp. 182–202.

[53c] ———, Untersuchungen über die Differentialgleichung der hypergeometrischen Reihe, *J. reine u. angew. Math.*, 56 (1859), pp. 149–165.

[54] JAHNKE, E., and EMDE, F., *Funktionen Tafeln*, Berlin (1933).

[55a] JORDAN, C., Sur la série de Fourier, *Compt. Rend. Acc. des Sci.*, 92 (1881), pp. 228–230.

[55b] ———, Cours d'Analyse de l'Ec. Polytec., I, 2nd ed., Paris, 1893, p. 99.

[56] KACZMARZ, S., and STEINHAUS, H., *Theorie der Orthogonalreihen*, Warsaw, 1935.

[57] KNOPP, K., *Theorie und Anwendung der unendlichen Reihen*, Berlin, 1924, Ch. XIII.

[58a] KOGBETLIANTZ, E., Sommation des séries et intégrales divergentes, par les moyennes arithmetiques et typiques, *Mem. des Sci. Math.*, LI (1931),

[58b] ———, Sur la sommabilité des séries d'Hermite, *Ann. Sci. Ec. Norm. Sup.* (3), 49 (1932), pp. 137–221. (Includes a full bibliography on expansions in series of the polynomials $L_n^{(\alpha)}(x)$, $H_n(x)$).

[58c] ———, Sur les moyennes arithmétiques des séries-noyaux de développements en séries d'Hermite et de Laguerre et sur celles de ces séries-noyaux derivées terme à terme, *J. Math. and Physics*, 14 (1935), pp. 37–99.

[58d] ———, Contribution à l'étude du saut d'une fonction donnée par son développement en série d'Hermite ou de Laguerre, *Trans. Am. Math. Soc.*, 38 (1935), pp. 10–47.

[59] KOWALEWSKI, G., *Einführung in die Determinatentheorie . . .*, Leipzig, 1909, pp. 320–337.

[60] KOWALLIK, U., Entwicklung einer willkürlichen Funktion nach Hermiteschen Orthogonalfunktionen, *Math. Z.*, 31 (1930), pp. 498–520.

[61] KRYLOFF, N., Sur quelques formules de l'interpolation géneralisée, *Bull. Sci. Math.*, 41 (1917), pp. 313–315.

[62] LAGRANGE, J., Solutions analytiques de quelques problemes sur les pyramides triangulaires, *Acc. Sci. Berlin*, 1773, Oeuvres, III, pp. 662–663,

[63a] LAGUERRE, E. DE, *Oeuvres*, I (Paris, 1898), pp. 428–437; Sur l'intégrale $\int_x^{+\infty} x^{-1} e^{-x}\, dx$, *Bull. Soc. math. France*, VII (1879), pp. 72–81.

[63b] ———, *Oeuvres*, I (Paris, 1898), p. 126; Sur les équations algébriques dont le premier membre satisfait à une équation linéaire de second ordre, *Compt. Rend. Ac. des Sci.*, 90 (1880), pp. 809–812.

[64a] LANDAU, E., Primzahlen. *I* (1909), p. 61.

[64b] ———, Über die Approximation einer stetigen Funktion durch eine ganze rationale Funktion, *Rend. Circ. Mat. Palermo*, 25 (1908), pp. 337–345.

[64c] ———, Über die Bedeutung einiger neuen Grenzwertsätze der Herren Hardy und Axer, *Prace Math. Fiz.*, XXI, Warsaw, 1910, p. 107.

[65a] LAPLACE, P. S., Mécanique Céleste, 1805, *Oeuvres* IV, Book X, p. 257.

[65b] ———, Mécanique Céleste, 1825, *Oeuvres* V, Book XI, Ch. II, p. 41.

[65c] ———, Mécanique Céleste, 1825, *Oeuvres* V, Book XI, Supplement 1, p. 473.

[65d] ———, VII Théorie analytique des probabilités, *Oeuvres* VII, 3rd ed., 1820, p. 105.

[65e] ———, *Oeuvres* XII, p. 368 (Mémoire sur les intégrales définies, *Acc. Sc.* (1), XI, 1810–1811.)

[66] LAURICELLA, G., Sulla chiusura dei sistemi di funzioni ortogonali, *Rend. R. Acc. Naz. Linzei* (5), XXI, (1912, 1st Sem.), pp. 675–685.

[67a] LEBESGUE, H., *Leçons sur les séries trigonométriques*, Paris, 1906.

[67b] ———, *Leçons sur l'intégration et la recherche des fonctions primitives*, 2nd ed., Paris, 1928, pp. 252–313.

[67c] ———, Recherches sur la convergence des séries de Fourier, *Math. Ann.*, 61 (1905), pp. 251–280.

[67d] ———, Sur les intégrales singulières, *Ann. Toulouse* (3), I (1909), p. 44.

[68] LEGENDRE, A. M., Sur l'attraction des Sphéroides, *Mém. Math. et Phys. présentés à l'Ac. r. des, sc. par divers savants*, X, (1785).

[69a] LÉVY, P., *Calcul des probabilités*, Paris, 1925, p. 61.

[69b] ———, ibid., p. 167.

[70] LITTLEWOOD, J. E., and HARDY, G. H., Notes on the theory of series (XVII); Some new convergence criteria for Fourier series, *J. London Math. Soc.*, 7 (1932), pp. 252–256.

[71] LOEWY, A., Der Stieltjessche Integralbegriff und seine Verwendung in der Versicherungsmathematik, *Bl. Vers. Math. u. ver. Gebiete*, II (1931), pp. 3—18, pp. 74–82.

[72] LORENZ, G., Sur la convergence forte des polynomes de Stieltjes-Landau, *Rec. Math.* (Moscow), I (43, (1936), pp. 553–555.

[73] MARKOFF, A. A., Sur une question de Mendeleieff, *Compt. Rend. Acad. Sci. St. Pétersbourg*, 62 (1889), pp. 1–21.

[74] MARKOFF, W. A., Sur les fonctions qui s'écartent le moins de zero, St. Pétersbourg, 1892, translated in German in *Math. Ann.*, 77 (1915), pp. 213–258.

[75] MEHLER, F. G., Notiz über Dirichlet'schen Integralausdrücke für die Kugelfunction $P^n(\cos\theta)$ und über eine analoge Integralform für die Cylinderfunction $J(z)$, *Math. Ann.*, 5 (1872), p. 142.

[76] MINKOWSKI, H., Géometrie der Zahlen, I, Leipzig, 1896, pp. 115–117.

[77] MOECKLIN, E., Asymptotische Entwicklungen der Laguerreschen Polynome, *Comm. Math. Helvetici*, 7 (1934), pp. 24–46.

[78] MOIVRE, A. DE, *Miscellanea analitica de Seriebus et Quadraturis*, London, 1730.

[79] NALLI, P., and ANDREOLI, G., Sui processi integrali di Stieltjes, *Ann. Mat. pura e appl.* (4), VII (1929), pp. 47–59.

[80] NASAROW, N., Über die Entwicklung einer beliebigen Funktion nach Laguerreschen Polynomen, *Math. Z.*, **33** (1931), pp. **481–487**.

[81] NEUMANN, F. E., *Beiträge zur Theorie der Kugelunctionen*, **1878**, p. **133**.

[82] NICOLETTI, O., Sulle condizioni iniziali che determinano gli integrali delle equazioni differenziali ordinarie, *Atti R. Acc. Sci. Torino*, **33** **(1897)**, pp. 746–748.

[83] OTTAVIANI, G., Sulla convergenza e sommabilita delle serie di Hermite; Disuguaglianze fondamentali per i polinomi di Laguerre e di Hermite, *Ann. R. Sc. Norm. Sup. Pisa*, (2), 7 (1938), p. **13**.

[84a] PALAMÁ, G., La trasformazione di Gauss e i polinomi di Hermite, *Rend. R. Acc. Naz. Lincei* (6), 25 (1937), pp. **356–361**.

[84b] ———, Sui polinomi di Legendre, di Laguerre, di Hermite, *Rend. R. Ist. Lombardo Sc. e Lettere*, 70 (1937), pp. 147–191.

[84c] ———, Sulla soluzione polinomiale della $(a_1x + a_0)y'' + (b_1x + b_0)y' - nb_1y = 0$, *Boll. Un. Mat. It.* (2), I (1939), p. 28.

[85] PAPPUS D'ALÉXANDRIE, *La Collection Mathématique*, P. Ver Eecke, Bruges, 1933, p. 258.

[86] PARSEVAL, A., Intégration générale et complète de deux équations importantes dans la mécanique des fluides, *Mém. présentés Inst. Sci.*, I (1805), pp. 524–545.

[87] PERRON, O., Über das Verhalten einer ausgearteten hypergeometrischen Reihe bei unbegrenztem Wachstum eines Parameters, *J. reine u. angew. Math.*, 151 (1921), pp. 63–78.

[88a] PICONE, M., *Lezioni di Analisi Infinitesimale*, Catania, 1923.

[88b] ———, *Appunti di Analise Superiore*, Naples, 1940, p. 46; p. 89; p. 113; p. 211 et seq., p. 259, p. 260, p. 265, pp. 507–560, pp. 775–793.

[88c] ———, Sul determinante di Gram, *Boll. Un. Mat. It.*, V (1926), pp. 81–84.

[88d] ———, Trattazione elementare dell approssimazione lineare in insiemi non limitati, *Giorn. Ist. It. Attuari*, V (1934), pp. 155–195.

[88e] ———, Sull'integrazione delle funzioni, *Rend. Mat. e delle sue appl.* (5), **3** (1942), p. 129.

[89a] POINCARE, H., *Théorie Analytique de la Propagation de la Chaleur*, Paris, 1895, pp. 36–53.

[89b] ———, *Calcul des probabilités*, Paris, 1912, p. 206.

[90a] POISSON, S. D., Mémoire sur la manière d'exprimer les fonctions par les séries de quantités périodiques, *J. Ec. Polytec.*, XVIII (XI), (1820), pp. 417–489.

[90b] ———, Addition au Mémoire precédent et au Mémoire sur la manière d'exprimer . . . , *J. Ec. Polytec.*, XIX (XII), (1823), pp. 145–162.

[90c] ———, Additions pour la connaissance des temps pour l'an 1829 et pour l' an 1831, *J. Ec. Polytec.* XVII (1835).

[90d] ———, *Théorie mathématique de la Chaleur*, Paris, 1835, p. 212.

[91] POLYA, G., Über den zentralen Grenzwertsatz der Wahrscheinlichkeitsrechnung, und das Momentenproblem, *Math. Z.*, 8 (1920), p. 173, p. 174, p. 175, p. 177.

[92] RIEMANN, B., Sur la possibilité de représenter une fonction par une série trigonométrique, *Oeuvre Math.*, 1896, p. 258.

[93a] RIESZ, F., Sur les systèmes orthogonaux des fonctions, *Comp. Rend. Acad. des Sci.*, 144 (1907), pp. 615–619.

[93b] ———, Über die Approximation einer Funktion durch Polynome, *Jahrb. deutsch. Math. Ver.*, 17 (1908), pp. 196–211.

[93c] ———, Démonstration nouvelle d'un théorème concernant les opérations fonctionnelles linéaires, *Ann. Sci. Ec. Norm. Sup.* (3), XXXI (1914), pp. 9–14.

[93d] ———, Untersuchungen über Systeme integrierbarer Funktionen, *Math. Ann.*, 69 (1910), p. 456.

[93e] ———, Su alcune disuguaglianze, *Boll. Un. Mat. It.* (1), 7 (1928), pp. 77–79.

[94a] RIESZ, M., Sur les séries de Dirichlet et les séries entières, *Compt. Rend. Acad. Sci.*, 149 (1909), pp. 309–312.

[94b] ———, Sur la sommation des séries de Fourier, *Acta Math. Sc. R. Un. Hung. Szeged*, I (1923), pp. 104–113.

[95] RODRIGUES, O., Mémoire sur l'attraction des spheroïdes, *Corr. Ec. Roy. Polytech.*, III (1816).

[95′] ROTACH, W., *Reihenentwicklungen einer willkürlichen Funktion nach Hermite'schen und Laguerre'schen Polynomen*, Inaug.-diss. Tech. Hochschule in Zürich, Geneva, 1925, pp. 1–33.

[96] RUNGE, C., Über eine besondere Art von Integralgleichungen, *Math. Ann.*, 75 (1914), pp. 130–132.

[97a] SAKS, S., *Théorie de l'Intégrale*, Warsaw, 1933, p. 49, p. 69, pp. 91–98.

[97b] ———, *Theory of the integral*, Warsaw, 1937, pp. 56–104.

[98a] SANSONE, G., *Lezioni di Analisi Matematica*, I, 9th ed., Padua, 1946, p. 143, p. 428, p. 487, II, 7th ed., Padua, 1948, p. 91, p. 110, p. 455.

[98b] ———, *Equazioni Differenziali nel campo reale*, I, 2nd. ed., Bologna, 1948, pp. 162–164, pp. 165–170, 200–203.

[98c] ———, Sugli zeri delle soluzioni polinomiali dell'equazione $(a_1x + a_0)y'' + (b_1x + b_0)y' - nb_1y = 0$, *Rend. Acc. Naz. Lincei*, (5), 15 (1932), pp. 125–130, 194–197.

[98d] ———, La chiusura dei sistemi orogonali di Legendre, di Laguerre e di Hermite rispetto alle funzioni di quadrato sommabile, *Giorn. Ist. It. Attuari*, IV (1933), pp. 71–82.

[98e] ———, Sul teorema di Parseval in intervalli infiniti, *Ann. R. Sc. Norm. Sup. Pisa* (2), IV (1935), pp. 35–41.

[98f] ———, Sulla sommabilità di Cesaro delle serie di Laplace, *Rend. R. Acc. Naz. Lincei* (6), 25 (1937), pp. 75–81.

[98g] ———, Due semplici limitazioni nel campo complesso delle funzioni associate ai polinomi di Tchebychef-Hermite, e del termine complementare della loro rappresentazione asintotica, *Math. Z.*, 52 (1950), pp. 593–598.

[98h] ———, La formula di approssimazione asintotica dei polinomi di Tchebychef-Laguerre col procedimento di J. V. Uspensky, *Math. Z.*, 53 (1950), pp. 97–105.

[99] SCHLÄFLI, L., *Über die zwei Heine'schen Kugelfunktionen*, Bern, 1881.

[100a] SHOHAT, J. A., Théorie générale des polynomes orthogonaux de Tchebychef, *Mém. des Sc. Math.*, 66, 1834, pp. 1—68.

[100b] ———, Hille, E., and Walsh, J. L., *A bibliography on orthogonal polynomials*, New York, 1940.

[101] Sibirani, F., Sulla rappresentazione approssimata delle funzioni, *Ann. Mat. pura e appl.*, (3), 15 (1909), pp. 203–221.

[102] Steiner, J., Sur le maximum et le minimum des figures dans le plan, sur la sphère, et dans l'espace en général, *J. reine u. angew. Math.*, 24 (1842), pp. 93–152, 189–250.

[103] Steinhaus, H., and Kaczmarz, S., *Theorie der Orthogonalreihen*, Warsaw, 1935.

[104a] Stieltjes, T. J. *Oeuvres complètes*, II, Groningen, 1918, pp. 45–47.

[104b] ———, ibid. p. 241.

[104c] ———, *ibid.*, pp. 234–252.

[104d] ———, *ibid.*, p. 469.

[104e] ———, *Corr. d'Hermite et de Stieltjes*, II, Paris, 1905, pp. 337–339; lettr. no. 385 de Stieltjes à Hermite (1893).

[105] Stirling, J., *Methodus differentialis; sive tractatus de summatione et interpolatione serierum infinitarum*, London, 1739, p. 137.

[106] Stone, M. H., Developments in Hermite polynomials, *Ann. Math.* (2), 29 (1927), pp. 1–13.

[107a] Szegö, G., *Orthogonal Polynomials*, Am. Math. Soc. Colloq. Publs., 23, 1933, p. 6, p. 116, pp. 126–127, p. 194, pp. 221–229.

[107b] ———, Ein Beitrag zur Theorie der Polynome von Laguerre und Jacobi, *Math. Z.*, 1 (1918), pp. 341–350.

[107c] ———, Beiträge zur Theorie der Laguerreschen Polynome, I: Entwicklungssätze, *Math. Z.*, 25 (1926), pp. 87–115.

[107d] ———, Über einige asymptotische Entwicklungen der Legendreschen Funktionen, *Proc. London Math. Soc.* (2), 36 (1934), pp. 427–450.

[107e] ———, Inequalities for the zeros of Legendre polynomials and related functions, *Trans. Am. Math. Soc.* 39 (1936), p. 1–17.

[108a] Tchebychef, P. L., *Oeuvres*, I, St. Petersburg, 1899, p. 138; p. 301, pp. 499–508.

[108b] ———, Théorie des mecanismes connus sous le nom de parallélogrammes, *Mém. Acad. Sci. St. Pétersbourg* (6), 7 (1854) pp. 539–568.

[108c] ———, Sur les questions de minima qui se rattachent à la représentation approximative des fonctions, *Mem. Acad. Sci. St. Pétersbourg*, 9 (1859), pp. 199–291.

[108d] Sur le développement des fonctions à une seule variable, *Bull. ph.-math, Acad. Imp. Sc. St. Pétersbourg*, I (1859), pp. 193–200.

[109] Titchmarsh, E. C., *The theory of functions*, 2nd. ed., Oxford, 1939, p. 168; pp. 233–235.

[110] Togliatti, E. G., Massimi e minimi, in L. Berzolari, G. Vivanti, D. Gigli, *Enciclopedia delle Matematiche Elementari*, Vol. II, Pt. II, (1938), pp. 1–71.

[111a] Tonelli, L., *Fondamenti di Calcolo delle Variazioni*, I, Bologna, 1921, p. 182, p. 183; II, Bologna, 1923, pp. 542–543.

[111b] ———, *Serie Trigonometriche*, Bologna, 1928; pp. 145–148; 186, 188, 189–190, 196, 246, 343, 436.

[111c] ———, I polinomi d'approssimazione di Tchebychef, *Ann. Mat. pura e appl.* (3), 15 (1908), pp. 47–119.

[111d] ———, Sulla rappresentazione analitica delle funzioni di piu variabili reali, *Rend. Circ. Mat. Palermo*, 29 (1910), pp. 1–36.

[111e] ———, Sopra alcuni polinomi approssimativi, *Ann. Mat. pura e appl.* (3), 25 (1916), (pp. 275–316), p. 287.

[111f] ———, Sulla convergenza delle serie di Fourier, *Rend. R. Acc. Naz. Lincei* (6), 2 (1925), pp. 85–91.

[112] Tonolo, A., Sulla chiusura del sistema $1/\sqrt{2\pi}$, $(\cos kx)/\sqrt{\pi}$, $(\sin kx)/\sqrt{\pi}$) *Boll. Un. Mat. It.*, 6 (1927), pp. 121–122.

[113a] Toscano, L., Formule limiti dei polinomi di Laguerre (with a bibliography) *Boll. Un. Mat. It.* (2), 1 (1939), pp. 337–339.

[113b] ———, Relazioni tra i polinomi di Laguerre e di Hermite, *Boll. Un. Mat. It.* (2), 2 (1940), pp. 460–466.

[113c] ———, Formule di addizione e moltiplicazione sui polinomi di Laguerre, *Atti R. Acc. Sc. Torino*, 76 (1940–41), pp. 417–432.

[114a] Tricomi, F., Serie ortogonali di funzioni, Corso di Analisi Superiore 1947–48, Torino, pp. 1–343.

[114b] ———, Les transformations de Fourier, Laplace et Gauss, et leurs applications au calcul des probabilités et à la statistique, *Ann. Sci. Inst. Poincaré*, 8 (1938), pp. 130–132.

[114c] ———, Generalizzazione di una formula asintotica sui polinomi di Laguerre e sue applicazioni, *Atti R. Acc. Sc. Torino*, 76 (1940–41), pp. 288–316.

[114d] ———, Sviluppo dei polinomi di Laguerre e di Hermite in serie di funzioni di Bessel, *Giorn. Ist. It. Attuari*, XII (1941), pp. 14–33.

[114e] ———, Sugli zeri di cui si conosce una rappresentazione asintotica, *Ann. Mat. pura e appl.*, (4), 26 (1947), pp. 283–300.

[114f] ———, Sul comportamento asintotico dell'n^{esimo} polinomio di Laguerre nell' intorno dell'ascissa $4n$, *Comm. Math. Helvetici*, 22 (1949), pp. 150–167.

[114g] ———, Sul comportamento asintotico dei polinomi di Laguerre, *Ann. Mat. pura e appl.*, (4), 28 (1949), pp. 263–289.

[115] Uspensky, J. V., On the development of arbitrary fonctions in series of Hermite and Laguerre polynomials, *Ann. Math.* (2), 28 (1927), pp. 593–619.

[116a] Vitali, G., *Geometria nello spazio Hilbertiano*, Bologna, 1929.

[116b] ———, Sopra la serie di funzioni analitiche, *Ann. Mat. pura e appl.* (3), 10 (1904), pp. 65–82.

[116c] ———, Sulla condizione di chiusura di un sistema di funzioni ortogonali, *Rend. R. Acc. Naz. Lincei* (5), 30 (1921), p. 498.

[117] Wallis, J., *Opera Matematica*, I Oxford, 1695, p. 469.

[118] Walsh, J. L., Hille, E., and Shohat, J. A., A bibliography on orthogonal polynomials, New York, 1940.

[119a] Weierstrass, K., Werke, III, Berlin, 1903, p. 5; p. 22.

[119b] ———, Über die analytische Darstellbarkeit sogenannter willkürlicher Functionen reeller Argumente, *Sitzber. Kgl. preuss. Akad. Wiss. Berlin*, 1885, pp. 633–639; pp. 789–805.

[120] Zalcwasser, Z., Sur le phenomène de Gibbs dans la théorie des séries de Fourier, des fonctions continues, *Fund. Math.* (Warsaw), 12 (1918), pp. 126–151.

[121a] Zygmund, A., Trigonometrical Series, Warsaw, 1935, pp. 1–331.

[121b] ———, *ibid.*, pp. 171–185.

INDEX

A CATALOG OF SELECTED

DOVER BOOKS

IN SCIENCE AND MATHEMATICS

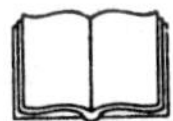

A CATALOG OF SELECTED
DOVER BOOKS
IN SCIENCE AND MATHEMATICS

QUALITATIVE THEORY OF DIFFERENTIAL EQUATIONS, V.V. Nemytskii and V.V. Stepanov. Classic graduate-level text by two prominent Soviet mathematicians covers classical differential equations as well as topological dynamics and erqodic theory. Bibliographies. 523pp. 5⅜ × 8½. 65954-2 Pa. $10.95

MATRICES AND LINEAR ALGEBRA, Hans Schneider and George Phillip Barker. Basic textbook covers theory of matrices and its applications to systems of linear equations and related topics such as determinants, eigenvalues and differential equations. Numerous exercises. 432pp. 5⅜ × 8½. 66014-1 Pa. $8.95

QUANTUM THEORY, David Bohm. This advanced undergraduate-level text presents the quantum theory in terms of qualitative and imaginative concepts, followed by specific applications worked out in mathematical detail. Preface. Index. 655pp. 5⅜ × 8½. 65969-0 Pa. $10.95

ATOMIC PHYSICS (8th edition), Max Born. Nobel laureate's lucid treatment of kinetic theory of gases, elementary particles, nuclear atom, wave-corpuscles, atomic structure and spectral lines, much more. Over 40 appendices, bibliography. 495pp. 5⅜ × 8½. 65984-4 Pa. $11.95

ELECTRONIC STRUCTURE AND THE PROPERTIES OF SOLIDS: The Physics of the Chemical Bond, Walter A. Harrison. Innovative text offers basic understanding of the electronic structure of covalent and ionic solids, simple metals, transition metals and their compounds. Problems. 1980 edition. 582pp. 6⅛ × 9¼. 66021-4 Pa. $14.95

BOUNDARY VALUE PROBLEMS OF HEAT CONDUCTION, M. Necati Özisik. Systematic, comprehensive treatment of modern mathematical methods of solving problems in heat conduction and diffusion. Numerous examples and problems. Selected references. Appendices. 505pp. 5⅜ × 8½. 65990-9 Pa. $11.95

A SHORT HISTORY OF CHEMISTRY (3rd edition), J.R. Partington. Classic exposition explores origins of chemistry, alchemy, early medical chemistry, nature of atmosphere, theory of valency, laws and structure of atomic theory, much more. 428pp. 5⅜ × 8½. (Available in U.S. only) 65977-1 Pa. $10.95

A HISTORY OF ASTRONOMY, A. Pannekoek. Well-balanced, carefully reasoned study covers such topics as Ptolemaic theory, work of Copernicus, Kepler, Newton, Eddington's work on stars, much more. Illustrated. References. 521pp. 5⅜ × 8½. 65994-1 Pa. $11.95

PRINCIPLES OF METEOROLOGICAL ANALYSIS, Walter J. Saucier. Highly respected, abundantly illustrated classic reviews atmospheric variables, hydrostatics, static stability, various analyses (scalar, cross-section, isobaric, isentropic, more). For intermediate meteorology students. 454pp. 6⅛ × 9¼. 65979-8 Pa. $12.95

RELATIVITY, THERMODYNAMICS AND COSMOLOGY, Richard C. Tolman. Landmark study extends thermodynamics to special, general relativity; also applications of relativistic mechanics, thermodynamics to cosmological models. 501pp. 5⅜ × 8½. 65383-8 Pa. $11.95

APPLIED ANALYSIS, Cornelius Lanczos. Classic work on analysis and design of finite processes for approximating solution of analytical problems. Algebraic equations, matrices, harmonic analysis, quadrature methods, much more. 559pp. 5⅜ × 8½. 65656-X Pa. $11.95

SPECIAL RELATIVITY FOR PHYSICISTS, G. Stephenson and C.W. Kilmister. Concise elegant account for nonspecialists. Lorentz transformation, optical and dynamical applications, more. Bibliography. 108pp. 5⅜ × 8½. 65519-9 Pa. $3.95

INTRODUCTION TO ANALYSIS, Maxwell Rosenlicht. Unusually clear, accessible coverage of set theory, real number system, metric spaces, continuous functions, Riemann integration, multiple integrals, more. Wide range of problems. Undergraduate level. Bibliography. 254pp. 5⅜ × 8½. 65038-3 Pa. $7.00

INTRODUCTION TO QUANTUM MECHANICS With Applications to Chemistry, Linus Pauling & E. Bright Wilson, Jr. Classic undergraduate text by Nobel Prize winner applies quantum mechanics to chemical and physical problems. Numerous tables and figures enhance the text. Chapter bibliographies. Appendices. Index. 468pp. 5⅜ × 8½. 64871-0 Pa. $9.95

ASYMPTOTIC EXPANSIONS OF INTEGRALS, Norman Bleistein & Richard A. Handelsman. Best introduction to important field with applications in a variety of scientific disciplines. New preface. Problems. Diagrams. Tables. Bibliography. Index. 448pp. 5⅜ × 8½. 65082-0 Pa. $10.95

MATHEMATICS APPLIED TO CONTINUUM MECHANICS, Lee A. Segel. Analyzes models of fluid flow and solid deformation. For upper-level math, science and engineering students. 608pp. 5⅜ × 8½. 65369-2 Pa. $12.95

ELEMENTS OF REAL ANALYSIS, David A. Sprecher. Classic text covers fundamental concepts, real number system, point sets, functions of a real variable, Fourier series, much more. Over 500 exercises. 352pp. 5⅜ × 8½. 65385-4 Pa. $8.95

PHYSICAL PRINCIPLES OF THE QUANTUM THEORY, Werner Heisenberg. Nobel Laureate discusses quantum theory, uncertainty, wave mechanics, work of Dirac, Schroedinger, Compton, Wilson, Einstein, etc. 184pp. 5⅜ × 8½. 60113-7 Pa. $4.95

INTRODUCTORY REAL ANALYSIS, A.N. Kolmogorov, S.V. Fomin. Translated by Richard A. Silverman. Self-contained, evenly paced introduction to real and functional analysis. Some 350 problems. 403pp. 5⅜ × 8½. 61226-0 Pa. $7.95

PROBLEMS AND SOLUTIONS IN QUANTUM CHEMISTRY AND PHYSICS, Charles S. Johnson, Jr. and Lee G. Pedersen. Unusually varied problems, detailed solutions in coverage of quantum mechanics, wave mechanics, angular momentum, molecular spectroscopy, scattering theory, more. 280 problems plus 139 supplementary exercises. 430pp. 6½ × 9¼. 65236-X Pa. $10.95

ASYMPTOTIC METHODS IN ANALYSIS, N.G. de Bruijn. An inexpensive, comprehensive guide to asymptotic methods—the pioneering work that teaches by explaining worked examples in detail. Index. 224pp. 5⅜ × 8½. 64221-6 Pa. $5.95

OPTICAL RESONANCE AND TWO-LEVEL ATOMS, L. Allen and J.H. Eberly. Clear, comprehensive introduction to basic principles behind all quantum optical resonance phenomena. 53 illustrations. Preface. Index. 256pp. 5⅜ × 8½. 65533-4 Pa. $6.95

COMPLEX VARIABLES, Francis J. Flanigan. Unusual approach, delaying complex algebra till harmonic functions have been analyzed from real variable viewpoint. Includes problems with answers. 364pp. 5⅜ × 8½. 61388-7 Pa. $7.95

ATOMIC SPECTRA AND ATOMIC STRUCTURE, Gerhard Herzberg. One of best introductions; especially for specialist in other fields. Treatment is physical rather than mathematical. 80 illustrations. 257pp. 5⅜ × 8½. 60115-3 Pa. $4.95

APPLIED COMPLEX VARIABLES, John W. Dettman. Step-by-step coverage of fundamentals of analytic function theory—plus lucid exposition of 5 important applications: Potential Theory; Ordinary Differential Equations; Fourier Transforms; Laplace Transforms; Asymptotic Expansions. 66 figures. Exercises at chapter ends. 512pp. 5⅜ × 8½. 64670-X Pa. $10.95

ULTRASONIC ABSORPTION: An Introduction to the Theory of Sound Absorption and Dispersion in Gases, Liquids and Solids, A.B. Bhatia. Standard reference in the field provides a clear, systematically organized introductory review of fundamental concepts for advanced graduate students, research workers. Numerous diagrams. Bibliography. 440pp. 5⅜ × 8½. 64917-2 Pa. $8.95

UNBOUNDED LINEAR OPERATORS: Theory and Applications, Seymour Goldberg. Classic presents systematic treatment of the theory of unbounded linear operators in normed linear spaces with applications to differential equations. Bibliography. 199pp. 5⅜ × 8½. 64830-3 Pa. $7.00

LIGHT SCATTERING BY SMALL PARTICLES, H.C. van de Hulst. Comprehensive treatment including full range of useful approximation methods for researchers in chemistry, meteorology and astronomy. 44 illustrations. 470pp. 5⅜ × 8½. 64228-3 Pa. $9.95

CONFORMAL MAPPING ON RIEMANN SURFACES, Harvey Cohn. Lucid, insightful book presents ideal coverage of subject. 334 exercises make book perfect for self-study. 55 figures. 352pp. 5⅜ × 8¼. 64025-6 Pa. $8.95

OPTICKS, Sir Isaac Newton. Newton's own experiments with spectroscopy, colors, lenses, reflection, refraction, etc., in language the layman can follow. Foreword by Albert Einstein. 532pp. 5⅜ × 8½. 60205-2 Pa. $8.95

GENERALIZED INTEGRAL TRANSFORMATIONS, A.H. Zemanian. Graduate-level study of recent generalizations of the Laplace, Mellin, Hankel, K. Weierstrass, convolution and other simple transformations. Bibliography. 320pp. 5⅜ × 8½. 65375-7 Pa. $7.95

THE FOUR-COLOR PROBLEM: Assaults and Conquest, Thomas L. Saaty and Paul G. Kainen. Engrossing, comprehensive account of the century-old combinatorial topological problem, its history and solution. Bibliographies. Index. 110 figures. 228pp. 5⅜ × 8½. 65092-8 Pa. $6.00

CATALYSIS IN CHEMISTRY AND ENZYMOLOGY, William P. Jencks. Exceptionally clear coverage of mechanisms for catalysis, forces in aqueous solution, carbonyl- and acyl-group reactions, practical kinetics, more. 864pp. 5⅜ × 8½. 65460-5 Pa. $18.95

PROBABILITY: An Introduction, Samuel Goldberg. Excellent basic text covers set theory, probability theory for finite sample spaces, binomial theorem, much more. 360 problems. Bibliographies. 322pp. 5⅜ × 8½. 65252-1 Pa. $7.95

LIGHTNING, Martin A. Uman. Revised, updated edition of classic work on the physics of lightning. Phenomena, terminology, measurement, photography, spectroscopy, thunder, more. Reviews recent research. Bibliography. Indices. 320pp. 5⅜ × 8¼. 64575-4 Pa. $7.95

PROBABILITY THEORY: A Concise Course, Y.A. Rozanov. Highly readable, self-contained introduction covers combination of events, dependent events, Bernoulli trials, etc. Translation by Richard Silverman. 148pp. 5⅜ × 8¼. 63544-9 Pa. $4.50

THE CEASELESS WIND: An Introduction to the Theory of Atmospheric Motion, John A. Dutton. Acclaimed text integrates disciplines of mathematics and physics for full understanding of dynamics of atmospheric motion. Over 400 problems. Index. 97 illustrations. 640pp. 6 × 9. 65096-0 Pa. $16.95

STATISTICS MANUAL, Edwin L. Crow, et al. Comprehensive, practical collection of classical and modern methods prepared by U.S. Naval Ordnance Test Station. Stress on use. Basics of statistics assumed. 288pp. 5⅜ × 8½. 60599-X Pa. $6.00

WIND WAVES: Their Generation and Propagation on the Ocean Surface, Blair Kinsman. Classic of oceanography offers detailed discussion of stochastic processes and power spectral analysis that revolutionized ocean wave theory. Rigorous, lucid. 676pp. 5⅜ × 8½. 64652-1 Pa. $14.95

STATISTICAL METHOD FROM THE VIEWPOINT OF QUALITY CONTROL, Walter A. Shewhart. Important text explains regulation of variables, uses of statistical control to achieve quality control in industry, agriculture, other areas. 192pp. 5⅜ × 8½. 65232-7 Pa. $6.00

THE INTERPRETATION OF GEOLOGICAL PHASE DIAGRAMS, Ernest G. Ehlers. Clear, concise text emphasizes diagrams of systems under fluid or containing pressure; also coverage of complex binary systems, hydrothermal melting, more. 288pp. 6½ × 9¼. 65389-7 Pa. $8.95

STATISTICAL ADJUSTMENT OF DATA, W. Edwards Deming. Introduction to basic concepts of statistics, curve fitting, least squares solution, conditions without parameter, conditions containing parameters. 26 exercises worked out. 271pp. 5⅜ × 8½. 64685-8 Pa. $7.95

TENSOR CALCULUS, J.L. Synge and A. Schild. Widely used introductory text covers spaces and tensors, basic operations in Riemannian space, non-Riemannian spaces, etc. 324pp. 5⅜ × 8¼. 63612-7 Pa. $7.00

A CONCISE HISTORY OF MATHEMATICS, Dirk J. Struik. The best brief history of mathematics. Stresses origins and covers every major figure from ancient Near East to 19th century. 41 illustrations. 195pp. 5⅜ × 8½. 60255-9 Pa. $7.95

A SHORT ACCOUNT OF THE HISTORY OF MATHEMATICS, W.W. Rouse Ball. One of clearest, most authoritative surveys from the Egyptians and Phoenicians through 19th-century figures such as Grassman, Galois, Riemann. Fourth edition. 522pp. 5⅜ × 8½. 20630-0 Pa. $9.95

HISTORY OF MATHEMATICS, David E. Smith. Non-technical survey from ancient Greece and Orient to late 19th century; evolution of arithmetic, geometry, trigonometry, calculating devices, algebra, the calculus. 362 illustrations. 1,355pp. 5⅜ × 8½. 20429-4, 20430-8 Pa., Two-vol. set $21.90

THE GEOMETRY OF RENÉ DESCARTES, René Descartes. The great work founded analytical geometry. Original French text, Descartes' own diagrams, together with definitive Smith-Latham translation. 244pp. 5⅜ × 8½. 60068-8 Pa. $6.00

THE ORIGINS OF THE INFINITESIMAL CALCULUS, Margaret E. Baron. Only fully detailed and documented account of crucial discipline: origins; development by Galileo, Kepler, Cavalieri; contributions of Newton, Leibniz, more. 304pp. 5⅜ × 8½. (Available in U.S. and Canada only) 65371-4 Pa. $7.95

THE HISTORY OF THE CALCULUS AND ITS CONCEPTUAL DEVELOPMENT, Carl B. Boyer. Origins in antiquity, medieval contributions, work of Newton, Leibniz, rigorous formulation. Treatment is verbal. 346pp. 5⅜ × 8½. 60509-4 Pa. $6.95

THE THIRTEEN BOOKS OF EUCLID'S ELEMENTS, translated with introduction and commentary by Sir Thomas L. Heath. Definitive edition. Textual and linguistic notes, mathematical analysis. 2500 years of critical commentary. Not abridged. 1,414pp. 5⅜ × 8½. 60088-2, 60089-0, 60090-4 Pa., Three-vol. set $26.85

A HISTORY OF VECTOR ANALYSIS: The Evolution of the Idea of a Vectorial System, Michael J. Crowe. The first large-scale study of the history of vector analysis, now the standard on the subject. Unabridged republication of the edition published by University of Notre Dame Press, 1967, with second preface by Michael C. Crowe. Index. 278pp. 5⅜ × 8½. 64955-5 Pa. $7.00

THE HISTORICAL ROOTS OF ELEMENTARY MATHEMATICS, Lucas N.H. Bunt, Phillip S. Jones, and Jack D. Bedient. Fundamental underpinnings of modern arithmetic, algebra, geometry and number systems derived from ancient civilizations. 320pp. 5⅜ × 8½. 25563-8 Pa. $7.95

CALCULUS REFRESHER FOR TECHNICAL PEOPLE, A. Albert Klaf. Covers important aspects of integral and differential calculus via 756 questions. 566 problems, most answered. 431pp. 5⅜ × 8½. 20370-0 Pa. $7.95

CATALOG OF DOVER BOOKS

CHALLENGING MATHEMATICAL PROBLEMS WITH ELEMENTARY SOLUTIONS, A.M. Yaglom and I.M. Yaglom. Over 170 challenging problems on probability theory, combinatorial analysis, points and lines, topology, convex polygons, many other topics. Solutions. Total of 445pp. 5⅜ × 8½. Two-vol. set.

Vol. I 65536-9 Pa. $5.95
Vol. II 65537-7 Pa. $5.95

FIFTY CHALLENGING PROBLEMS IN PROBABILITY WITH SOLUTIONS, Frederick Mosteller. Remarkable puzzlers, graded in difficulty, illustrate elementary and advanced aspects of probability. Detailed solutions. 88pp. 5⅜ × 8½.
65355-2 Pa. $3.95

EXPERIMENTS IN TOPOLOGY, Stephen Barr. Classic, lively explanation of one of the byways of mathematics. Klein bottles, Moebius strips, projective planes, map coloring, problem of the Koenigsberg bridges, much more, described with clarity and wit. 43 figures. 210pp. 5⅜ × 8½. 25933-1 Pa. $4.95

RELATIVITY IN ILLUSTRATIONS, Jacob T. Schwartz. Clear non-technical treatment makes relativity more accessible than ever before. Over 60 drawings illustrate concepts more clearly than text alone. Only high school geometry needed. Bibliography. 128pp. 6⅛ × 9¼. 25965-X Pa. $5.95

AN INTRODUCTION TO ORDINARY DIFFERENTIAL EQUATIONS, Earl A. Coddington. A thorough and systematic first course in elementary differential equations for undergraduates in mathematics and science, with many exercises and problems (with answers). Index. 304pp. 5⅜ × 8¼. 65942-9 Pa. $7.95

FOURIER SERIES AND ORTHOGONAL FUNCTIONS, Harry F. Davis. An incisive text combining theory and practical example to introduce Fourier series, orthogonal functions and applications of the Fourier method to boundary-value problems. 570 exercises. Answers and notes. 416pp. 5⅜ × 8½. 65973-9 Pa. $8.95

THE THOERY OF BRANCHING PROCESSES, Theodore E. Harris. First systematic, comprehensive treatment of branching (i.e. multiplicative) processes and their applications. Galton-Watson model, Markov branching processes, electron-photon cascade, many other topics. Rigorous proofs. Bibliography. 240pp. 5⅜ × 8½. 65952-6 Pa. $6.95

AN INTRODUCTION TO ALGEBRAIC STRUCTURES, Joseph Landin. Superb self-contained text covers "abstract algebra": sets and numbers, theory of groups, theory of rings, much more. Numerous well-chosen examples, exercises. 247pp. 5⅜ × 8½. 65940-2 Pa. $6.95

GAMES AND DECISIONS: Introduction and Critical Survey, R. Duncan Luce and Howard Raiffa. Superb non-technical introduction to game theory, primarily applied to social sciences. Utility theory, zero-sum games, n-person games, decision-making, much more. Bibliography. 509pp. 5⅜ × 8½. 65943-7 Pa. $10.95

Prices subject to change without notice.

Available at your book dealer or write for free Mathematics and Science Catalog to Dept. GI, Dover Publications, Inc., 31 East 2nd St., Mineola, N.Y. 11501. Dover publishes more than 175 books each year on science, elementary and advanced mathematics, biology, music, art, literary history, social sciences and other areas.